The Materials Revolution

The Materials Revolution

Superconductors, New Materials and the Japanese Challenge

Edited by Tom Forester

Basil Blackwell

First published 1988

Basil Blackwell Ltd
108 Cowley Road, Oxford, OX4 1JF, UK

British Library Cataloguing in Publication Data

The Materials revolution : superconductors,
new materials, and the Japanese challenge.
1. Materials. Technological innovation
I. Forester, Tom
620.1'1

ISBN 0-631-16699-8
ISBN 0-631-16701-3 Pbk

Printed in the United States of America

Contents

III Materials and the Economy

IV Materials Innovation and Substitution

V New Frontiers in Materials

Preface

Three megatechnologies will dominate the last decade of the twentieth century: information technology, biotechnology, and new materials. The Japanese recognized this years ago—that's why they targeted all three for special R&D effort. Only now is the United States beginning to wake up to what's been going on.

In recent years the information technology revolution has generated whole libraries of books, while biotechnology has spurred a flurry of publications. New materials, on the other hand, have not been so widely written about, and their importance is less well understood by the general public. Indeed, their existence is barely touched upon in many university and college courses that should deal with such matters.

With this volume I shall attempt to correct this imbalance in our knowledge of the triumvirate of new technologies and to demonstrate that new materials play a crucial, enabling role in technological progress because developments in materials technology are the *precursor* to innovation in many other areas. And the arrival of high-temperature ceramic superconductors in early 1987 has given the whole subject of new materials an added topicality.

For assistance and encouragement in this project, I thank Frank Satlow of MIT Press and his team of four anonymous referees; Michael Marien, editor of *Future Survey*; and Don Lamberton of the University of Queensland. Griffith University kindly provided me with the time and the facilities necessary for the preparation of the manuscript.

Acknowledgments

The editor and the publisher are grateful to the following:

American Ceramic Society, for W. David Kingery, "Looking to the Future," reprinted from W. D. Kingery (ed.), *Ceramics and Civilization: High-Technology Ceramics, Past, Present, and Future,* vol. 3, pp. 371–381. Copyright © 1986 American Ceramic Society. Reprinted by permission of the American Ceramic Society.

Elsevier Sequoia S.A., for Melvin Kranzberg and Cyril Stanley Smith, "Materials in History and Society," reprinted from *Materials Science and Engineering* 37, 1 (1979). Reprinted with permission.

Fortune, for Gene Bylinsky, "What's Sexier and Speedier Than Silicon," reprinted from *Fortune,* June 24, 1985. Copyright © 1985 Time Inc. All rights reserved.

High Technology Business, for Gordon Graff, "Ceramics Take on Tough Tasks," reprinted from *High Technology,* December 1983 and for Gordon Graff, "High-Performance Plastics," reprinted from *High Technology,* October 1986. Reprinted with permission, *High Technology* magazine, Copyright © 1983 and 1986 by Infotechnology Publishing Corporation, 214 Lewis Wharf, Boston, MA 02110.

International Journal of Materials and Product Technology, for E. D. Hondros, "Materials, Year 2000—A Perspective," reprinted from *International Journal of Materials and Product Technology* 1, 1 (1986), and for Gene Gregory, "New Materials Technology in Japan," *International Journal of Materials and Product Technology* 2, 1 (1987). Copyright © 1986 and 1987.

Penguin Books Ltd. for "Materials Processing in Space," reprinted from Peter Marsh, *The Space Business,* Penguin Books, 1985. Copyright © Peter Marsh, 1985.

Pinter Publishers Ltd. for George F. Ray, "Innovation in Materials," reprinted from Roy Macleod (ed.), *Technology and the Human Prospect,* Frances Pinter, London, 1986. Reprinted by permission of Pinter Publishers, copyright © 1986.

Resources for the Future Inc., for John E. Tilton, "Materials Substitution: Lessons from Tin-Using Industries," reprinted from John E. Tilton (ed.), "Materials Substitution: Lessons from the Tin-Using Industries," Resources for the Future, Washington, DC, 1983. Copyright © 1983.

Science, for James M. Broadus, "Seabed Materials," reprinted from *Science* 235, February 20, 1987, pp. 853–860, with permission. Copyright © 1987 by the American Association for the Advancement of Science (AAAS), Washington, DC.

Scientific American, for Eric D. Larson, Marc H. Ross, and Robert H. Williams, "Beyond the Era of Materials," reprinted from *Scientific American* (June 1986), pp. 24–31, and for Joel P. Clark and Merton C. Flemings, "Advanced Materials and the Economy," reprinted from *Scientific American* (October 1986), pp. 43–49. Both reprinted with permission. Copyright © 1986 Scientific American, Inc. All rights reserved.

Technology Review, for Les C. Gunderson and Donald B. Keck, "Optical Fibers: Where Light Outperforms Electrons," reprinted from *Technology Review* (May–June 1983); Joel P. Clark and Frank R. Field III, "How Critical Are Critical Materials?" reprinted from *Technology Review* (August–September 1985); Doug Stewart, "Skylines of Fabric," reprinted from *Technology Review* (January 1987); Thomas W. Eagar, "The Real Challenge in Materials Engineering," reprinted from *Technology Review* (February–March 1987). All reprinted with permission from *Technology Review,* copyright © 1983, 1985, and 1987.

The Boston Globe, for Alison Bass, "Superconductivity Research: Pace Slows, Reality Catches Up," reprinted from *The Boston Globe,* November 9, 1987. Reprinted courtesy of The Boston Globe.

The Economist, for "Not-So-Superconductors," reprinted from *The Economist,* June 13, 1987. Reprinted with permission.

Time, for Michael D. Lemonick et al., "Superconductors!" reprinted from *Time* magazine, May 11, 1987. Copyright © 1987 Time Inc. All rignts reserved. Reprinted by permission from *Time.*

Whole Earth Review, for "Nanotechnics and Civilization," reprinted from *Whole Earth Review,* no. 54 (Spring 1987). Reprinted with permission from Whole Earth Review, 27 Gate Five Road, Sausalito, CA 94965. Annual subscription: $20.

The Materials Revolution

Introduction

Superconductors

On March 18, 1987, a scientific "happening" occurred in the United States, one that greatly excited members of the scientific community and the world at large. In retrospect it may be seen to be a major landmark on the path of technological progress. Dubbed the "Woodstock of physics" by the media, the annual meeting of the American Physical Society at the New York Hilton took a remarkable turn. The Society had planned to hold just a small workshop on a relatively unpublicized area of science—the search for superconductivity, or materials that conduct electricity without resistance. But such was the interest in 1987 that not only was the session switched to the hotel's vast Sutton Ballroom, but by the time the doors were opened, 5,000 people were clamoring to get into a hall that seated 1,200. The excited scientists had come to hear news of dramatic breakthroughs in the discovery of new materials—materials that become superconductors at temperatures far higher than anything previously achieved. Being able to transmit electricity without loss would, among other things, make possible much cheaper energy, superfast "flying" trains and would spark a further revolution in electronics, the "core" technology of today.

Way back in 1911 the Dutch scientist, Heike Kamerlingh Onnes, first observed that some metals like mercury became superconductive when cooled to near the absolute zero temperature of minus 273 degrees Celsius (minus 459 degrees Fahrenheit, or zero on the Kelvin scale preferred by scientists). This is the point at which atoms cease moving around. But such low temperatures could only be achieved with the generous application of costly liquid helium, which really ruled out superconductors for most practical purposes. In fact, superconductors stayed on the scientific back-burner for decades, and progress in the search for superconductivity at warmer temperatures was painfully slow. "Room temperature" superconductors seemed light-years away.

Then, in a series of stunning breakthroughs beginning in late 1985, scientists started to discover new materials that became superconductors at hitherto unheard-of temperatures. First, researchers at IBM's laboratory in Zurich, Switzerland, led by Karl Alex Müller and Johannes Georg Bednorz, came across an oxide of copper that became superconductive at an amazingly warm 35 K. In response, research teams all over the world in 1986 began a series of similar experiments with different ceramic oxides based on a variety of materials and rare earths. In February 1987 a group at the University of Houston, Texas, headed by Ching-Wu "Paul" Chu, switched a rare earth, yttrium, for the lanthanum Müller had used in his recipe and found a compound that became superconductive at 98 K, a temperature well above that at which inexpensive nitrogen can be used for cooling. Breaking the barrier from helium to nitrogen created a sensation, thereby ensuring Chu's place in the history books. As one scientist put it, "Chu ran the four-minute mile in superconductivity."

New developments have since come thick and fast. In the United States venture capitalists have rushed to create start-up companies in a bid to cash in on superconductors, no doubt with the aid of lucrative government contracts. With a probable Nobel Prize at stake, there has been a wave of feverish activity in labs around the world, as scientists have vied with each other to discover ever-warmer superconductor materials. In June 1987, for example, scientists in the United States, Japan, the U.S.S.R., India, and China reported that they had found ceramic oxides that lost all resistance to electricity at close to room temperature. In September 1987 it was announced that Dr. Ihara of the Japanese Electrotechnical Laboratory at Ibaraki had actually found a ceramic material that became superconductive *at* room temperature—or 65° C. Researchers at IBM's Watson Research Center in Yorktown Heights, New York, announced that they had made the world's first superconducting quantum interference devices, or SQUIDs, using the new superconducting materials in the form of thin films. And a British team at the ICI company claimed a further world "first" with their development of a ceramic superconducting wire which is said to be strong, flexible, and able to carry considerable electrical current. In October 1987 news came from Scandinavia that Müller and Bednorz had won the Nobel Prize for Physics—less than two years after their initial discovery of high-temperature superconductivity had been published.

Governments rushed to boost superconductor research, in the belief that superiority in superconductors was vital to the national interest. The U.S. government launched a $1.6 million program through the National Science Foundation to fund research at three major university centers, and President Ronald Reagan announced he would ask Congress to build a $4.5 billion Superconducting Super-Collider (SSC), which would be the world's largest high-energy particle accelerator. Some twenty-six states immediately started lobbying furiously in attempts to get the project. In August 1987 the U.S. government launched a further eleven-point superconductor initiative, comprising research grants, advisory groups, information sharing, regulatory reform, and $150 million in funding from the Department of Defense. At a conference in Washington to which foreigners were not invited, the President spoke of the breakthroughs in superconductivity bringing us to the threshold of a new age: "We are increasingly moving from an age of things to an age of thoughts, to an age of mind over matter," he said. We are facing "a revolution of shattered paradigms and long-held certainties."

Meanwhile in Japan the Ministry of International Trade and Industry (MITI) moved to establish a new materials research lab at Nagoya, and the private sector Nomura Research Institute stepped in with its own superconductor plan. In September MITI asked the Finance Minister to boost government backing of superconductor research from under $5 million in the current fiscal year to $40 million in the next. But most Japanese corporate efforts focused on early commercialization, and there was a rush to register patents. In Britain the Science and Engineering Research Council (SERC) launched a four-point program of R&D assistance, while in France,

West Germany, Australia, and Canada there were similar moves to "catch the wave" before it was too late. Even so, the British magazine *New Scientist* (June 8, 1987) warned that superconductors were being oversold "in an unwitting attempt to squeeze money out of governments," and it would be wise for governments to sit back and think awhile rather than panic and throw millions at the subject. At the same time sceptical voices were being raised about many aspects of superconductivity and its prospects—scepticism that is reflected in chapter 2 of this volume.

Superconductors hold out the prospect of a new generation of microchips and computers that would be much more powerful and would never need cooling. Some are already talking about a "Third Age of Electronics" based on superconductors (after vacuum tubes and transistors) and even the creation of an "Oxide Valley" to supercede Silicon Valley in California. Transmitting electricity without loss would obviously save millions of dollars and revolutionize the electricity industry by doing away with a lot of costly transmission equipment and by making power plants more efficient. Thanks to the so-called "Meissner effect"—a phenomenon by which a superconductor excludes any magnetic field that comes near to it—superconductors also make possible the economic construction of magnetic levitation trains which could revolutionize ground transportation. Medical scanners could be vastly improved because they too rely on powerful superconducting magnets.

Superconductors may in addition transform science itself by enabling scientists to smash atoms more easily. Ultimately, this might lead to practical nuclear fusion, and thus inexhaustible, cheap energy supplies. Nobody underestimates the practical difficulties and the time scale involved in translating the latest breakthroughs into everyday products. Scientists don't even properly understand how the things work yet. But as Arno Penzias, vice-president for research at Bell Labs, put it, "The recent advances in the field of superconductivity are almost without comparison." There seems little doubt that superconductors are here to stay and that in time they will have a tremendous impact on technology and on society—and in ways that cannot be predicted. Some have likened the arrival of superconductors in the late 1980s to that of the transistor in the 1950s, but Jack Kilby, co-inventor of the integrated circuit, says that this is, if anything, an understatement: "This is much broader. It could impact almost anything."

A Materials Revolution

Yet advances in superconductors are but one (albeit well-publicized) example of progress in ceramics research, which in turn is but a part of a much wider revolution taking place in new materials and materials processing. The new ceramics, which have little in common with china teacups, are being used increasingly in engines and electronics. They are light, they never wear out, and they can withstand enormous temperatures. New high-performance plastics and composites based on carbon fibers look likely to transform, for example, the automobile and aerospace industries. Already some cars have plastic body parts, and all-plastic cars and planes—held together by the new superglues—could be common in the next century. More sophisticated semiconductors made from silicon and new materials like gallium arsenide promise further quantum leaps in computing power, while advances in fiber optics are transforming the telecommunications industry and influencing other areas such as medicine. New metal alloys—made by processes such as "rapid solidification" and "powder metallurgy"—will have a major impact right across the range of manufacturing industries. Even humble cement has gone high-tech.

Despite these exciting developments the public still knows comparatively little about the work of materials scientists and physicists working in this area. Materials research has for years been very much the Cinderella of the sciences, yet progress in materials research is crucial to overcoming such problems as the finiteness of the world's resources and possible shortages of strategic materials. Innovation in materials to a great extent determines the pace of technological advance in many key industries—especially the computer industry—which in turn greatly influences productivity, capital formation, the demand for labor, and the overall rate of economic growth. Our standard of living today has been largely determined by past discoveries of "new" materials, and our future prosperity will in large part depend on the fruits of contemporary research into even newer materials and new materials production processes.

Of course materials of one kind or another have been vitally important throughout human history. Do we not talk of the Stone Age, the Bronze Age, and the Iron Age, implying that a *material* was the defining technology of an era? Likewise historians have argued that the invention of hydraulic cement by the Romans played an important part in the growth of the Roman Empire, the invention of gunpowder clearly had an explosive effect on the technique of warfare, and the adoption of another "new" material—glass—was a shattering blow to the birds and insects that used to seek shelter in medieval homes. New technologies like printing were predicated by advances in materials such as paper, and society was never the same again after iron was first smelted. Indeed, the Industrial Revolution itself was based on new ways of processing the materials coal, iron, and steel. Since then, progress in the more efficient use of the "traditional" manufacturing metals, iron and steel, has continued: for example, the power-to-weight ratio of an 1810 locomotive was 1,000 kilograms per horsepower, by 1900 it was about 100; by 1950 it was down to 25, and by 1980 it was 14 (Eric D. Larson et al., chapter 6 in this volume).

In recent years materials research has been coming up with a string of useful money-saving innovations such as plastic film, plastic bottles, the radial tire, and the two-piece aluminum can. But the big difference between earlier centuries and today is that in the past the human race has tended to adapt naturally occurring materials and minerals for its use. Now science and technology is giving humans the ability to *design* the materials they require. It will become increasingly possible to predict the properties of materials before they are even made and to modify the recipe to get the desired result to suit a particular application. These so-called "advanced materials" begin life in the mind of the scientist in the laboratory—they aren't ripped out of the ground in some far-off continent and transported at enormous cost across the seas. The shift to designer materials will thus benefit nations that are consumers of raw materials and disadvantage the traditional producer nations, especially those in the third world. The new materials technology could therefore represent an entirely new way of going about things, and as such it will present a major new challenge not only to managers, designers, and entrepreneurs but to governments worldwide.

Ceramics and Plastics

High-tech ceramics are a particularly important part of the materials revolution. Japan's MITI put ceramics at the top of its list of "next generation industries" when it launched a ten-year, $120 million program of ceramics research in 1981. These distant cousins of crockery are actually highly purified artificial substances based on minerals like alumina, titania, and sand, which are mixed and fired to make materials like aluminium oxide, silicon carbide, and silicon nitride. The new ceramics are harder, stronger, lighter, and more durable than many metals; they don't rust and they can withstand much higher temperatures. Eventually, they will require little or no machining. For example, sialon (trade name: Syalon), a ceramic alloy of silicon, aluminium, oxygen, and nitrogen, is reputedly as hard as diamonds, stronger than steel, as light as aluminium, and tough enough to withstand almost anything. With the prospect of traditional metals and metal-based engineering declining in favor of ceramics, U.S. analysts fear that if America fails to keep up with the Japanese in ceramics research, it could miss out on a key growth industry of the 1990s.

Already advanced ceramics are showing up in scissors, knives, fishhooks, batteries, sensors, stoves, heat exchangers, artificial limbs, and dentures. The main industries affected by developments in ceramics will be electronics, telecommunications, automobiles, aerospace, energy, medicine, and the military. But the two most important applications are ceramic engines and electronic devices. Because metals have pretty much reached their limits, researchers in Japan and the United States are racing to develop ceramic engines and ceramic parts for engines. Nissan already sells a version of the 300 ZX sports car with a ceramic turbocharger rotor, and both Isuzu and Toyota are talking about part-ceramic diesel engines by 1990. Being able to withstand temperatures up to 1,500° C without cooling or lubrication means that ceramic engines can boost fuel efficiency by 30 to 40 percent. Electronics actually accounts for 80 percent of all ceramics sold at present: ceramics are being used in capacitors, chips, transformers, horizontal chip carriers, and piezoelectric devices (highly sensitive ceramic parts that generate small voltages when stimulated) that have a variety of electronic and audio applications. Over the horizon are optical computers made of ceramic materials.

It would be wrong, however, to imply that ceramics have no disadvantages. They are in fact still difficult to mold and manufacture, and they are very brittle. This means that they are prone to crack or shatter. It is taking scientists much longer than expected to see how ceramic materials stand up to the stresses and strains of actual use, particularly in engines. For instance, the "thermal shock" of a sudden change in temperature can create micro-cracks that can in turn trigger catastrophic failure—not much fun in an aero-engine. Solutions—like improved atomic structures and better monitoring devices—are being developed, but it all takes time.

Plastics have been around since 1907, when chemist Leo Baekeland, an immigrant to the United States from Belgium, produced the first truly synthetic plastic, which he called Bakelite. Despite being a pejorative term, plastics have become more and more popular throughout the century. Indeed, in 1979 the annual volume of plastics produced in the United States exceeded that of steel for the first time. Now new generations of high-performance plastics and of advanced composites (or fancy versions of fiberglass) are set to transform products and production processes across a range of industries. Because they are entirely artificial, plastics and composites for specific uses can be tailor-made by scientists from the molecules upward. The new plastics are not only being used increasingly in cars, planes, chips, and memory discs as a substitute for metals or glass; in recent years we have also seen the invention of plastics that conduct electricity, that change color with variations in temperature or voltage, that can be washed away with water, and that disappear after sixty days in sunlight without spoiling the environment.

The new plastics can often be cheaper, lighter, and stronger than metals. They can lower manufacturing costs because they use less energy, they are self-coloring, they can utilize snap-fit assembly, and they can reduce a large number of parts into one molding. Plastics can be made to have good thermal and electrical insulation properties, and they don't easily corrode. Furthermore they can decrease noise, vibration, friction, and wear. Their disadvantages include high initial investment costs (switching to plastics involves entirely new plant) and an inability to withstand heat or certain chemicals. Plastics are also difficult to join, they can't be recovered and melted-down like metal after they've outlived their usefulness, and the finish on plastic parts is often not as good as that on metals. Even so, and despite the heat problem, a U.S. company and a Canadian company have actually developed plastic auto engines that are lighter, quieter, and cooler than conventional engines.

Of course plastics are already used in a whole range of applications, from domestic kettles and contact lenses to PVC pipes and gutters and safety glass. Recently, we have seen the arrival of the plastic, squeezable ketchup bottle, a plastic version of the neon sign, the plastic Army helmet (made of Kevlar), and the Kevlar-based bullet-proof vest. But the two main areas to watch are cars and planes. Cars like the *Corvette* and the Pontiac *Fiero* have plastic body panels and GM claims that 20 percent of its cars and 50 percent of its vans will have all-plastic bodies by 1990. The main aim of the auto manufacturers is to reduce weight and thus boost fuel economy. In addition plastics in cars don't rust, they don't dent so easily, and they allow for greater flexibility both in the style of the product (more custom versions of models) and in the production process itself (part of the general move to FMS, or flexible manufacturing systems). The forces holding things back in the auto industry are inertia, high investment costs, the high cost of repairs, and sheer prejudice against plastics.

In the aerospace industry advanced composites—fibers of metals or ceramics embedded in plastic—promise to revolutionize plane-making. The older carbon fiber (as found in tennis racquets) has been used for a long time in numerous aircraft parts. But the new composites now make possible the construction of stiff, lightweight structures suitable for use in aircraft wings, tails, and even whole airframes. Such materials were behind recent successes with human-powered flight and the nonstop round-the-world trip by the *Voyager* aircraft in 1987. Thirty percent of the airframe and wings of the AV-8B *Harrier* jump-jet are made of composite materials. And despite the collapse of the *Lear Fan* plastic plane project in 1985, lightweight turboprop aircraft made almost entirely from plastics could become common in the late 1990s.

Semiconductors and Cement

New materials lie behind past, present, and promised advances in semiconductor technology, the very basis of the computer revolution. Transistors, integrated circuits, and the chip itself only came about after materials scientists learned how to process silicon from common sand. Silicon turned out to be the ideal material on which to etch the minute electronic circuits that are built up layer by layer to create the entire central processing unit of a computer on a chip no bigger than a fingernail. Now new materials processes such as plasma etching, ion implantation, and molecular beam epitaxy are being utilized to further miniaturize chip circuitry, in order to create megabit or so-called "superchips" which will be faster, more complex, and more powerful than even today's chips.

What's more, entirely new materials like gallium arsenide—formed by combining gallium with arsenic, the favorite Victorian poison—are threatening to replace silicon as the main constituent of semiconductors. Gallium arsenide chips have achieved faster operating speeds than silicon chips, they consume less power, they are radiation-resistant, and they can operate at higher temperatures (making them an obvious choice for military applications). However, many claims made for gallium arsenide devices in the past have not been realized, and they are still very expensive and difficult to make. In addition new types of semiconductors such as CCDs (charge-coupled devices) or imager chips, "smart" chips, wafer-scale chips, superlattices, and ballistic transistors promise to grow in importance, as do so-called Josephson junctions which switch on and off 1,000 times faster than silicon transistors. The first commercial versions of these latter, long-promised, superconducting devices were unveiled in February 1987, and they could help pave the way to superconducting chips proper.

In the related sphere of photonics—in which light pulses (representing information) are generated by lasers and transmitted by fiber optic cables—advances in materials technology are transforming telecommunications and promising to make optical computing a reality. Optical fibers are tiny strands of pure glass no wider than a human hair that can carry thousands of telephone conversations or other digitized data in the form of extremely fast streams of light pulses. Many such strands are bundled together into a cable, which is typically one-fifth of the size of conventional copper cable. Optical fibers have a much higher capacity than copper cable, they are faster, easier to handle, and they are getting cheaper. It's therefore hardly surprising that optical fiber is quickly becoming the preferred medium for transmitting voice and data, especially over long distances, and nations are rushing to install fiber optic cables between major cities and under the oceans.

Other exciting new materials include Space Age metal alloys that are lighter, stronger, and easier to use than conventional metals like aluminium. The new alloys are made by powder metallurgy—the forming of parts by pressing and heating metallic powders in molds. This technique reduces the amount of machining involved in manufacture. Rapid solidification or quenching also rearranges the atoms during cooling to create alloys that are much stronger than conventional metals, and they also have a much lower resistance to electricity. For example, *SunRaycer,* the solar-powered car developed by General Motors which easily won the 1987 solar car race across Australia, has a so-called Magnaquench electric motor made in this way. Being constructed of stronger material means that it is lighter, smaller, and therefore more efficient than other electric motors. In the aerospace industry there is great interest in powder metallurgy and rapid solification. Companies like Alcoa are engaged in major research efforts to find new alloys of aluminium of use to planemakers. Other techniques for handling the new alloys include superplastic forming (allowing parts to be formed to their final shapes), direct sheet casting, and squeeze casting. Scientists have also come up with "shape memory alloys"—metals that can be bent and twisted but revert to their original shape when heated or cooled to a predetermined temperature. Applications include springs, valves . . . and the underwire female bra, which tends to get mangled in the washing machine. At room temperature this superbra made by a Japanese company offers maximum support because it always returns to its original bodily form.

Cement may have been a "new" technology to the Greeks and the Romans, but few people realize that humble sidewalk cement is undergoing a revolution today. A new generation of high-tech cements—blended with metals, plastics, and ceramics and set by heating or chemical processes—is raising the possibility of concrete bottle tops, springs, and brake shoes, and even cement loudspeakers and cement armor-plating for tanks. So-called "macro-defect-free (MDF) cement, developed in England, is unusually strong, stiff, soundproof, and resistant to chemicals and can be machined like metal. At the same time scientists in the United States have been taking a closer look at the superior cements used by the Egyptians to build the pyramids—prompting one wag to comment that we can now make cement just like the Mummies used to make.

Finally, mention should be made of the new high-performance toughened adhesives or "superglues" developed in recent years. These new glues can be used for joining plastics, metals, and ceramics. In the auto industry, it is suggested that superglues could replace spot-welding as the traditional form of bonding materials because adhesive-bonded joints are allegedly stronger, stiffer, and longer-lasting. Just recently in the United States doctors started using a type of superglue to stick detached retinas in the human eye back into place, saving the sight of patients when other approaches had failed.

Economic and Social Implications

"Science-Push" and "Market-Pull"

Advanced materials will have a major impact on manufacturing industries, services, and indeed the economy and society as a whole. But it's often not so much a case of science-push *impacting* on the rest of the economy as a case of market-pull in key sectors forcing the pace of development in materials science itself. Thus, to take an obvious example, it is the desire on the part of computer and telecommunications manufacturers to increase the functional power of information and communications systems (and the desire of customers to buy them) that has stimulated the remarkable series of innovations in materials such as semiconductors, ceramics, and plastics and in materials processing techniques such as lithography, ion implantation, and thin-film coating. To take another example, the desire of ship-owning nations like Norway, Japan, and West Germany to hold onto their merchant fleets and to stay at the forefront of maritime technology via their "Ship of the Future" projects has created a demand for corrosion-resistant and maintenance-free deck equipment made of ceramics, plastics, and metal alloys that can withstand the salty rigors of life at sea.

In the auto industry the driving forces behind innovation in materials have been more complex: in the 1950s rising demand for motor vehicles put pressure on the supply of "raw" materials, but it was increased foreign (mainly Japanese) competition in the early 1970s that provided the trigger for change in the United States. The oil price shock of 1973 was swiftly followed by new federal regulations governing fuel economy, and soon after came new regulations on safety and exhaust emission. U.S. manufacturers first began a frantic attempt to boost fuel efficiency by reducing vehicle weight—the "downsizing" trend. But this ongoing effort led in turn to a total reexamination of the process by which cars are designed, developed, and manufactured—not just to reduce costs but to improve quality and reliability.

Much of this reexamination involved new materials. Indeed, planning the production of a car these days involves complex computations of the relative merits of particular materials and production processes. This means trade-offs between cost, weight, ease of production, and even production-run planned (tooling) costs and component-integration (reducing the number of components). Changing consumer preferences in durability (reduced corrosion), performance (improved aerodynamics), and style have created a further wave of materials innovation in the auto industry. In fact, the combined impact of all these forces has been to dramatically change the materials content of cars to the extent that many late 1980s cars bear little resemblance to late 1950s models.

In the field of aerospace a similar combination of forces has been at work. The desire to reduce the weight of aircraft in order to boost fuel economy and to reduce costs has been the main consideration. Aircraft manufacturers have also been keen to cut fabrication costs in order to remain competitive. The reduction of engine noise in order to conform to new government regulations has become a top priority as environmentalists and community groups have succeeded in getting noise restrictions imposed on airports in urban areas. Jet engines are now quieter than ever, yet the thrust delivered by your average aero-engine has increased sixfold over the past thirty years. In the wider aerospace business, governments have set successive objectives for the performance of military aircraft, the construction of satellites and missiles has required the development of many new materials, and materials like Teflon (as in frying pans) were specifically invented to assist in the Space Race.

In 1986 the Reagan administration in the United States set goals for subsonic, supersonic, and hypersonic flight. In the subsonic area a new generation of fuel-efficient and quiet turboprop or prop-fan aircraft is envisaged, supersonic flight will see big improvements in long-distance efficiency, and hypersonic vehicles like the planned aerospace plane are expected to fly from New York to Tokyo in two hours. In order to meet these goals, further advances in materials technology will be necessary. In particular, advanced alloys of aluminium and titanium, alloy steels, and advanced composites will have to be developed to achieve further savings of fuel and weight and to cope with the very high and very low temperatures experienced in transatmospheric flight. The aerospace plane will require quantum leaps in propulsion, aerodynamics, and avionics (electronic control systems) and the experiences with Space Shuttle successes (and disasters) should be helpful here.

Energy and materials are, in a sense, the lifeblood of advanced industrial societies, and the energy industry has come a long way since man relied upon muscles, wood, water, and wind. Every aspect of energy production and distribution involves the use of appropriate materials—such as materials that conduct electricity and materials that resist heat or corrosion. In fact, materials are becoming more important in energy production because energy is being generated in more complex ways such as solar power, nuclear fission, and coal liquefaction. A basic ingredient of economic growth is cheap energy, and the development of new materials and improvements in the use of existing ones have a direct impact on the cost of energy. New materials can reduce transmission losses, and superconductors might eliminate them altogether. New materials can reduce friction losses, and materials science is behind recent advances in solar photovoltaic cells. New materials can improve the reliability and safety of energy production. The never-ending demand for cheaper, cleaner, more reliable energy will doubtless bring about further innovation in materials.

Advances in medicine and health care are very significant from the materials point of view. The desire to prolong life and relieve suffering is creating a revolution in so-called "biomaterials"—artificial tissues and organs made of plastics, ceramics, glasses, and composites that can augment or totally replace bodily tissues and organs. Polyurethanes, silicones, rubbers, carbons, and titanium are being used in the cardiovascular system as man-made heart valves, artificial arteries, synthetic blood, and entire artificial hearts. Bone implants and replacement hip joints are now widely available, thanks to new ceramics, polymers, and alloys, and bones can be joined more easily with the aid of bioglasses and calcium phosphate ceramics. Just recently, scientists in West Germany implanted artificial bone, made from the tiny skeletons of algae, into the jaws of laboratory animals. The muscular system is also benefiting from tendons and ligaments made of polymers and carbon fibers, artificial lens implantation promises to save the sight of many patients, and burn victims are benefiting from current attempts to develop synthetic substitutes for human skin.

A Global Battle

Recent advances in materials have fired up a global race for world supremacy in advanced materials, the leading contenders being the United States and Japan, with Europe and the U.S.S.R. some way behind. From the time of the National Academy of Sciences' massive study of the field of materials science and engineering (the COMSAT report) in the early 1970s, successive reports to the U.S. government have highlighted the need for the United States to stay ahead in materials research. Writing in *Science* in 1984, Richard Reynolds, new director of the Defence Sciences Office in the Department of Defense, warned: ". . . the United States is losing its international competitive edge with respect to the technology base necessary to support materials self-sufficiency in the manufacture of . . . strategic materials." In 1985 the National Research Council reported on the growing Japanese R&D effort in ceramics, and in 1986 an Office of Technology Assessment (OTA) study warned that the United States was in danger of falling behind the Japanese in developing commercial (as opposed to military) applications of new materials. As if in response, the Midwest Technology Development Institute, set up by the governors of nine midwestern states, formed the Advanced Ceramics and Composites Partnership to help boost the advanced materials content of locally made products. As well as recent decisions to pour money into superconductor research, the

U.S. government has been quietly spending large sums of money—chiefly through the Department of Defense and NASA—on research into advanced materials for military and space applications. NASA also has a long-standing interest in the processing of materials in space (where zero gravity makes for better molecular structures), and the Strategic Defense Initiative (SDI)—or "Star Wars"—project has provided a further boost in funding.

The Japanese government long ago recognized the vital importance of new materials technology. Back in 1981 MITI launched a program of research in six areas: advanced ceramics; high-performance plastics; composites, especially carbon fiber composites; electrically conductive polymers; advanced alloys; and synthetic membranes. As Gene Gregory explains in his contribution to this volume, the goal of this project is to produce materials that are light but strong, functional at high temperatures, of superior electrical and magnetic properties, made from readily available resources, and energy efficient to make and use.

There is moreover a greater awareness among the scientific community in Japan of the way in which most scientific advances depend upon prior advances in materials. Every five years, for instance, Japan's Science and Technology Agency surveys around 2,000 Japanese scientists for their predictions of when major scientific advances might be expected—particular favorites being knowledge about and cures for cancer, ability to forecast earthquakes, economically viable deep-sea mining, clarification of the ageing process, replacement of human organs, abundant nuclear and solar energy, and further advances in information technology. But as Hondros points out in chapter 3, 14 out of 22 categories of prediction in the 1982 survey involved breakthroughs that depended crucially on a materials precursor. Advanced materials, he says, are more widely seen as the enabling factor behind other scientific and technological advances. The 1987 survey demonstrates an even more remarkable Japanese obsession with materials: the commercialization of ceramic superconductors, the construction of superconducting maglev trains, and the widespread use of biomaterials figure prominently in the list. Having successfully targeted industry after industry—from transistor radios and motorbikes to microchips and supercomputers—for takeover in the postwar decades, the

Japanese seem determined to maintain their dominant position with the aid of new materials.

In Europe the European Commission (EC) in 1986 launched the European Research on Advanced Materials (EURAM) program to help generate a European production capacity in order to prevent European industry from falling behind Japan and the United States. Meanwhile, the U.K. government had received in early 1985 the report of its Materials Advisory Group, chaired by John Collyear ("the Collyear committee"), which recommended a $180 million program of research to keep Britain abreast of new developments. But aside from minor initiatives like a $3 million ceramic engine program and the formation of an Advanced Composites Group, the Collyear committee's report was largely ignored. With the notable exception of some leading-edge companies, poor old Europe in general looks like it's being left behind in the advanced materials race.

In addition to the simple desire to stay one jump ahead of the perceived competition, an important objective of nations in supporting materials R&D is to reduce their consumption of, and dependence on, "strategic" materials—which become "critical" materials if imported from hostile countries, temporarily inaccessible countries, or countries with unstable political regimes. As Joel P. Clark and Frank R. Field III show in chapter 15, the main materials in question are chromium, manganese, platinum, and cobalt, and they tend to come from places like South Africa, Turkey, the Philippines, Zaire, and Zimbabwe. Clark and Field argue that for various reasons (like large stockpiles), the United States has little to fear from disruptions in the supply of these critical materials, but nevertheless U.S. dependence can be further reduced by improving materials conservation, exploiting new mineral resources such as the low-grade ores and seabed materials, and especially by expanding government-sponsored R&D efforts to develop alternative new materials.

No Limits to Growth

In fact, there is growing evidence that demand for all basic materials—including those of strategic importance—is falling, or at least is not keeping pace with economic growth. Contrary to what was being predicted fifteen years ago by the Club of Rome and the "Limits to Growth" lobby, the world today has a large surplus of food and raw materials. According to the Organization for Economic Cooperation and Development (OECD), the prices of agricultural products and metals in late 1987 were at their lowest in real terms for more than thirty years. The OECD argued that technological change was the biggest single factor behind falling prices, because it had both expanded supply by improving productivity and reduced demand by making cheaper substitutes available—the development of ceramics, composites, and fiber optics was apparently unforeseen by the Malthusians of the early 1970s. As George F. Ray pointed out in a 1980 paper (see bibliography), history shows that predicted materials shortages rarely materialize because substitutes and new ways of processing materials are invariably developed.

In chapter 6 Eric D. Larson et al. go further and argue that the industrial nations are witnessing a fundamental and probably irreversible historic shift from the Materials Age characterized by low-tech products with a high materials content to an Information Age characterized by high-tech products with a low materials content. Identifying a "cycle" in the demand for materials, they say that four factors are responsible for the declining consumption of basic materials: the substitution of one material for another, which has slowed the growth of demand for particular materials; design changes in products that have increased the efficiency of materials use; the saturation of traditional markets that expanded rapidly during the era of materials; and the growth of new markets which tend to involve products that have a relatively low materials content. This "profound shift" in the pattern of demand for materials will inevitably mean a further shift in wealth and power away from traditional producer nations in favor of the advanced industrial nations.

Despite all the euphoria over superconductors and new materials—and presidential talk about "New Ages"—it is important to appreciate the time span involved in the materials revolution. Coal and iron took centuries to catch on and it took plastics seventy years to overtake steel in terms of production volume. George F. Ray in chapter 14 cautions that the diffusion of innovations in materials "often takes an unusually long time," and Thomas W. Eagar (chapter 13) gives a whole number of reasons why we shouldn't expect to see dramatic changes overnight. The transition of a new material from the laboratory to the production line can often take fifteen years. Getting it incorporated widely into everyday commercial products can take much longer. And in the case of superconductors, even pioneer Paul Chu has warned: "The superconductor payoff will be great. But it will take time."

Nevertheless, it is also time for the world to wake up to the amazing implications of superconductors and the materials revolution—both good and bad. The manufacture of new materials could present serious health hazards to workers and cause environmental pollution unless the production processes are adequately monitored. With any new substance there are bound to be new and unforeseen problems and side effects. And is anyone out there studying the management implications of new materials technology? The available evidence suggests that many industrial designers and managers are largely ignorant about the latest developments. This book may help to redress the balance by informing and alerting all concerned to the potential and dangers of the materials revolution.

References

Materials for Economic Growth. Special issue of *Scientific American* 255, 4 (October 1986).

New Structural Materials Technologies: Opportunities for the Use of Advanced Ceramics and Composites. Office of Technology Assessment, U.S. Congress, Washington, DC, 1986.

Peter A. Psaras and H. Dale Langford (eds.). *Advancing Materials Research.* National Academy Press, Washington, DC, 1987.

Michael D. Lemonick with Thomas McCarroll, J. Madeleine Nash, and Dennis Wyss. "Superconductors! The Startling Breakthrough That Could Change Our World." *Time,* May 11, 1987.

Gene Gregory. "New Materials Technology in Japan." *International Journal of Materials and Product Technology* 2, 1, (1987).

Tom Forester. *High-Tech Society.* MIT Press, Cambridge, MA, 1987.

Jeffrey Zygmont. "Plastic Cars Hit the Road." *High Technology* (December 1985).

John W. Dizard. "The Amazing Ceramic Engine Draws Closer." *Fortune,* July 25, 1983.

Gordon Graff. "High-Performance Plastics." *High Technology* (October 1986).

Nancy W. Stauffer. "New Fangled Cement." *Technology Review* (July 1986).

T. A. Heppenheimer. "Making Planes From Powder," *High Technology* (September 1986).

Eric D. Larson, Marc H. Ross, and Robert H. Williams. "Beyond the Era of Materials." *Scientific American* 254, 6 (June 1986).

The Superconductor Story

I

Superconductors! The Startling Breakthrough That Could Change Our World

Michael D. Lemonick with Thomas McCarroll, J. Madeleine Nash, and Dennis Wyss

1

Although the first of the recent breakthroughs in superconductivity occurred in December 1985, it was a full year before the media—including the science press—began writing about them (see bibliography). After the New York Hilton meeting in March 1987, superconductors became hot news, and just about everyone did the story. In my opinion, the best popular account appeared in Time *magazine on May 11, 1987, and here it is. Written by journalist Michael D. Lemonick, this article provides an excellent description of the search for superconductivity, right from the early days up to the present.*

They began lining up outside the New York Hilton's Sutton Ballroom at 5:30 in the afternoon; by the time the doors opened at 6:45, recalls Physicist Randy Simon, a member of TRW's Space and Technology Group, "it was a little bit frightening. There was a surge forward, and I was in front. I walked into the room, but it wasn't under my own power." Recalls Stanford Physicist William Little: "I've never seen anything like it. Physicists are a fairly quiet lot, so to see them elbowing and fighting each other to get into the room was truly remarkable."

Thus began a session of the American Physical Society's annual meeting that was so turbulent, so emotional, and so joyous that the prestigious journal *Science* felt compelled to describe it as a "happening." AT&T Bell Laboratories Physicist Michael Schluter went even further, calling it the "Woodstock of physics." Indeed, at times it resembled a rock concert more than a scientific conference. Three thousand physicists tried to jam themselves into less than half that number of seats set up in the ballroom; the rest either watched from outside on television monitors or, to the dismay of the local fire marshal, crowded the aisles. For nearly eight hours, until after 3 a.m., the assembled scientists listened intently to one five-minute presentation after another, often cheering the speakers enthusiastically. Many lingered until dawn, eagerly discussing what they had heard and seen.

What stirred all the excitement at that tumultuous meeting in March was a discovery that could change the world, a startling breakthrough in achieving an esoteric phenomenon long relegated to the backwaters of science: superconductivity. That discovery, most scientists believe, could lead to incredible savings in energy: trains that speed across the countryside at hundreds of miles per hour on a cushion of magnetism; practical electric cars; powerful, yet smaller computers and particle accelerators; safer reactors operating on nuclear fusion rather than fission and a host of other rewards still undreamed of.

Superconductivity is aptly named. It involves a remarkable transition that occurs in many metals when they are cooled to temperatures within several degrees of absolute zero, or, as scientists prefer to designate it, 0 Kelvin. Absolute zero, equivalent to −460° F or −273° C, represents a total absence of heat; it is the coldest temperature conceivable. As the metals approach this frigid limit, they suddenly lose all their electrical resistance and become superconductors. This enables them to carry currents without the loss of any energy and in some cases to generate immensely powerful magnetic fields. Scientists have recognized for years that the implications of this phenomenon could be enormous, but one stubborn obstacle has stood in their way: reaching and maintaining the temperatures necessary for superconductivity in these metals is difficult and in most instances prohibitively expensive.

Now, in a series of rapid-fire discoveries, researchers around the world have begun concocting a different class of materials that become superconductors at significantly higher temperatures—levels that, while still beyond the reach of a kitchen refrigerator, are easier and less costly to attain. These achievements have had an electrifying effect on a subject that just a year ago would have elicited yawns from physicists. Indeed, hardly a week has passed since the New York City meeting without reports from competing scientists of new superconducting materials and ever higher temperature ranges. An effect that once could be detected only with sophisticated equipment has become a common sideshow at conferences: a sample of one of the new materials is placed in a dish of liquid nitrogen, and a magnet placed above it. Since superconductors repel magnetic fields, a phenomenon called the Meissner effect, the magnet remains suspended in midair.

Fun and games aside, though, the competition is growing more intense. Researchers around the world are canceling vacations, ignoring their families, moving cots into their labs and subsisting on takeout food and microwave popcorn. "We've been working since right after Christmas," says Physicist J. T. Chen of Wayne State University in Detroit. "We do experiments almost every day. Sometimes we sleep only three or four hours. Maybe it was like this when the transistor was invented, but in my personal experience this is unique." Says Japanese Chemist Kohji Kishio: "The race is for the Nobel Prize."

The world's leading industrial nations are in a race of another kind. Quick to recognize the commercial potential of the new development, Japan's Ministry of International Trade and Industry plans to subsidize private-sector research, and will establish a center in Nagoya to test equipment made from superconducting materials. In Washington, the Department of Energy has decided to double this year's research support for superconductors to $40 million; it is also compiling a computerized data base that will enable American scientists to keep up to date on fast-breaking superconductor research results, and will co-sponsor a White House conference on superconductivity this summer. "It's a monumental subject," says Energy Secretary John Herrington. "It ranks up there with the laser." In the Senate, Minnesota Republican David Durenberger has co-sponsored a bill calling on the President to form a national commission to coordinate superconductivity research and development. Says Durenberger: "We cannot stand idly by while Japan targets another industry for industrial supremacy." Last week the National Science Foundation announced $1.6 million in grants to help keep the U.S. competitive in superconductivity research.

The superlatives continue to roll in. "In terms of the societal impact, this could well be the breakthrough of the 1980s in the sense that the transistor was the breakthrough of the 1950s," says Alan Schriesheim, director of Argonne National Laboratory near Chicago. Indeed, scientists hardly know where to start in describing the bonanza that superconductors could yield.

Take the transmission of electricity, for example. As much as 20 percent of the energy sent through high-tension lines is now lost in the form of heat generated as the current encounters resistance in the copper wire. If the electricity could be sent through superconducting cable, however, not a kilowatt-second of energy would be lost, thus saving the utilities (and presumably consumers) billions of dollars. Furthermore, at least in theory, all of a large city's electrical energy needs could be supplied through a handful of underground cables.

The elimination of heat caused by electrical resistance could have a profound effect on the design and performance of computers. In their never-ending quest to produce smaller and faster computers, designers try to cram more and more circuits into chips and ever more chips into a tiny space. But they are limited in their scaling-down endeavors by heat; even the tiny currents in computer circuits generate enough cumulative heat to damage components if they are too tightly packed. Today's personal computers could not operate without vents or internal fans to dissipate the heat. Now, with practical superconducting circuitry on the horizon, computer designers may soon see the way clear for even more remarkable miniaturization. . . .

In still other applications, the intense magnetic fields that might someday be generated by the new superconductors should benefit any device that now uses electromagnetism in its operation—medical diagnostic imaging machines, magnetically levitated trains, fusion-energy generators—and will undoubtedly spawn a host of new machines. Electric motors could increase in power and shrink in size.

But these are just the most obvious examples. Scientists like Robert Schrieffer, who shared the 1972 Nobel Prize in Physics for the first successful theory of how superconductivity works, believe its most dramatic applications have yet to be conceived. "When transistors were first invented, we knew they'd replace tubes,"

Schrieffer says. "But no one had any idea there would someday be large-scale integrated circuits." Robert Cava of Bell Labs agrees. "We don't know where this will lead," he says. "It's exciting—and I guess frightening at the same time."

From the time that Dutch Physicist Heike Kamerlingh Onnes discovered superconductivity in 1911 until the recent rash of breakthroughs, there was only one way to produce the phenomenon: by bathing the appropriate metals—and later, certain metallic alloys—in liquid helium. This exotic substance is produced by lowering the temperature of rare and costly helium gas to 4.2 K (−452°F), at which point it liquefies. But the process is expensive and requires considerable energy. Furthermore, unless the liquid helium is tightly sealed in a heavily insulated container, it quickly warms and vaporizes away. Thus the practical use of superconductors has been limited to a few devices—an experimental Japanese magnetically levitated train, a few giant particle accelerators, and medicine's magnetic-resonance imaging machines—that operate with intense magnetic fields.

But in the past year and a half physicists have stumbled on an unusual class of ceramic compounds that change everything. They too must be cooled to become superconductors, but only to a temperature of 98 K (−283°F). And that suddenly brings superconductivity into the range of the practical; liquid helium can be replaced as a coolant by liquid nitrogen, which makes the transition from a gas at the easily produced temperature of 77 K (−320°F). Moreover liquid nitrogen is cheaper by the quart than milk and so long-lasting that scientists carry it around in ordinary thermos bottles. Also the ceramics may be able to generate even more intense magnetic fields than metallic superconductors. Thus, if these new substances can be turned into practical devices—and most scientists believe they can—technology will be transformed. Declares Arno Penzias, vice-president for research at Bell Labs: "The recent advances in the field of superconductivity are almost without comparison."

Success and celebrity have been a long time in coming to the field of superconductivity. "Until recently," says John Ketterson, a physicist at Northwestern University, "people were glum. There hadn't been a breakthrough in a long time. Funding was drying up. This has sent everyone back into the field with a new burst of enthusiasm." Although Kamerlingh Onnes envisioned early on that his discovery might pave the way for extremely powerful, compact electromagnets, he and other experimenters were stymied by a strange phenomenon: as soon as enough current was flowing through the then known superconductors (lead, tin, and mercury, among others) to generate significant magnetic fields, the metal lost their superconductivity.

It was not until the 1950s that scientists discovered alloys, such as niobium tin and niobium titanium, that keep their superconductivity in the presence of intensely strong magnetic fields. And it was not until the 1960s and 1970s that the manufacture of large superconducting magnets became standardized. But progress toward the other goal of superconductivity researchers, pushing the phenomenon into a practical temperature range, was even slower. By 1973 some 62 years after Kamerlingh Onnes had found superconductivity in mercury at 4.2 K, scientists had upped the temperature to only 23 K, using an alloy of niobium and germanium. After 1973: no improvement.

That was the situation in 1983 when Karl Alex Müller, a physicist at the IBM Zurich Research Laboratory in Switzerland, decided to pursue an approach to superconductivity that had met with limited success in the past. Instead of using the kind of metallic alloys that held the existing record, he turned his attention to the metallic oxides (compounds of metals and oxygen) known as ceramics. Some theorists had suggested ceramics as potential superconductors even though they were poor conductors at room temperatures. In fact, ceramics are often used as insulators—for example, on high-voltage electric-transmission lines.

Müller and his colleague, Johannes Georg Bednorz, tinkered with hundreds of different oxide compounds over the next few years, varying quantities and ingredients like alchemists in search of the philosopher's stone. Finally, in December 1985 they came across a compound of barium, lanthanum, copper, and oxygen that seemed promising. When Bednorz tested the compound, he was startled to see signs of superconductivity at an unprecedented 35 K, by far the highest temperature at which anyone had observed the phenomenon. Could this result be correct? Aware of some hastily made superconductivity claims that later could not be reproduced, the IBM team proceeded cautiously, painstakingly repeating their experiments. In April 1986 Müller and Bednorz finally submitted the findings to the German journal *Zeitschrift für Physik,* which published it five months later.

As Müller had anticipated, other physicists were skeptical. For one thing, the IBM scientists had lacked the sensitive equipment to test for the Meissner effect, the surest proof of superconductivity, and thus could not confirm it in their report. More important, in a field where improvements of a few degrees were reason for celebration, this great a temperature leap seemed unlikely. Douglas Finnemore, a physicist at Iowa State University, admits that he was among the doubters. "Our group read the paper," he says. "We held a meeting and decided there was nothing to it."

Not everyone was so quick to dismiss the discovery. Scientists from the University of Tokyo took a look at the substance. Says Müller: "The Japanese weren't smiling, and they confirmed it. Then the United States sat up." By the end of the year, confirmation had come from China and the United States, and suddenly a nearly moribund branch of physics was the hottest thing around. Large industrial and government laboratories jumped in; so did major universities. At Bell Labs a team led by Bertram Batlogg and Ceramist Cava had launched their own program of alchemical tinkering. Soon they had manufactured a similar compound that became a superconductor at 38 K, one-upping their archrivals at IBM. "That's when the hysteria started." says Cava. "The place was abuzz with excitement."

But Bell Labs too was soon to be upstaged. For among those who had given early credence to the news from Zurich was a small, modestly equipped team of researchers headed by Paul C. W. Chu of the University of Houston. Chu had been studying superconductivity since 1965; now he and his group, including scientists from the University of Alabama, quickly reproduced the IBM results and moved on to their own experiments.

Since the Houston lab had special equipment for testing materials at high pressure, Chu wondered what would happen if he pressurized the IBM compound. "Using known theories," he says, "you don't expect the transition temperature to go up rapidly under pressure, but it shot up like a rocket. It suggested to us that there might be some new mechanism involved." That unexpected result, says

Chu, played right into what he considers his group's strong suit: "We feel we have an advantage over some other groups because we are not confined to conventional thinking. We think wildly." Chu found that the compound remained a superconductor up to 52 K (−366°F) when subjected to from 10,000 to 12,000 times normal atmospheric pressure.

Forcing the pressure higher than that had no effect; it was time for more wild thinking. Chu reasoned that the high pressure worked because it squashed the compound's molecular structure and that this somehow boosted its superconducting temperature. Since more pressure did no good. Chu decided to compress the molecules in a different way—from within. He replaced the barium with strontium, which is similar chemically but has a smaller atomic structure. Sure enough, the temperature rose again, to 54 K, then stopped. So he turned to calcium, an element with even smaller atoms. This time the temperature dropped. It appeared to be a dead end.

Now Chu's team tried lanthanum, the rare-earth[1] component of the IBM compound. Maw-Kuen Wu, head of the team's Alabama unit and a former graduate student of Chu's, replaced the lanthanum with another rare-earth element, yttrium.

The new substance showed so much promise that Chu filed a patent application on January 12. That promise was soon fulfilled. At the end of the month, after subjecting their creation to a series of heat and chemical treatments, Wu and his assistants began chilling a bit of the compound, by dousing it with liquid nitrogen, and sending an electric current through it. To their amazement, the sample's resistance began to drop sharply at a towering 93 K. Recalls Wu: "We were so excited and so nervous that our hands were shaking. At first we were suspicious that it was an error." But a few days later he and Chu duplicated the feat in Houston and even bettered it by 5°.

The accomplishment of Chu and his team did nothing to dampen their competitors' enthusiasm. Indeed, the effect was just the reverse. In order to protect his patent, Chu refused to disclose the exact composition of his new material before the formal report was published in the March 2 *Physical Review Letters,* but other scientists thought they could easily guess its makeup and went to work.

At the University of Illinois, Physicist Donald Ginsberg raced out to buy an air mattress and an alarm clock, anticipating a spate of all-nighters. At IBM's Almaden Research Center in San Jose, scientists successfully duplicated the compound, analyzed its crystal structure, and passed the information on to the company's labs in Yorktown Heights, NY, where their colleagues were able to make thin films of the substance literally overnight. At the University of California, Berkeley, a group that included Theoretical Physicist Marvin Cohen, who had been among those predicting superconductivity in the oxides two decades ago, reproduced the 98 K record, then started trying to beat it. "I'm a standard American scientist," says Cohen. "My definition of research is to discover the secrets of nature—before anyone else."

In short, says Douglas Scalapino, of the University of California at Santa Barbara, recent developments are something like the breaking of the four-minute mile. Beforehand, it had been considered nearly impossible; afterward, "you could go to any track meet and some guy was breaking it." The activity, says Cava, "is more exciting than a supernova. Astrophysicists can watch it, but when it happens, it happens and it's gone. In superconductivity, the events are still going on, and the physics is just beginning to pour in."

So are the scientific papers. Says Metallurgist Frank Fradin, director of Argonne's materials science division, who is also an associate editor of *Physical Review Letters:* "As of three weeks ago, we had 98 papers submitted on the subject, and only a small fraction of them will ever get published. Progress is so rapid that a result of two to three weeks ago is already out of date. We've had to institute a whole new system to speed up the publication process." One important discovery: at least a dozen different compounds, all subtly different from the one Chu found, appear to act as high-temperature superconductors.

Although scientists know the chemical composition of the new class of superconductors, they are less certain about how they work, True, a theory exists that explains low-temperature superconductivity. It is known as BCS, from the initials of Author John Bardeen and his colleagues Leon Cooper and Robert Schrieffer, who shared the 1972 Nobel Prize for Physics for their effort. But BCS may not apply to the strange goings-on at higher temperatures.

Ordinary conductivity, the measure of a material's ability to transmit electrical current, is determined by events that take place at the atomic level. Atoms consist of a tiny dense nucleus that contains positively charged protons and chargeless neutrons. Around the nucleus whirl the negatively charged electrons, residing in shells with shapes determined by the electrons' energy levels.

In many atoms, particularly those of metallic conductors, the outer shell has a number of empty slots, and the electrons that it does contain are not bound as tightly to it as those in the inner shells. Just as the sun's gravitational pull is weaker on distant Pluto than on nearby Mercury, the hold of an atomic nucleus is also weaker on electrons in the outermost layers.

So when an electric current—which is simply a stream of moving electrons—flows in a conductor, electrons move from empty slot to empty slot in the outer shells of the atoms. A material like rubber, on the other hand, is an insulator: it consists largely of atoms with completely filled, stable outer shells. Thus when voltage is applied, electrons have no empty slots to move into, and no current flows.

But even the best of ordinary conductors have some resistance to the flow of electrical current. The reason: as current passes through, some of the electrons collide with other electrons, thus dissipating their energy in the form of heat. According to the BCS theory, these collisions are avoided in superconductivity. "What causes a material to become superconducting is a phase change,"[2] explains Bardeen, now a professor emeritus at the University of Illinois. "You can think of it as electrons condensing into a new state." That state involves the pairing of electrons and a kind of group discipline.

Iowa's Finnemore compares the movement of the electrons in a superconductor to a crowd moving across a football field. "If they act as individual particles," he explains, "they will bump into each other and scatter. That's the equivalent of electrical resistance. But suppose someone starts counting cadence, and everyone locks arms and marches in step. Then, even if one person falls into a chuckhole, he won't fall because his neighbors hold him up." Thus in a superconductor electrons move unhindered.

Although the BCS theory works well near absolute zero, some physicists think it will have to be modified or even scrapped as an explanation for the behavior of higher-temperature superconductors. According to Bardeen, his theory can explain superconductivity up to around 40 K. But at 90 K, he says, "I think it's highly unlikely. We no doubt are going to need a new mechanism." In fact, says Schrieffer, "superconductivity may turn out to have as many causes as the common cold."

Confusion at the level of theory has put no damper on the orgy of speculation about potential applications. Some ideas involve upgrading existing superconducting technology; others push marginal technology into the realm of the profitable; still others raise the prospect of entirely new uses of the phenomenon.

Giant particle accelerators are one target for possible upgrading. Currently the most powerful such devices use conventional superconducting electromagnets. If high-temperature superconducting magnets can be developed, millions of dollars could annually be saved in electrical and liquid-helium bills.

Electromagnets are also crucial to fusion energy, which depends on fusing atoms (the same process that powers the sun), rather than splitting them. Key to one promising fusion process, which is under development in several countries, is a "bottle" composed not of any material substance but of powerful magnetic fields, generated at great expense by conventional electromagnets. Such fields are the only envelopes that can contain and squeeze atoms together at the hundred-million-degree temperature required to initiate fusion. But superconducting magnets, especially warm-temperature ones, could produce more intense fields at less expense and thus could "help make fusion power possible and practical," says Harold Furth, director of Princeton University's Plasma Physics Laboratory.

In medicine, superconducting magnets are at the heart of magnetic resonance-imaging machines. The magnets' powerful fields first align the atoms of the body. Then a pulse of radio waves knocks them momentarily out of alignment. When the atoms return to their previous attitudes, they emit radiation that produces detailed images of the body's soft tissues. MRI machines in use today are enormous (6 ft by 8 ft by 10 ft), largely because of the more than $100,000 worth of bulky insulation required to preserve the liquid helium coolant, which costs an additional $30,000 annually. The improved economics of the new superconductors, says Walter Robb, of General Electric's Research and Development Center, should eventually enable medical institutions to install many more MRI machines, which are invaluable for diagnosing disorders like brain tumors.

High-temperature superconducting magnets may become important in the maglev, or magnetically levitated, trains under development in Japan and West Germany. And scientists at Japan's Mercantile Marine University in Kobe have already developed a working scale model of a ship with a propulsion system based on magnetism. Physicist Yoshiro Saji sends current through the seawater from an on-board electric generator via ship-bottom electrodes. A superconducting magnet, also on board, creates a strong magnetic field. As the electromagnetic field produced by the electric current pushes against the field of the magnet, the ship moves forward. Saji has already moved up his timetable and hopes to complete a 100-ton "mag-ship" within four years. "Thanks to the new materials," he says, "magnets will be lighter and easier to handle. Once we can replace liquid helium with liquid nitrogen, the whole process of outfitting the ship will be simplified. It's a fantastic development."

On a smaller scale, superconductors have already been used to create superfast electronic switches called Josephson junctions (after Nobel Laureate Brian Josephson, the British physicist who discovered the principle on which they are based), which until now could operate only at liquid-helium temperatures. For both technical and economic reasons, IBM abandoned its Josephson junction project in 1983. But IBM Physicist Sadeg Faris quit the company, obtained licenses for the technology and formed Hypres, Inc., which has begun marketing its first Josephson junction product—a high-speed oscilloscope. Says Faris: "The new materials are at a primitive stage, but we're anxious to exploit them to bring down costs and improve speed." Since switches are a limiting factor in computer speed, an economical Josephson junction could prove invaluable.

At Westinghouse, scientists are working on the idea of using superconductors for electric-power production. Today's nonsuperconducting generators produce electricity by spinning wire-wrapped rotors in a magnetic field; their output is typically some 300 megawatts a generator. If the field were generated even by conventional superconducting electromagnets, says Research Director

John Hulm, the output could be doubled. The benefits would be even greater with high-temperature superconductors.

And then there are the daydreams: giant underground loops of superconducting cable that can store vast amounts of electricity for later use; cars that run on tiny, powerful electric motors, drawing current from superconducting storage devices. But even the daydreams are taken at least somewhat seriously. At Ford, for example, a study group has been assembled to rethink the feasibility of the electric car in light of the recent advances in superconductivity. Says IBM Physicist John Baglin: "The question is not 'How can we take this material and do something everyone has wanted to do?' but 'How can we do something that no one has yet imagined?'" Some tongue-in-cheek suggestions overheard at a superconductor meeting: superconducting ballroom floors and rinks that would enable dancers and skaters literally to float through their motions.

All the applications, though, depend on bringing the technology out of the lab, and despite the bubbly confidence of many scientists, obstacles remain. One is the need to form the new materials into usable shapes. Although metals bend, anyone who has dropped a dinner plate knows that ceramics do not. And a flexible material has a big advantage over a brittle one if it is to be coiled around an electromagnet. Says Osamu Horigami, chief researcher at Toshiba's Energy Science and Technology Laboratory: "To get a magnet or coil or even a wire we could use with complete confidence could take another five years." Agrees Hulm: "It will take extraordinary engineering to solve the brittleness problem."

IBM scientists may already have a partial answer: they announced last week that the new compounds can be "spray-painted" onto complex forms, where they solidify. Says IBM Scientist Jerome Cuomo, who described the technique at the American Ceramic Society conference in Pittsburgh: "This opens the door wider than ever to the fabrication of useful objects made of superconducting materials."

More fundamental is the fact that though the new ceramics remain superconductors at high temperatures and can withstand intense magnetic fields, they can as yet carry only about a hundredth of the current capacity of conventional superconductors. And because the amount of current flowing through the magnetic field de-

Trains That Can Levitate

One item is on everyone's list of potential benefits of high-temperature superconductors: maglevs, or magnetically levitated superfast trains. It is a safe prediction, since the new materials give promise of electromagnets far more powerful and economical than those in use today. And it is the electromagnet that lifts and propels existing maglevs in Japan, West Germany, and Britain.

As long ago as 1979 an unmanned Japan Railways Group prototype fitted with low-temperature superconducting electromagnets hit 321 mph on a test track; a version carrying three passengers made it to 249 mph earlier this year. That beats any conventional rival, including Japan's celebrated bullet train, which goes as fast as 149 mph, and the French TGV, which provides the world's fastest regularly scheduled rail service, at speeds of up to 186 mph.

Japan's maglev is faster because instead of pounding along a set of rails, it floats 4 inches above a guideway on a cushion of magnetic force; there is no friction to slow it down, no fear of derailment on a section of bent track. This maglev has wheels, but the only times it uses them are while picking up speed before lift-off and while slowing down after landing.

The principle behind the maglev is simple: opposite magnetic poles attract each other; like poles repel. In Japan's version, eight superconducting electromagnets are built into the sides of each train car, and thousands of metal coils are set into the floor of the guideway. When the train is in motion, the electromagnets on the train induce electric currents in the guideway coils, which then themselves become electromagnets. As power is increased, the opposing sets of magnets repel each other and lift the train into the air. Two other rows of electromagnets, one on each wall of the U-shaped guideway, repeatedly reverse polarity to push or pull on the coach's magnets and thus move the train forward.

In planning the train, Japanese engineers chose superconducting magnets because for a given input of electricity they generate more intense magnetic fields—and thus greater lifting and propulsion power—than conventional electromagnets. The drawback: the liquid-helium coolant needed for the superconducting magnets is expensive, and a heavy compressor is required in each coach to reliquefy the evaporating helium. That is why maglev engineers are excited by the idea of the new high-temperature superconductors, which would use considerably less expensive liquid nitrogen as a coolant and require far smaller compressors. The developments of the past few months, says Research Chief Kazuo Sawada, who has been in on the project from the beginning, are a "promising sign."

In West Germany, on the other hand, the new superconductors are of little interest to maglev engineers, who abandoned superconducting magnets in 1979. They opted to use conventional electromagnets instead. The German system is based on magnetic attraction, not repulsion. The magnets are on assemblies attached to the cars' undercarriages that curve around and under the crossbar of a T-shaped track. When the magnets are energized, they pull themselves up toward the crossbar's metallic underside and the car is lifted into the air; magnets in the track provide propulsion. Which technique is better? Both have advantages. The German maglev is simpler and less expensive to operate. But so far the Japanese trains are about 100 mph faster.

termines its strength, scientists are concerned that a quick fix may not be in sight. Warns GE's Robb: "What we need now is a second invention that would modify copper oxides to allow high currents to flow at high temperatures. There's a fifty-fifty chance that second invention will ever be made."

Finally, there is a human problem that could hinder progress in the suddenly vigorous field of superconductivity: the increasing unwillingness of scientists to exchange information about their experiments. At the Woodstock of physics meeting, for example, some were miffed when Stanford researchers, following their presentation, refused to divulge further details of their research; they had been advised by patent attorneys to reveal as little as possible until their work was legally protected. The competition extends beyond legal rights. Two weeks after Chu's record-breaking temperature was announced, the Berkeley team independently came up with the same superconducting compound. They immediately mailed a report of their results to *Physics Letters,* hoping it would be received before Chu's paper was published. Reason: they wanted to establish that they had not merely copied his work.

Still, there are hints that some of the physical barriers, at least, are starting to fall. At the March meeting, scientists were already showing rings and flexible tapes made of high-temperature superconductors; by the end of the month, teams at IBM, Bell Labs, Toshiba, Argonne and a handful of other places were developing wire-thin ceramic rods. Says Toshiba's Horigami: "We weren't even sure this was possible. When we finally had a wire that could potentially be coiled, there was absolutely no way to measure our sense of triumph." Argonne Ceramist Roger Poeppel now talks of building a furnace ten feet long to fire his group's wire almost continuously as it is extruded. "We think it will be flexible enough to twist into cable," he says, "and cable is the building block for magnetic coils and electrical transmission lines. With two miles of wire, we'll make a superconducting magnet. To get a practical device is now the race."

Later, in April, scientists at Stanford and IBM announced that they had made thin films of the new substances, important for computer applications. The spotlight then shifted to IBM Researchers Robert Laibowitz and Roger Koch, who reported that they had made their own thin film into a working gadget called a SQUID (for superconducting quantum interference device). Such tools are already used in low-temperature versions to measure extremely faint magnetic fields. They are also employed by physicists in the search for elusive gravity waves and magnetic monopoles, predicted by some theories but not yet observed. Medical researchers use SQUIDs to detect the minute fields generated by electrical activity within the brain. High-temperature SQUIDs should make all these searches a little easier.

Other scientists are seeking a better understanding of why the ceramics become superconductors. Many labs have taken pictures of the materials with electron microscopes, pulsed beams of neutrons, X rays, and ultrasound. A team of Bell Labs and Arizona State scientists has produced electron-microscope photographs that show defects in the compound's crystalline structure. Says Team Leader Abbas Ourmazd: "We don't quite understand what role the defects play, but it raises some provocative questions. Is it the perfect material that is superconducting? Or is it the defects? If it turns out that it is the defects, then we will want to control them and increase their density and put them in intentionally."

Most intriguing of all are reports that the temperature record set by Chu and since matched by dozens of other researchers has already been surpassed. Some physicists have even reported superconductivity-related effects—though not true superconductivity—at the torrid heights of 240 K, or −27°F, which is warmer than many wintry nights in North Dakota.

Those results suggest an intriguing possibility. Says Bell Labs' Penzias: "Transition temperatures have increased by a factor of four in the past year. If temperatures are raised by another factor of four in the same period of time, we'll have room-temperature superconductors in less than a year." Adds IBM's Praveen Chaudhari: "All the mental barriers are gone. No one is asking how high it will go anymore." If room-temperature superconductivity is achieved, whether in a year or in a score of years, its impact will be incalculable. The need for refrigerators and insulation, even for liquid nitrogen, will be gone. And the costs of this still futuristic technology could drop more dramatically than anyone expects. Says IBM's Paul Grant: "We're looking. Everyone is." Adds IBM's William Gallagher: "We shouldn't let our imaginations be constrained by things we now know about. We're just not able to imagine the things you can do."

Notes

1. The so-called rare earths, a group of seventeen chemical elements, are not rare at all; yttrium, for example, is thought to be more abundant than lead. These elements were mislabeled because they were first found in truly rare minerals.

2. The most familiar example of a phase change is the transformation, at 32°F, of water from a fluid into crystalline ice.

Are Superconductors Really That Super?

Two reports from *The Economist* and *The Boston Globe*

2

After the hyperbole of the Time *article, we balance the picture with two down-to-earth accounts that set out to debunk the superconductor revolution. While* The Economist *emphasizes that superconductors have been around for years and that high-temperature superconductivity is a "comparatively minor breakthrough," Alison Bass of* The Boston Globe *concentrates on the lack of a theory to explain how superconductivity works. Both highlight the problem of getting electrical current through superconductive material and question the widely predicted application of superconductors in electric power lines, magnetic levitation trains, and computers.*

Not-So-Superconductors

Near the absolute zero of temperature, $-273°C$, some materials lose all electrical resistance and thus waste no energy when conducting a current—that is, they superconduct. What makes the new type of superconductor special is that it keeps superconducting at up to 98° above absolute zero: four times higher than most scientists thought possible.

The material discovered by Paul Chu and his colleagues earlier this year is a sooty-black ceramic compound of the elements barium, yttrium, copper, and oxygen. At the end of May, Chu announced that another closely related (and still secret) ceramic compound showed signs of superconducting at 225° above absolute zero. New records are claimed almost every week. Researchers describe the new superconductors not only as a scientific marvel but also as the basis of a technological revolution.

They are right about it being a marvel. The sudden jump in temperature is a surprise because it seems to defy the standard theory of superconductivity, which is often described as one of the great successes of theoretical physics. Physicists are now reexamining an older, rival account in the hope of explaining high-temperature superconductivity.

The claim that it represents a technological revolution is more debatable. Superconductivity has held out great promise for quite a while. Yet few of its applications have ventured out of the labora-

How Electrons Mate

It took forty-five years, at a rate of about three failed theories a year, to find an apparently workable explanation for superconductivity. In 1957 John Bardeen, Leon Cooper, and John Schrieffer, all then at the University of Illinois, developed a theory known as BCS that seemed to account for all the quirks of superconductors. Now the BCS theory needs modifying too, and others are coming into vogue, including one that is almost as old but was until recently unfashionable.

To see why the BCS theory is not the whole story, look first at how it works. At the level of atoms, heat energy in a crystal simply consists of atoms jiggling around fixed positions within the atomic lattice of the crystal. Passing electrons collide with the jiggling atoms and lose energy. Such collisions are the main source of electrical resistance in normal conductors.

However, under certain conditions, electrons join together in pairs, called "Cooper pairs," and behave oddly. All electrons are negatively charged, so it might seem suprising that two like charges can attract. The explanation is that an electron distorts the crystal structure of the atoms around it and attracts a partner, in much the same way as two people sleeping in a soft bed will tend to roll into the middle because of the depressions they make in the mattress.

A Cooper pair of electrons is best thought of as a single entity, belonging to the category of elementary particles known as "bosons." Strangely, a lone electron belongs in another category, that of "fermions." Bosons and fermions behave very differently. Fermions are individualists—a single electron may be involved in a collision with atoms, quite independently of what other electons in the same crystal are doing. Bosons are groupies, and do it together or not at all. Jiggling atoms cannot deflect all the Cooper pairs in a crystal at once. Thus, once the pairs are set in motion by a voltage, the Cooper pairs travel along, oblivious to the atoms.

This works only so long as the atoms are not jiggling too violently—that is, so long as the crystal is cold. If the temperature is too high, the Cooper pairs are wrenched apart, and the electrons no longer behave as bosons, so they keep bashing into atoms.

In order to raise the critical temperature at which a material becomes superconducting, the force with which the Cooper pairs are held together must be increased. This means increasing the distortions that passing electrons produce in the lattice. There are at least two ways to do this:

- The chemist's approach is to weaken the chemical bonds of the crystal by changing its constituent atoms. The atoms are then more easily displaced by the electrons, and the critical temperature for superconductivity becomes higher. But there are limits to this technique. Eventually, the crystal is held together so weakly that it changes to a more stable structure that does not superconduct.

- The physicist's approach is to switch one chemical element in the crystal with another of the element's isotopes. Although two isotopes of an element are chemically identical, they have different masses. In theory, if the new isotope is lighter, the distortion of the lattice is larger, the Cooper pair is more strongly bound together, and so the critical temperature for superconductivity should be higher.

This April [1987] Bertram Batlogg and his colleagues at AT&T's Bell laboratories in New Jersey tested the isotope effect by replacing an oxygen isotope in one of the new superconductors with a more massive one. The BCS theory led them to expect a decrease in the critical temperature of their superconductor of over three degrees, but they found no change at all. This was the finishing stroke for BCS. It became clear that the theory needs some big changes.

For more than twenty years, William Little of Stanford University in California has championed a controversial change to superconductor theory that might fit the bill, and also help scientists to find room-temperature superconductors. His theory is inspired by biology.

In the 1960s Little suggested that superconductivity might be achieved in organic molecules, which are long and have "side-chains," or branches, sticking out of them. Instead of nudging, and thus polarizing, atoms as it passes through a crystal lattice, an electron traveling down the backbone of an organic molecule could polarize side-chains of the molecule by displacing electrons in them. The polarized chains would then serve to attract another electron, and a Cooper pair would be formed. Since electrons are far lighter than atoms, superconductivity in an organic molecule should be possible at much higher temperatures.

Researchers are now busy studying several organic compounds in the hope of proving Little right. Superconductivity has been found in some of them but, so far, only at extremely low temperatures. Does the new family of superconductors, based on barium-yttrium-copper-oxide, fit the theory? Although not an organic compound, it does suit Little's model. Copper-oxide side-chains stick out from each plane of atoms, making the new superconductors somewhat like organic molecules. And Little's theory neatly accounts for the puzzling result of Batloggs' experiment (because the mass of atoms, on Little's theory, does not affect the strength of the bond between a Cooper pair). Until others have had time to cook up their own theories, Little's twenty-year-old idea has the field.

tory. Fewer still have achieved commercial success. High-temperature superconductors make a big difference. They can use liquid nitrogen at 77° above absolute zero as a fluid coolant rather than the liquid helium at 4° absolute that other superconductors must rely on. Nitrogen is abundant and cheap to cool. Helium is scarce and expensive to keep cold.

First, consider what superconductors do best: small, lightweight superconductor coils can produce intense magnetic fields. Copper coil electromagnets can produce magnetic fields of 4 tesla and more (one tesla is about 200,000 times the earth's own magnetic field) but need huge amounts of energy to do so. The water needed to keep such a magnet from melting would amount to the flow of a large river. Compare this with a 4-tesla superconductor magnet. Without its heat insulation, the magnet is about the size of a coffee pot. When running, the magnet consumes next to no power, so only the cost of helium refrigeration needs to be reckoned with.

It is not hard to see, then, why the word "revolution" comes to researchers' lips. The catch is that the revolution occurred in 1960 when American researchers developed a compound of niobium and tin that was ideal for powerful superconducting magnets. That compound and its relatives have since been at the heart of nearly every practical superconductor project.

What happened to superconductors in the 1960s was more radical than what is happening now. It was not merely that a new technology displaced an old one. The superconductor magnets of the 1960s go far beyond the limits that the old technology had met: 10-tesla magnets are now built almost as a matter of routine. Like the laser and the integrated circuit, superconducting magnets made the apparently impossible suddenly seem trivial.

High-temperature superconductors are a comparatively minor breakthrough. Scientists are not claiming that the new material can do things that the older superconductors cannot. Only that it can do them much more cheaply. Even that is questionable.

High-temperature superconductors offer big savings for those applications in which cooling the magnet is the main expense. Medical scanners that use 3- or 4-tesla superconducting magnets are expensive. The initial cost of the helium, with its bulky insulation, is about $100,000; the running costs of the helium refrigeration are around $50,000 a year. Nitrogen-cooled magnets could slash the capital and running expenses to one-tenth or less of their current cost, and allow scanners to be made much smaller.

But few other applications give high-temperature superconductors such a clear advantage because cooling costs are not usually the limiting factor. Here are some sobering examples in which, contrary to some common claims, superconductors do not make a revolutionary difference.

For the past ten years Eric Forsyth and his colleagues at Brookhaven National Laboratory in New York have studied the feasibility of superconducting underground cables for the transmission of electricity. Superconductors offer two advantages over conventional underground cables. First they can carry 3 gigawatts of power—six times more than normal underground cables—and as much as the highest capacity overhead lines. Second, they can span distances of up to 100 miles, some five times longer than would otherwise be possible underground, and comparable with the longest distances overhead lines can reach.

But, contemporary myth not withstanding, superconductor cables do not transfer power without losing a single watt. Although a superconductor offers no resistance to a direct current, it loses part of the energy it transfers when the current is alternating—and alternating current (AC) is what the utilities provide.

Put simply, AC power losses happen because an alternating current generates radio waves that are absorbed by the insulating material in the cable. This happens whether the cable is superconducting or not. In the best cable design to come out of the Brookhaven studies, the AC losses were close to those of an overhead line of similar capacity. But the cost of helium cooling doubles the cables' losses. So even reducing cooling costs will not make underground cables suddenly look the best bet.

Magnetic levitation trains (maglevs) are another instance in which the advantages of superconductors are unclear. Japanese National Railways has built a test maglev using superconductor magnets. A West German firm, Transrapid International, has built one using conventional iron-core magnets. The Japanese train boasts the highest speeds—over 500 kilometers an hour—but also has the bumpiest ride. The Japanese model could certainly be made lighter and cheaper to run by switching to liquid nitrogen cooling. But a technical hitch remains—that of accelerating the train on wheels until the magnetic field induced by the superconductors in the metal guide rails is enough for takeoff.

That complication makes the train and track more expensive. The Transrapid, by contrast, can hover at a halt. Also the superconductor maglev generates much stronger magnetic fields than its competitor. These fields may disrupt electronic devices, such as a passenger's wristwatch or, more seriously, his pacemaker.

High Hopes for High-Tech

However they are built, one problem for maglevs could be lack of demand. Building maglevs alongside Japan's loss-making bullet trains would look like overkill. Last year, transport ministers from France, West Germany, Britain, Holland, and Belgium agreed in principle to an ambitious intercity high-speed-link project. But the competition for those lines will probably be between conventional high-speed trains, such as France's Grande Vitesse, and West Germany's Intercity Express. Sceptics say that maglevs have missed their chance by about a century. As they see it, when rail transport was an expanding, competitive industry, new technology generated excitement. It is now a mainly contracting, nationalized industry, in need of better management rather than expensive new toys. On that reasoning, superconductors should have brighter prospects in high-tech applications, such as computers.

A few years ago superconducting computers were the futurologists' favorite remedy for the problems that bedeviled semiconductors. The idea was to replace semiconductor transistors with a superconducting version called a "Josephson junction," after its inventor, Brian Josephson of Cambridge University.

The role of a transistor is to amplify a small voltage signal into a large current change, thus acting as an on–off switch that can represent the ones and zeros of binary arithmetic that every computer uses. The advantage of a Josephson junction is that it responds to a tiny voltage signal. The signal needed to flip it is a hundred times weaker than that required by the best semiconducting transistors. This is helpful because the time it takes to switch a transistor is roughly proportional to the voltage signal, and the heat produced by the transistor increases as the square of that voltage. Thus a one hundredfold reduction in the voltage signal should mean speedy computers that can be crammed into minute spaces without the risk of melting. Or so the theory goes.

In 1983 IBM closed down its Josephson junction project. Others cut their efforts substantially—with the notable exception of some Japanese companies. Two technical problems prompted the shift away from superconductors. First, the difficulty of mass-producing junctions that could be cooled repeatedly to 4° above absolute zero. Second, the slow, but sure, progress of semiconductor technology threatened to whittle away the advantages of Josephson junctions, and make helium-cooled computers not worth the bother. But nitrogen-cooled superconductors should change all that, say those who believe that Silicon Valley will be replaced by Barium-Yttrium-Copper-Oxide Valley. Cooling with nitrogen does make a difference, but that difference is not entirely for the better.

To understand why, consider the real secret behind the Josephson junction. The fact that it can be triggered by such a small voltage has little to do with superconductivity, and much to do with the temperature of liquid helium. At 4° above absolute zero, the thermal "background noise" in a circuit, which is produced by electrons bouncing around, is very low. A tiny voltage signal registers like a sneeze in a library. At 77° absolute—the temperature of liquid nitrogen—that signal is more like a sneeze on a crowded street. It cannot be heard.

A simple rule of thumb says that the voltage signal required to switch a transistor is proportional to the absolute temperature. Working at 77° absolute, rather than 4°, means forfeiting speed and increasing heat output. What little theoretical advantage remains for Josephson junction computers threatens, once more, to be eaten away by the plodding advance of semiconductor technology.

Josephson junctions have other uses, though. An intriguing one is the superconducting quantum interference device (or SQUID) which exploits the Josephson junction's high sensitivity to magnetic fields. SQUIDs are used as magnetometers, and are some 1,000 times more sensitive to changes in a magnetic field than are other instruments. IBM researchers have demonstrated a high-temperature SQUID. Although working at liquid nitrogen temperature degrades the performance of a SQUID, the fact that it can be packed in much less cumbersome insulation would be a clear advantage for many applications, including mineral prospecting and submarine tracking.

High-temperature superconductors may yet find their way into the microelectronics industry, but probably in a less-than-grand role: by transmitting the signals that semiconductor transistors send to one another. Superconductor wiring between transistors would generate less heat and boost the speed at which the transistors can switch. Theodore Geballe of Stanford University in California, whose group was one of the first to demonstrate a Josephson junction using high-temperature superconductors, reckons that hybrid superconductors-semiconductor circuits could boost computer speeds about fivefold. If that is the case, the technical hurdles can be jumped.

High-temperature superconductors are not yet out of the laboratory. Some researchers have already produced wires and tapes of the stuff—no mean feat, considering that the new ceramic superconductors are brittle, whereas a wire must be flexible. But cautious scientists say that it could take years to make wires good enough for industrial use. The main stumbling block seems to be the maximum current that a superconductor will carry before it reverts to normal conductivity. At the moment the biggest current that researchers have managed to push through a wire measuring 1 millimeter across is around 10 amperes. For powerful magnets, that figure needs to be boosted about 1,000 times.

The source of the problem is not yet apparent. A wealth of experience with conventional superconductors suggests that small nonsuperconducting defects in the wire are partially blocking the current and causing resistance when the current is raised too high. For the superconductor wiring on a silicon chip, those defects are bad news. Such wires would be only a few billionths of a meter across. Single defects, which the current could easily bypass in a thicker wire, would block the current entirely in a tiny wire.

In May IBM's Watson Research Center in Yorktown Heights, New York, announced that its scientists had made some headway with this problem. By growing a thin single crystal layer of the superconductor, researchers were able to achieve reasonable currents. Unfortunately, the key to IBM's success lies in the perfection of the material. The jumble of minute crystallites that makes up a wire is bound to have more defects than a single crystal.

Such pessimistic observations may enrage scientists, especially those in America who are trying to convince Congress that superconductors are the battlefield for the next technology fight with the Japanese. But nothing will stop hundreds of fertile minds from looking for new applications of superconductivity. Perhaps they will soon find them, and make this article look out of date. But let it be clear that the superconductor revolution has not quite happened yet.

Superconductivity Research: Pace Slows, Reality Catches Up

Alison Bass

The party is over—at least for now. The avalanche of dazzling advances that began last year in the once-sleepy field of superconductivity has slowed almost as abruptly as it began, say scientists, and the pace of discovery has returned to a much more typical scientific crawl.

At the same time, researchers are facing up to some harsh realities about what the newly discovered superconducting ceramics can actually do. They say there may be inherent limitations in the materials that caused so much excitement last spring—limitations that may make large-scale applications of these materials prohibitively expensive.

Some scientists even question whether such highly touted applications as extremely efficient power line transmission, levitating trains, and cheaper high-energy atom smashers will ever come to fruition.

"This is not the ideal material," Simon Foner, chief scientist at the National Magnet Laboratory at Massachusetts Institute of Technology, said in an interview. "It turns out these materials have some major problems."

The new reality may well be apparent at a major conference on high-temperature superconductivity that opens today [November 9, 1987] in Boston. Several scientists who plan to attend described in interviews last week their concern with the current pace of progress.

From *The Boston Globe,* November 9, 1987. Reprinted courtesy of *The Boston Globe.*

"It's extremely frustrating for those of us who have been working in the field for the last ten months," says Robert J. Cava, a physicist at Bell Laboratories. "But you have to keep an optimistic attitude."

In laboratories throughout the world, the search for stable materials that are superconducting at higher temperatures seems stalled at 100 K (280° below zero Fahrenheit). Although some teams have reported tantalizing glimpses of superconductivity at up to room temperature, none of those effects has lasted or been reproduced by other labs.

Furthermore scientists still have not found a basic theory that explains why these materials become superconducting. Without a theory, the search for even better superconductors remains a hit-or-miss process.

Researchers, however, remain hopeful that superconductivity at even higher temperatures will eventually be found, in either the new ceramic compounds or an entirely new class of still unknown materials.

"Theoretically, room temperature is not impossible," says Philip W. Anderson, a Nobel laureate theoretical physicist at Princeton University. "But I expect the next advance to take place in a totally different material."

The ceramic compounds do continue to show promise in some smaller, high-technology applications such as ultrafast microcomputers and extremely sensitive magnetic sensors for use in geological prospecting. They may also find a niche in high-speed microwave transmission devices used to detect enemy missiles and submarines.

And some scientists say the "best applications" for these materials have yet to be discovered. "Once clever people start taking the existence of superconductivity at a reasonable temperature for granted, they will come up with applications we haven't dreamed of yet," says Cava.

The first major advance in high-temperature superconductivity came in January 1986 when two researchers in Switzerland noticed a sudden drop in electrical resistance when they ran a current through a ceramic compound that had been cooled only to 35 K (−396°F). That was well above the temperature required for any previously known material to become superconducting, and the achievement resulted in a Nobel Prize for the two scientists last month.

However, news of their breakthrough did not reach the scientific and popular press until almost a year later. Then, in the first heady months of this year, scientists in the United States, Japan, and China succeeded in pushing the temperatures of these superconducting compounds up to 93 K (−292°F). At those temperatures, the materials could be cooled with liquid nitrogen, a gas that is much cheaper and easier to handle than the liquid helium required to chill conventional metal superconductors.

With the breakthroughs, many scientists began talking about a host of astonishing applications, from ultrafast computers to extremely cheap power transmission to powerful magnets that could propel magnetically levitated (maglev) trains.

Shortly afterward scientists at IBM made another major breakthrough—they developed an extremely thin film or coating of the new material that could be applied to silicon chips and carry sufficient electrical current to be useful in high-speed computers.

More recently, researchers at Cornell University, the University of Rochester, IBM, and MIT's Lincoln Laboratories have found that superconducting thin films can transmit electronic signals at high speeds without excessive distortion, which may make it possible to use the materials to transmit microwave data for short distances. Such high-speed transmission systems may be particularly useful in military radar systems, where cost is less of a concern and high speeds are vital.

Despite the absence of a basic theory to account for this high-temperature superconductivity, scientists are also fast compiling a large body of information about the structure of the materials and how they work.

"In the last year, everyone has changed what he or she's doing and started to work on these new materials," says Marc Kastner, an MIT physics professor. "And as a result we probably know as much about these new materials as we do about silicon, and that took thirty years."

But until scientists understand *exactly* why they work, it may be difficult to find mixtures of compounds that are more effective than the ones now being explored. Although twenty or thirty theoretical hypotheses exist, none has been confirmed experimentally and thus accepted as the standard explanation for high-temperature superconductivity.

"To get a superconductor, you may have to mix together exactly the right ratio of metals and oxides," Cava says. The Swiss researchers tried more than 200 materials before "hitting" the right combination of the elements lanthanum, barium, copper, and oxygen to achieve superconductivity.

In the search for practical applications, the biggest obstacle so far has been that these materials cease to be superconducting as soon as researchers try to pass anything more than a very small current through them. Until researchers find a way to transmit significantly higher currents, they will not be useful for many applications.

Scientists have achieved higher currents with the thin films used to carry electrical signals on tiny computer chips and other electronic devices. But even here they face serious technical problems.

Researchers, for instance, are having trouble making thin-film technology that can be used in extremely sensitive magnetic detectors known as SQUIDS. In building an essential part of a SQUID—the "tunnel junction"—two layers of thin film must be deposited and separated by a thin oxide barrier.

"The problem is that to get the high-transit temperatures [for superconductivity], you have to apply a heat treatment to this junction, and in the process the oxide barrier is destroyed," explains John Clarke, a physicist at the University of California at Berkeley. "This is an extremely serious problem and until it's resolved, you're not going to see very sensitive SQUIDS."

Even if these technical problems are solved, thin-film technology is very expensive and would not be practical for large-scale uses. It would be prohibitively expensive, for instance, to coat miles of telephone wire with thin films in order to transmit data over long distances.

"Thin films are extremely expensive. They're high technology, and in practice it may be too difficult to make thin films and roll them up into wires," says Princeton's Anderson. "It's like making solar cells cost-effective. We've had [solar-cell] technology for quite a while, but we still haven't made it cost-effective."

MIT's Kastner agrees. "Part of the excitement with these materials in the beginning was because the fabrication cost of ceramics is so low," he says. "It's like putting pasta through a pasta maker; it's very low-tech. Keep in mind, they make toilet bowls out of ceramic."

Some Still Hold Hope

Some scientists, however, think the hiatus is temporary.

"People were spoiled by the unbelievable rate of progress in the first few months of this year, and now we're getting into a more normal rate of scientific progress," says Cava of Bell Labs. "People have to be patient. We're up against some difficult problems and they will take time to fix."

Richard Withers, a researcher at MIT's Lincoln Labs, adds: "We have a lot of work to do, but I wouldn't say it's impossible to have bulk materials with high critical currents."

But even if scientists do create room-temperature superconductors or learn to make ceramic superconductors that can carry high currents, some scientists doubt they will change the way we live. The existence of this technology alone, for instance, would not determine whether the United States builds high-speed trains that fly above the tracks at 300 miles an hour or atom-smashers that are 53 miles in circumference and cost close to $6 billion—like the superconducting supercollider now envisioned.

"The United States has yet to build a [nonlevitating] high-speed bullet train between New York and Washington," although the idea has been batted around for a long time, MIT's Foner notes. "This kind of project is simply not a public priority in this country. People don't travel by train." The availability of high-temperature superconductors, which would be a small part of the cost of a high-speed train, would not make this project any more of a priority, he says.

Similarly, the cost of replacing all the overhead power lines in the country with underground superconducting cable would be much greater than the amount of money that could be saved over time by energy-efficient superconductors.

"Power lines are already relatively efficient, but you don't read that in the newspapers," Foner says. He and other specialists say transmission losses run about 5 to 10 percent nationwide—not 20 or 30 percent as has been widely reported. "There may well be opportunities for new applications. But in terms of economics, using superconductors in long-distance transmission lines is the least feasible of all the various applications I've read about."

To Foner and others, the excitement with the ceramic superconductors is not so much in the applications—but in the science.

"From a scientific point of view, it's spectacularly exciting," Kastner says. "It means there's a new force that we've never seen before."

What Makes Them Work May Make Them Fail

Scientists at Princeton and MIT have come up with one plausible explanation for why an obscure class of ceramics can carry electricity with no loss of energy at unusually high temperatures.

But they kind of wish they were wrong.

If their theory is correct—it's one of a number of competing hypotheses—the very property that makes these materials superconducting would prevent them from carrying high electrical currents. And that would be very bad news indeed.

"Almost all the applications depend on getting high currents through these materials," says Marc Kastner, one of the physicists at the Massachusetts Institute of Technology who discovered this particular property in the new ceramics. "So there may be a real problem here."

In the laboratory, scientists have found they cannot run large electrical currents through these new ceramic compounds, and create powerful magnetic fields, without losing the materials' unprecedented ability to superconduct at temperatures up to 100 K ($-280°$ F).

The new ceramics are crystalline in structure, meaning they are made up of many tiny crystals. Each crystal consists of a latticework of atoms, which are surrounded by "free" electrons. A material's electrical resistance is, essentially, a measure of how much the lattice blocks the transit of these free electrons, which carry electrical current.

In most materials the electrons move every which way, colliding with each other and with the lattice and losing energy in the process. But in superconducting compounds, the electrons pair up and move together in collision-free planes without dissipating energy.

Researchers have understood for years why these electrons flow so smoothly in conventional metallic superconductors. But many believe that the force that causes electrons to move evenly through ordinary superconductors is not strong enough to keep electrons from colliding at much higher temperatures.

It is a law of nature that all atoms begin vibrating at temperatures above absolute zero. At high enough temperatures the movements of these atoms become great enough to knock some of the electrons out of their smooth planes of passage.

Given nature's constraints, many scientists felt they needed to look for a stronger mechanism or force to explain the phenomenon of superconductivity at relatively high temperatures. Theoretical physicist Philip W. Anderson of Princeton University says he has predicted in a mathematical theory just what this new force might be, and Kastner, Robert Birgeneau, and Philip Picone of MIT think they have found solid evidence that the force exists.

The MIT physicists observed that the electrons inside each of these ceramic crystals spin, like a top, in alternating directions that change every trillionth of a second. Electrons in ordinary superconductors also spin, but the direction of spin is random, and does not create a magnetic field. That's why there is no "permanent" magnetism in ordinary superconductors; you can pass an electrical or magnetic current through them, but without that current, the material is not magnetic.

However, in the new ceramic materials, adjacent electrons spin in opposite directions, and that alternating pattern of spin produces a strong magnetic force that holds the electrons together in pairs, enabling them to move in collision-free planes through the material. In measuring the electrons' spin, Birgeneau and Kastner found that this force is strong enough to hold the electrons together at higher temperatures.

But this force also allows the electrons to move in only one or two directions in parallel planes. This property is known as 2-D (two-dimensional) magnetism, and it is not found in ordinary metal superconductors, or for that matter, in any other known material. However, because ceramics are made up of many different crystals, the direction of the planes in one crystal is not always parallel to the direction of the planes in other adjoining crystals. And that's the problem.

In a sense these ceramics can be likened to a patchwork quilt made up of patches with patterns running in all different directions. When one piece or crystal doesn't match up with its neighbor, the electrons cannot easily move from one crystal to the next.

When a large current is passed through the material, the electrons jumping between crystals end up bumping into each other and losing much of their energy. The material develops electrical resistance and loses its ability to superconduct. (Conventional superconductors have three-dimensional electrons, which is why they can carry high currents but only at low temperatures.)

"We hope to find entirely new materials that have 3-D electrons and are still superconducting. And perhaps there will be ways of making ceramic crystals that are oriented so all the planes are parallel to each other," Kastner says. "But there may be fundamental limits to what can be overcome."

Materials and Society

II

Materials, Year 2000

E. D. Hondros

3

In this scene-setting article, the author emphasizes the crucial, enabling role of materials as a precursor to major technological innovations expected in the near future. After noting increased government awareness of the importance of materials, Hondros considers evidence on the world reserves of important industrial materials. He is cautiously optimistic about future supplies, steering a middle course between the "Cornucopian school" and the "school of Armageddon." Even so, he advocates greater conservation, substitution, reclamation and recycling. The author is the director of the Joint Research Center of the Patten Establishment in the Netherlands.

Introduction: The Criticality of Materials

In this overview of the future of materials we begin by considering the results of a long-range technology forecast of a particularly alert people, the Japanese, who recently carried out a survey through their Science and Technology Agency to discover the measures that will be necessary to cope with future requirements in energy, industry, natural resources, and the environment. This was part of a systematic study of the conditions necessary for satisfying expectations for a stable and prosperous future. Many of the general issues addressed and solutions proposed relate to Japan, though any country will develop strategic priorities that satisfy principally its parochial

From *The International Journal of Materials and Product Technology,* vol. 1, no. 1, 1986.

needs (for example, Japan recognizes the enormous importance of being able to predict earthquakes reliably). In this forecast, fifteen fields of science and technology were evaluated, ranging from education, environmental matters, and health to the microelectronics revolution. Many distinguished men and women of learning were canvassed for their views on the development of technology in the future.

The results of this study are presented in table 3.1. Here, forecasts that relate to health improvement such as cancer prevention rank the highest, followed by those that relate to energy resources. Each of the fifteen categories produced a list of specific technology requirements, and the table gives a summary of the principal science and technology innovations with a high expectation of realization. The vast amount of information collected allows a chronological display of the future developments. Some of these items in the table refer to materials per se; however, in a great many of the forecasts the *materials factor* is implied—these items have been marked with an asterisk. In other words, to realize the major developments indicated in the table, there will be a concurrent need to develop materials with specific functions—optical, electromagnetic, chemical, biological, thermal, and electronic. For example, the first half of the 1990s will see important developments in materials for replacing human organs (prosthetic devices). In the same period there will be major developments in high strength, heat-resistant engineering ceramics as engine components. In the latter half of the 1990s we expect to see the development of macromolecular materials with electrically conducting properties.

It is quite remarkable, then, that in most of the categories in this forecast, technologies are cited that depend on materials developments—these materials-related technologies are destined to be crucial in innovations to the year 2000 and beyond. Materials are the *enabling factor*—be it equipment for deep-sea mining, fusion reactors, ersatz human organs, or large-area solar conversion cells—the critical step in each realization will be the materials precursor.

Conventional and New Materials

Despite the rapidly growing interest in new materials, it is important to remember that the more conventional materials and materials

Table 3.1
Technology development forecast (Japan)

2006	Forecasting earthquakes within one month
2005	*Superconductors with critical temperature of liquified nitrogen (77 K)
2004	Means of converting cancerous cells into normal cells
2003	*Steel production from nuclear energy
2002	*Submersible cargo ships; aircraft and automobiles using hydrogen fuel
2001	Drugs for arteriosclerosis
2000	Large-scale environmental purification technologies
1999	Chemical agents for treating clotted cancer growths
1998	*Direct aluminium refining technology (direct reduction) Earthquake prediction through seabed crust activity
1997	*Deep-sea extraction of metallic nodules *Large-scale commercial nuclear fuel reprocessing plant
1996	*Engineering laboratory in space *Three-dimensional memory devices
1995	*Deep-sea (several hundred meters) drilling technologies *Disposal of high-level radioactive wastes by solidification techniques
1994	*Super LSI—10^9 devices per chip
1993	*Supercomputer (high-speed devices); high-efficiency thermo-electric conversion
1992	*Advanced robots for complex working environments
1991	Long-range meteorological forecasting
1990	*Large-area amorphous silicon solar cell
1989	Satellite prospecting (minerals, fishing, agricultural)

Source: *Science and Technology in Japan,* April–June 1983.

technologies will be with us into the next century. It may be self-evident, but it is worth emphasizing that steels, other alloys, and cement will be required in the future in greater quantities for advanced applications and in the less sophisticated aspects of man's daily life, such as in building materials. Thus a critical aspect of materials that we will consider here is the future availability of the many conventional materials in the face of rapidly increasing consumer demand and the widespread depletion of the richer ore bodies and other natural resources.

The feature that distinguishes materials from the other technologies that will define our future society is that materials are not an end product in themselves. Their value lies in that they permit an engineered product to fulfill its functions. In this sense, materials have a fundamental underpinning role in shaping industries; they provide the framework for the mechanical, chemical, building, process engineering, and transport sectors. It follows that any considerations of materials needs and innovations should relate to the functions of the final engineered product.

The term "material" denotes any physical stuff that is used by man to make things he needs. This covers a wide spectrum of materials types and usages, including machines, tools, and houses. In table 3.2 we list the major classes of materials, provide some examples, and note the principal properties with some examples of applications. The first four classes refer to the materials that have the highest promise of value-added improvement, that is, a high information content and a high latent economic gearing factor. These *engineering materials* are commonly distinguished from the generally large volume production materials such as cement and timber used in the construction industries.

Awareness of the Importance of Materials

The widespread recognition of the fundamental importance of materials to society is not new. Throughout history man has always needed minerals and materials, as illustrated in the anthropologist's characterization of man as a tool-using animal. Primitive man used tools and articles fashioned from the natural materials lying about him, such as bone, fiber, hide, wood, and stone. Archaeologists have recorded the effort that early man put into the search for suit-

Table 3.2
Use of materials

Classes	Some examples	Principal properties or characteristics	Examples of usage
Metal and alloys	Steels, superalloys, light alloys	Strength toughness	Automobiles, aircraft, pressure vessels
Ceramics	Alumina, silicon nitride, metal carbides	Temperature and corrosion resistance, high hardness	Furnace refractories, cutting tools, engine components
Plastics	Polymers, rubbers, polyurethanes	Strength, corrosion resistance, low density	Pipes, panels, process plant
Composites	Fiber-reinforced plastics, metals or ceramics	High toughness; high strength/ low weight	Aircraft and other transport components
Construction materials	Building stone, cement	Durable, plentiful supply, cheap	Buildings, roads, bridges
Timber	Wood, wood composites	Ease of fabrication, strength	Furniture, buildings
Fiber	Cotton, nylon, glass	Diverse; manufacturing and handling ease	Textiles, fiber/plastic composites
Paper	Paper, paperboard	Diverse	Printing, decorating

able deposits of flint which had desirable properties such as the ability to be shaped into instruments for cutting and shaping. Yet for millennia other comparable materials such as obsidian were neglected. This is a mystery that many writers have commented on. The interactions between early societies and the materials used is itself a fascinating field of inquiry: one school of thought places materials as a prime motivating force in the emergence of civilizations. The sagas of exploration undertaken by the early Phoenicians in search of tin and the documented (as well as debated) evidence that the decline of early Middle Eastern civilizations was due to a decline in the tin content of the bronze used for implements are examples of the central role of materials in the dominance and stability of early civilizations.

Today this unique position of engineering materials as an enabling technology—enabling future developments in microelectronics, information technology, transport and energy systems—has been widely recognized in all the industrially developed countries. One of today's sages in the materials science and engineering field, Morris Cohen, has remarked, "Materials are now being recognized as one of the basic resources of mankind, ranked along with living space, food, energy, and human knowledge . . . materials are seen to form a vast, connective web that ties nations and economies and peoples not only to one another, but also directly to the very stuff of nature. This huge materials complex is now called *the materials cycle*."

Partly as a consequence of such seminal perception, in the United States a National Materials and Minerals Program was recently submitted to Congress; it highlights the critical importance of minerals and materials to the national economy, defense, and standard of living and recommends urgent government intervention. At the same time Japan, with sharpened motivation arising from having to import almost all of its raw materials, has galvanized its institutions into action. Japan's position has been made quite clear; in a report, a U.K. scientific observer noted that following the analysis of developments of current project areas, "Japanese government officials have realized the importance of materials research as the *key technology for future progress*." The same note is being struck in many places and at different levels. Recently, the EEC has been discussing materials-based cooperative research programs on, for example, "Basic Technological Research" and "Materials in Transport." Other international structures such as OECD and NATO have pinpointed materials research as one of the pivotal areas for collaborative research. Again, the summit meeting at Versailles in June 1982 proposed a program on "Advanced Materials and Standards" as one of a number of projects of a "Technology, Growth, and Employment" plan that aims to create a climate that fosters advanced technology, including an open and competitive trading system and facilitating the cross-border flow of technology. In the United Kingdom a Materials Advisory Group has recently been formed to advise ministers of materials-related opportunities for industry.

The industrialized countries now recognize the inevitability of a shift toward knowledge-intensive technologies in order to underpin the changing complexion of their industries. As traditional, materials-based manufacturing industries such as foundry steel products become more mature, and as it becomes more difficult to procure advantages through improved productivity (even with application of the highest technical skills), such industries will transfer to countries where the unit labor cost provides the advantage. There are examples of this in all countries, including Japan. This highlights the crucial need for developed countries to embrace the new technologies and to venture into new products with high added value that incorporate advanced materials. In the fields of mechanical and electronic engineering, a prerequisite is the availability of materials with novel characteristics or a combination of unique properties—for example, light, strong, polymer composites, corrosion-resistant plastics, or temperature-resistant, toughened ceramics—materials for up-market products.

This widespread awareness of the critical position of materials has exacerbated concern about the dependence of industrial nations on the supply of primary materials from third world countries. Table 3.3 shows as a percentage of consumption the amounts of some principal metals imported in the form of ores and concentrates. For the world's major trading blocks, this dependence on external supplies of raw materials is most critical for Japan. Although the position of the United States is fair in comparison with the other free-market trading blocks, it has nevertheless led to considerable local concern especially in regard to the more vulnerable metals such

Table 3.3
Import dependence of raw materials by trading areas
(percent of consumption)

	EEC	United States	Japan	COMECON
Aluminium	61	85	100	28
Copper	90	14	94	4
Chromium	100	91	98	2
Cobalt	100	98	100	68
Iron ore	79	36	99	5
Lead	76	13	78	3
Nickel	100	70	100	13
Manganese	100	98	98	3
Tin	87	83	98	
Titanium	100			
Tungsten	99	52		
Zinc	91	57	74	9

Sources: Morgan (1980) and Crowson (1980).

as chromium, cobalt, tungsten, and manganese and fears that politico-economic movements in the third world could endanger supplies. Another important factor is that mineral producers, especially advanced countries such as Australia and Canada, will desire to increase the value-added component of their raw products and will insist on future exports of ores in the form of costly upgraded forms.

Patterns of Materials Consumption

With increasing economic growth throughout the world over the past fifty years, there has been a staggering increase in the consumption of raw materials. There is concern that continued economic growth may mean that the demand for raw materials will not be met. The current scale of world production of materials is truly impressive. This touches on two central questions of this paper: Will existing trends in consumption carry through to the year 2000? And will the world's resources be sufficient to meet these needs? By analyzing data on materials consumption in the United States throughout its history and taking into account important factors such as technological evolution, population growth, consumer expectations, and other social factors, we may approach an understanding of general historical trends—the experience of the United States may serve as a paradigm for the developed world and help us to make projections.

The production of major materials in the United States from the early nineteenth century to the present day is shown in figure 3.1. This illustrates graphically the immense growth in output of materials in the past century together with the growth in the U.S. population over the same period. During the nineteenth century, with a sevenfold increase in population and similar growth in gross national product (GNP), the production of iron and steel increased by a factor of 250. In the following half century, however, with the population doubling and the GNP rising even more rapidly, iron and steel production had increased only sixfold. This growth rate then leveled off to approximately that of the population. Similar behavior holds for lead and copper. This pattern seems to be indicative of a general phenomenon—namely that during the early phase

Figure 3.1 Growth in usage of industrial materials in the United States compared with population growth (source: Radcliffe 1976)

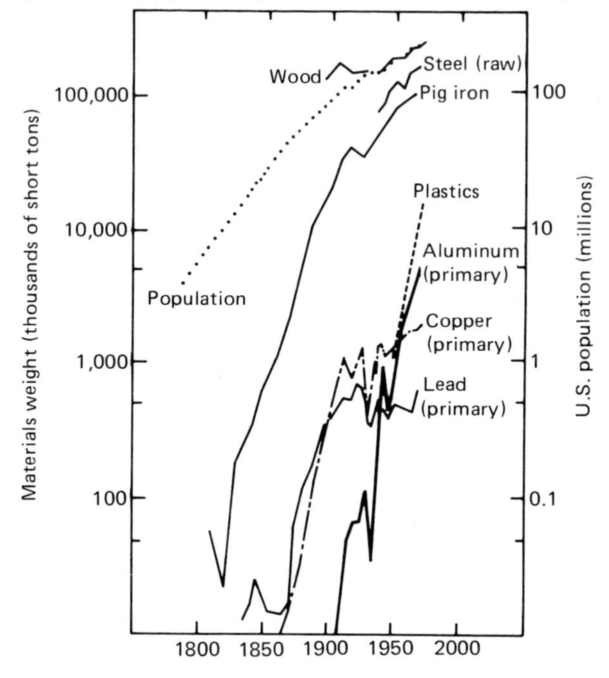

of industrialization, materials production grows more rapidly than population, and at a later stage of maturity it declines to a rate close to that of the population. This rapid growth period lasted eighty years for iron and steel and some forty years for copper and lead. On the other hand, aluminum (which is essentially a twentieth-century metal) and plastics are still in a period of rapid growth compared with that of the population. In fact, statistics from a recent UNIDO study show that world production of synthetic materials, such as plastics, fibers, and rubber, has increased fourfold during each decade up to 1970.

The Energy Materials Factor

Energy plays a central role in all aspects of materials, from the winning of the raw product to the manufacture of goods. Energy and materials are inextricably linked: the harnessing of primary energy to generate power to drive engines requires materials with specialized properties. Conventional and new power generating systems demand appropriate materials, as in coal gasification and liquefaction, oil shale, and tar sand developments, in magnetohydrodynamics (MHD), and in geothermal and nuclear systems. Indeed, the MHD method, which is potentially one of the most efficient ways of generating power, has not succeeded in practice because of the limitation of materials that must withstand the intense thermal and chemical attack of highly corrosive gases.

Conversely, and of equal importance, a plentiful supply of cheap energy is a necessary condition for making materials available at low cost. Much has been written of the energy problem, and we do not intend here to go over all the arguments. We emphasize, however, the crucial dependence of materials on energy—all our expectations of a future with an abundant supply of the necessary industrial materials at a competitive price will come to naught if the cost of energy becomes prohibitively high. The underlying thinking here can best be summarized in the words of Cottrell (1973), "the strictly materials problems of society do in the end boil down mainly to problems of energy. Given enough energy, we can virtually realize the old alchemist dream of making everything from anything."

Nowhere is the fundamental interdependence between energy

and materials illustrated better than in the minerals extraction technologies. Table 3.4 indicates that there are ample reserves of ore for the principal metals. However, the richer ore grades are being worked first, and in metals such as copper, increasingly poorer ore grades must be exploited. It was noted by Kellogg (1979) that the threat to the future of materials resources is not so much ultimate exhaustion; it is the prospect that we may reach a stage in which we can no longer afford the cost, particularly in energy requirements, for extraction from very low grade ores—a notion which he depicted as the "tyranny of ore grade." Figure 3.2 shows on a log-log scale the energy required to recover a ton of metal on the grade of ore. Each of the three bands relates to a particular extraction process; the lowest band refers to alluvial ores such as beach sands which are simply dredged from surface deposits and therefore require the least energy; the middle band refers to the common rocks which require energy-intensive processes such as crushing and grinding; the uppermost band represents ores that must be recovered by chemical leaching or pyrometallurgical processes such as for the nickel laterites. For a given ore grade the extraction processes from bottom to top require an approximately tenfold increase in energy—Kellogg depicted this as the "tyranny of ore types." The slopes of all three bands indicate the sharply increasing energy requirements as the

Table 3.4
World cobalt mine production and reserve base (short tons of cobalt content)

		Reserve base	
Country	Mine production	Quantity	Grade of ore (percentage)
Zaire	17,000	1,300,000	0.25–0.45
Zambia	5,000	400,000	0.09–0.4
Canada	2,000	30,000	0.03–0.11
Australia	1,700	50,000	0.08–0.12
Finland	1,000	20,000	0.2
Philippines	1,000	200,000	0.03–0.12
Other market economy countries	1,450	580,000	0.01–1.2
Central economy countries	4,000	450,000	0.05–0.15
World total	33,150	3,400,000	

Source: Adapted from Kruger (1983).

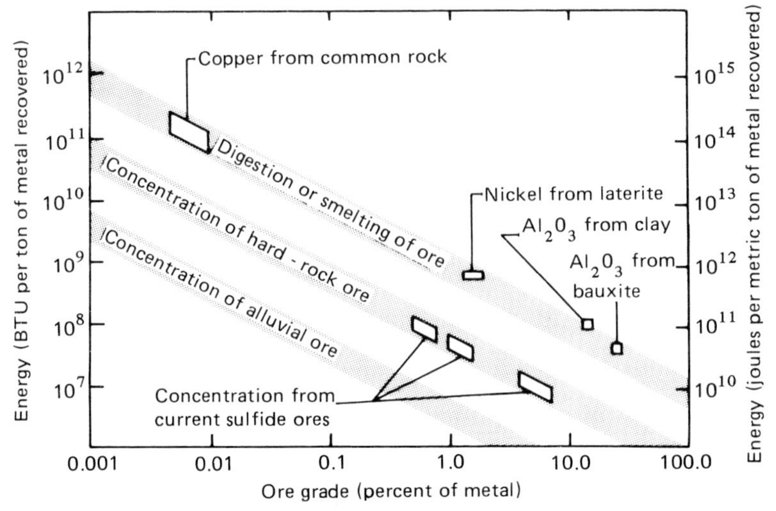

Figure 3.2 Energy dependence of metal winning on ore grade and extraction process (source: Kellogg 1979)

ores become leaner. The current levels of sulphide ores from which copper and nickel are extracted are indicated in the middle band, and similarly the position of nickel from laterite ores is indicated in the top band. Here, for example, it can be seen that as the grade of sulphide ore becomes impoverished, the tyranny of ore grade will determine whether it would be economically sound to jump onto the top band and recover nickel from richer concentrations of laterite ores by leaching processes.

Future Availability—Alarm and Euphoria

We pose the central question of "Materials 2000": Can availability be assured?

Over the past dozen years or so, a number of commentators have asked whether the present growth in demand in goods and services can continue and what constitute the limits to growth. The great public debates we have seen in recent years as illustrated in the deliberations of the Club of Rome (Meadows 1972) have focused on the impact on the environment and depletion of natural resources as the limiting factor. This depletion in raw materials is seen by many to be *the* limit to growth. Many popular books published in recent years present this alarmist vision of declining natural resources, a slow death of industrialization and the irreversible pollution of the environment. This view of a finite, exploited, and exhausted planet, the "Spaceship Earth" syndrome, has caught the imagination of the public. There is nothing new in this type of analysis: it is merely a restatement of the thesis proposed by Malthus some 150 years ago in relation to the earth's limited ability to feed a rapidly growing population. These views have led to a supply pessimism echoed in learned circles and in particular among the mass media. However, as one distinguished materials observer noted, "the sociologist can convince the public that we are running out of mineral resources easier than the geologist or minerals economist can allay public fear in this respect primarily because the media is more attracted to the depletion view" (Smith 1983).

On the other side of the opinion spectrum there exists a totally opposite school of thought that propounds the thesis that there will never be a shortage of raw materials: the application of materials conservation measures, recycling procedures (especially legislated enforcement of recycling), and the mining of raw materials from low quality ores (from seawater as well as from seabed nodules) will ensure a veritable cornucopia in the future. The free-market economists have added their support in this respect: as a given raw material becomes more expensive because of difficulty of extraction, inevitably less will be used, and the imagination of scientists will surely work miracles and find substitutes for the scarcer materials.

This confusing polarization of attitudes between, on the one hand, the Cornucopian school and, on the other, the school of Armageddon serves at least to concentrate the mind. Futurology which involves apocalyptic visions such as the gross spoliation of the earth and contrasts the barbaric overconsumerism of some advanced societies with the poverty and malnutrition in other societies engenders emotions that color conclusions. Our analysis in this paper supports a view that given man's control of his excesses, the unleashing of his scientific imagination, and geopolitical stability—not unreasonable assumptions—the prospects for the future are of cautious optimism.

Current Consumption and World Reserves of Materials

A single incident in 1978 provoked considerable alarm among the materials-using industrial countries and focused attention on the stability of materials supplies. In that year, Katangan rebels invaded the cobalt-producing area of Shaba province in Zaire. As table 3.4 indicates, Zaire is by far the biggest single producer of cobalt in the world, and following the occupation of this province by the rebels for a period of only weeks, the cobalt output was paralyzed. This incident, which triggered off a dramatic cobalt shortage and high price rises among the market economy countries, produced in its wake concern and debate on the vulnerability of western economies. In the United States this resulted in people drawing entirely opposed conclusions on national materials policies: on the one hand, some argued that the free-market system with minimum interference from central government would continue to meet the needs of industry; on the other, that there was an urgent need for federal intervention to guarantee an adequate supply of materials that are critical to the U.S. economy. Informed analysts such as Kruger (1983) concluded that much of the panic was due to over-reaction and that industry could reduce its import dependency through private stockpiling, substitution, and recycling. One interesting example of substitution provoked by the high cobalt prices at that time is that the producers of magnets used in hi-fi systems and loud speakers turned from cobalt-containing magnets to ferrite magnets.

The debate on potential supply disruption has often focused on the two critical metals, cobalt and chromium. There is an argument that these critical materials derive from geopolitically unstable countries and that politically motivated cartels could create shortages. One argument against the cartel syndrome is that minerals-producing countries, especially underdeveloped countries, must export their minerals, which constitute a good proportion of their GNP. Again, as a glance at table 3.4 will show, despite Zaire's predominance as a producer of cobalt, the metal comes from politically diverse countries. Many experts believe that market forces will outweigh all other factors, and if supplies continue to be disrupted, other countries will be encouraged to invest in mines in order to promote production even from low concentration ores.

In general, industrialized countries are now sufficiently equipped to cope with a potential materials disruption, and certainly they are now more sensitized to this possibility. The situation is not one of acute despair: a planned reaction to any impending materials disruption would include the use of government stockpiles; the use of alternative foreign suppliers; the introduction of substitute elements; the planned reduction of critical materials consumption by the introduction of better design, processing, and manufacturing procedures; the elimination of nonessential uses and the acceleration of domestic production.

From table 3.4 two facts are clear: first, although half the world's supply comes from one country, Zaire, there is a fair number of countries producing significant quantities; second, at the present rate of production the world reserve base is more than adequate to support the present rate of consumption well into the next century. Nevertheless, the fact that by far the world's biggest source of high concentration cobalt ore is Zaire may continue to color the debate on strategic supplies.

Chromium is another metal that is essential to modern industry and about which there has been much debate in recent years. We recall that it has vital uses in stainless steels, in tool and heat-resistant steels, superalloys, and in the chemical industry. In the aerospace industry chromium is essential for the nickel base superalloys used in turbine engines. However, the total chromium consumption in the aerospace industry in the United States is only several percent of total chromium production, the largest consumer being the automobile industry, through the use of stainless steel catalytic converters.

The chromium issue has been analyzed in considerable detail in the United States, and this should reflect the broad trend in other industrialized countries. Thus as a potential remedy to dependence on chromium importation, it has been estimated that the present 12 percent of domestic chromium that is being supplied by recycling from stainless steel scrap could be easily improved by another third through a concentrated drive. Furthermore some 25 percent of the current consumption of chromium could be saved by using existing substitution techniques in applications such as electroplating steel.

In addition to cobalt and chromium we consider the other principal industrial materials in table 3.5, which gives a summary of data from a recent study (1983) by Kruger. Here we present world mine production for each material with an estimate of the reserve base. For each entry the message is clear and quantified according to the best modeling procedures—at the present rate of world consumption, the world reserves are totally adequate to sustain present levels

Table 3.5
Critical materials: world mine production and reserve base

Material	Mine production	Reserve base
Antimony (short tons)	62,400	5,000,000
Asbestos (thousand metric tons)	4,000	104,000
Aluminium (thousand metric tons of bauxite)	75,800	22,400,000
Bismuth (thousand pounds of bismuth content)	6,700	222,000
Cadmium (metric tons of cadmium content)	17,100	795,000
Chromium (thousand short tons of chromite)	9,700	3,700,000
Cobalt (short tons of cobalt content)	33,150	3,400,000
Copper (thousand metric tons of copper content)	7,780	511,000
Diamond (million carat)	28	870
Lead (thousand metric tons)	3,450	146,000
Manganese (thousand short tons)	23,000	5,400,000
Mercury (76 pound flasks)	189,500	4,410,000
Nickel (thousand short tons of nickel)	576,000	61,800,000
Platinum (troy ounces)	6,400,000	1,200,000,000
Tantalum (pounds of metal content)	910,000	48,000,000
Tin (metric tons of metal content)	238,000	10,000,000
Titanium (thousand short tons of ilmenite and rutile)	4,930	947,000
Tungsten (thousand pounds of tungsten content)	94,200	6,400,000
Zinc (thousand metric tons of zinc)	6,160	243,000

Source: Adapted from Kruger (1983).

of consumption well into the next century. There is no indication that the world is heading toward a supply shortage in any of the principal materials. Localized political events in mineral-producing areas of the world could lead to sudden, but short-lived, deficiencies in some of the strategic metals. Policy planning in the industrialized countries must center around this possibility and the need for a deficiency buffer plan. Such a plan should contain such diverse actions as extending an adequate stockpile of the strategic metals over perhaps six months; conservation in the usage of these critical elements which should be the focus of supported R&D in areas such as substitution technology; encouragement of recycling in materials usage patterns.

Future Consumption Trends

The important question arises: To what extent will the demand for materials throughout the world increase in the future? A number of analytical methodologies have been used to broach this forecasting problem, and a particularly promising approach is that used earlier by Landsberg et al. (1963) which integrates the intensity of usage with demand aspects.

Malenbaum (1977) takes this up for a number of critical metals, and presents the outlook for demand in terms of two component elements: the intensity of use (IU) and the total output of final goods and services (GDP). A principal finding in Malenbaum's analysis is that the IU of most materials in most regions has been declining in recent decades, with the exception of aluminium. The forces giving rise to a declining pattern of intensity of use, especially in the richer countries, originate partly from technological developments, partly from substitutions among the raw materials, and partly from shifts in the types of final products that consumers expect.

By the year 2000 the intensity of use of most materials will have decreased to somewhere between 10 and 20 percent of recent levels. In general, this change in the *demand* prospect is a factor that mitigates the *supply* limitation prospect. It represents a conscious slackening of man's consuming need for raw materials. The other factor is the growth in GDP. For the period from now to 2000 it is expected that the GDP growth rate will be some 20 percent below that in the previous quarter century. Nevertheless, with the forecast population growth by the year 2000, it is expected that the *world per capita GDP* will exceed the 1970 to 1975 level by some 50 percent. Typical results of this analysis for the three elements chromium, tungsten, and cobalt are shown in figure 3.3 for the intensity of use per unit of GDP against time. This indicates at least saturation in intensity of use and even a significant reduction.

By combining the forecast IU factor with the per capita GDP and the total population, projections have been made for the future demand for raw materials by the year 2000. Data are plotted for some important metals in figure 3.4. By the year 2000 it is forecast that the world will be consuming two to three times more raw materials per year than it did in 1975. Indeed, most ratios are closer to two. The optimistic forecast obtained by these methods contrasts with previous forecasts which reported equivalent ratios of between three and five.

This analysis offers a hopeful vision of man and society working together toward reducing the rates of materials consumption, of a deliberate and conscious effort to limit the consumptive appetite of man. Analyses such as these are a salutory change from the doomsday doctrines that suggest man's helplessness in the face of resource exhaustion or of the deterioration of the environment. The technico-politico infrastructure of advanced countries should be fo-

Figure 3.3 Changes in intensity of use per unit GDP with time for a selection of metals (source: data from Malenbaum 1977). The left-hand ordinate scale refers to chrome ore only.

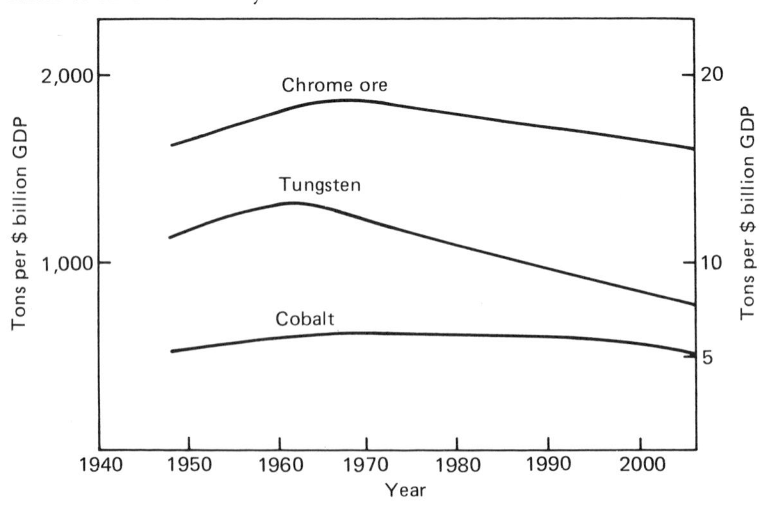

cused on maximizing the social, political, and technological factors that underline this view.

The Potential of Science and Technology in Securing a Sound Materials Future

We examine now the proposition: to what extent can man, by harnessing his imagination with advances in the science and technology of materials, secure a sound future for materials? The current broad technological trend toward materials that have a greater information content per unit weight is surely in the right direction. Through such scientific developments we may design lighter components, thereby saving energy and materials cost; we may design complex engineering structures to operate longer and be more reliable, that is, avoid overdesign; we may reduce the quantity of scarce materials used—thus, for example, understanding the specific function of chromium in a corrosion-resistant application may result in a steel component coated with a layer of a chromium alloy rather than a large volume of chromium. This approach reflects a strong move among competing advanced countries toward the development of products with *high added value*—these in turn will require materials with greater sophistication of content.

Figure 3.4 Forecast of future world requirement of three important metals (source: data from Malenbaum 1977)

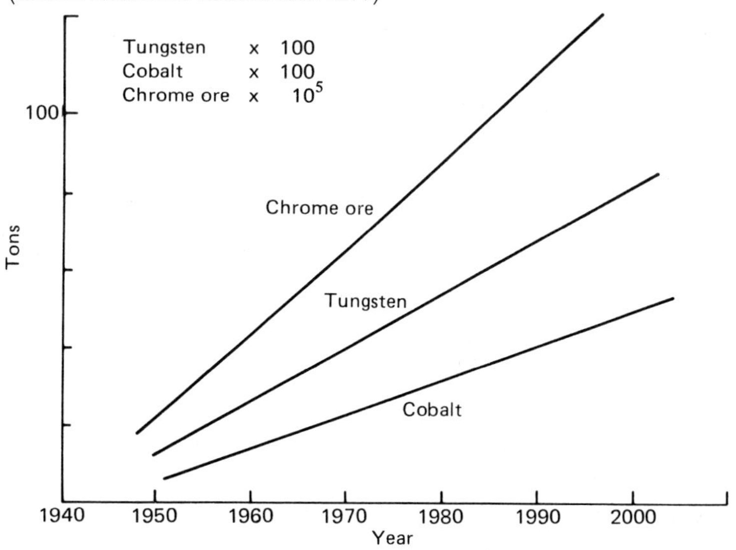

This important concept is illustrated in figure 3.5, which shows the relationship between the quantity of material in a product and the information content of the material (Altenpohl 1979). The vector extending from each point indicates the trend for the future. The expectation for the automobile sector is shown in the diagram; a more recent analysis carried out by the EEC in a collaborative research program on materials in transport applications indicates similar trends to the year 2000 and notes in particular the introduction of new materials such as composites in the body and fittings, and structural ceramics in the engines. The expectations for aeronautical materials are schematized in figure 3.6 (Balazard 1983), which shows clearly the progressive introduction of composite materials, for the sheathing and the primary structure, generally at the expense of aluminium alloys. The use of special steels and titanium in highly specific applications will remain fairly steady; however, there is growing expert opinion that there will be an increasing use of ceramics in turbines, heat exchangers, coatings, and thermal insulation.

A future materials strategy aimed at improving the value content of manufactured products, and in a broader sense combating the wasteful consumption of materials that we have known in the past decades, will consist of a mix of a number of ingredients which fall into the broad categories of materials substitution, reclamation, conservation, and performance.

We summarize certain requirements in the specific functions of the materials as follows:

Opportunities lie in developing near-net-shape forming *manufacturing techniques* that minimize the metal removed during processing and the associated waste scrap value, including processes such as sintering, casting, and precision forging. Computerized methods will be used for improved *product design* as well as for better control during manufacturing. There are opportunities in strip casting techniques combined with in-line rolling, in super-plastic forming operations, and in hot and cold isostatic pressing.

In the broad category of *improved performance* we include things such as enhanced reliability and resistance to failure. In addition to actual performance, there is scope for improvements in *durability* of materials or their life extension. Here the ubiquitous problems of losses through corrosion, erosion, and wear will be addressed

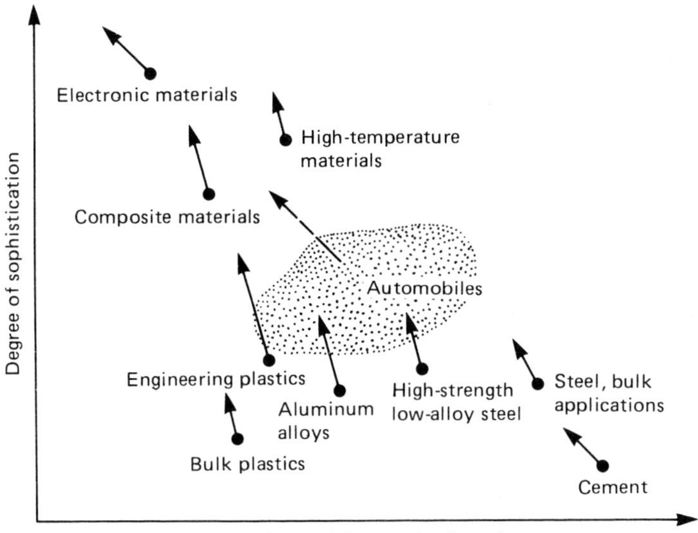

Figure 3.5 Heuristic model relating the quantity of material in a product to its materials information content (source: after Altenpohl 1979)

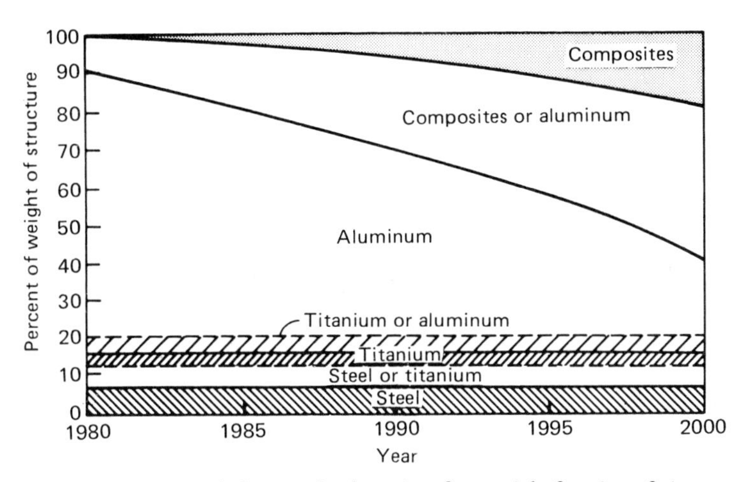

Figure 3.6 Expected changes in the mix of materials for aircraft (source: Balazard 1983)

through both the application of existing knowledge and the introduction of new protective technologies such as coatings and novel microstructures involving rapid solidification metallurgy. In this connection, we shall witness the introduction of new ceramic materials that will substitute for many of the metallic materials that lack the heat, corrosion, and wear resistance of ceramics. Already there have been major developments in "sialons" which have high strength at high temperatures, and which are substituting for cemented carbides in cutting tips used for machining hard superalloys.

The introduction of such *advanced materials* to replace certain strategic materials will in turn catalyze major technological innovations. These include composite materials such as metals, polymers, and ceramics reinforced with a variety of fibers; structural ceramics; structural polymers; rapidly solidified, microcrystalline, and glassy metals; and innovations in surface engineering, in particular tailored coatings designed to procure certain property advantages. Figure 3.7 gives an indication of the exciting potential applications of fine ceramics in a wide range of future technologies and products (*Science and Technology in Japan,* October–December 1983) and how all the specific primary functions of metal compounds—thermal, optical, electronic, magnetic, and mechanical—required in new applications are interrelated.

The Materials Conservation Ethic

The preceding discussion indicates that there is considerable potential in exploiting modern materials science and technological understanding to ensure a safe materials supply for the future. We must develop a *materials conservation ethic* based on respect for the total materials cycle. In this concept we embrace all aspects—extraction of the raw product from the earth, transportation of the ore, the energy costs in extracting metal from the ore, modification of the raw material for specialized alloys aimed for certain functions, and then the shaped materials as they are incorporated into larger assemblies—the final engineered product. After its useful life has terminated, this product may be recovered and made to re-enter the materials cycle at some point further back. A systems analysis suggests an optimum utilization of materials, yielding benefit in terms of performance, materials conservation, energy consumption, and a

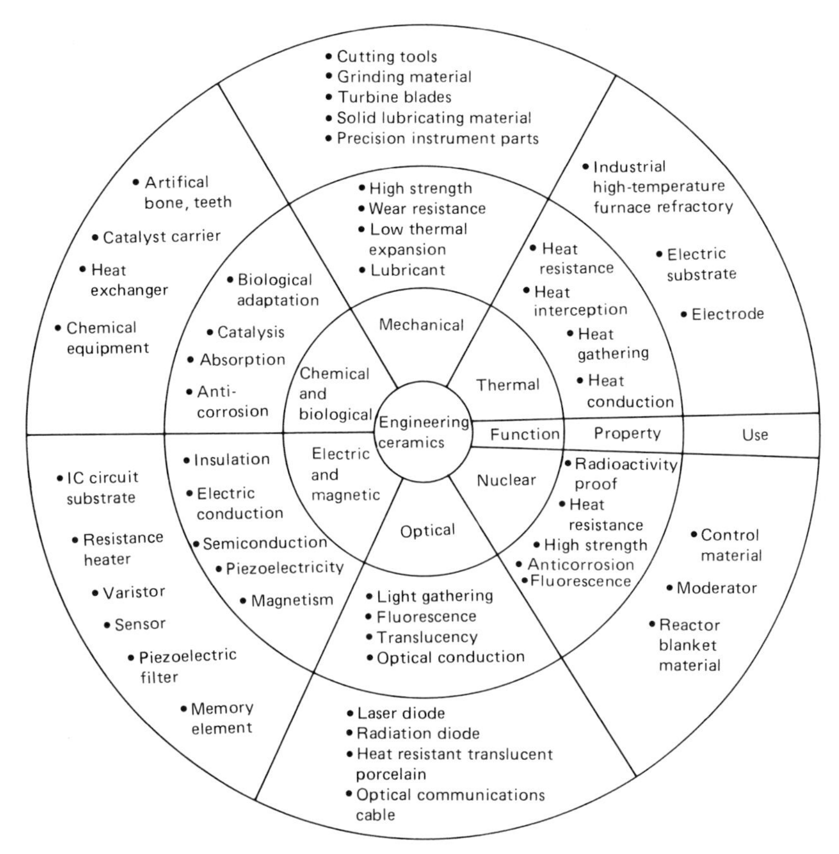

Figure 3.7 The potential of fine ceramics (source: *Science and Technology in Japan,* October-December 1983)

cleaner environment. Already we are seeing some successes with this approach: advanced coatings for industrial tools that enhance the tool life, base metals that replace platinum group metals in hybrid circuits, advanced wear-resistant bearing alloys, and cobalt-free alloy systems for the hard facings of engine exhaust valves. At all stages of the global materials cycle—extraction, production, processing, and manufacturing—opportunities exist for conserving valuable materials.

Finally, as part of a materials conservation ethic, we should encourage the development of a new concept of design. A methodology encompassing *recyclability* as one of the ingredients of design should emerge. Thus legislation or positive incentives might induce engineers to design machines with a built-in ease of dismantling so as to allow the separation of the various components of known constitution. Recovering alloys could result in secondary and tertiary alloy products that may not be adequate for the original property requirements but may be suitable for other applications. Already we are learning from some interesting examples of this design approach: thus recently a carbon steel has been introduced that has a markedly improved resistance to atmospheric corrosion. This results from the increased level of normally deleterious elements such as tin and copper, introduced through the use of scrap from urban waste.

References

D. G. Altenpohl. 1979. *Materials and Society* 3: 315.

J. Balazard. 1983. *Materiaux-Energie*. Vol. 27. Agence Française pour la Maitrise de l'Energie, rue Louis Vicat 75015, Paris, p. 301.

M. Cohen. 1979. "Materials and Man Linked with Nature." Alpha Sigma Mu Lecture. *ASM News* (July): 4–6.

A. Cottrell. 1973. *Metall. Trans. A.* 4: 405.

P. C. F. Crowson. 1980. *Future Metal Strategy.* The Metals Society, London, p. 9.

H. H. Kellogg. 1979. *Materials and Society* 3: 141.

L. G. Kruger. 1983. Report No. 83–171 SPR. Congressional Research Service. The Library of Congress.

H. L. Landsberg, L. L. Fischman, and J. L. Fisher. 1963. *Resources in America.* Resources for the Future, Washington, D.C.

W. Malenbaum. 1977. *World Demand for Raw Materials in 1985 and 2000.* National Science Foundation, Washington, D.C.

D. H. Meadows et al. 1972. *The Limits to Growth.* Pan Books, London.

J. D. Morgan. 1980. *Materials and Society* 4: 211.

S. V. Radcliffe. 1976. *Science* 191: 700.

"Technology Development Forecast up to 2010 in Japan." *Science and Technology in Japan,* April–June 1983, p. 24.

"High Performance Fine Ceramics in Japan." *Science and Technology in Japan,* October–December 1983, p. 16.

R. L. Smith. 1983. *Metall. Trans. A.* 14:2199.

Materials in History and Society

Melvin Kranzberg and Cyril Stanley Smith

4

An all-important historical perspective on materials in society is provided by this fascinating chronology which takes us from the earliest times through the Stone, Bronze, and Iron Ages to Roman times, famous for cement, and the Middle Ages, which gave us printing and gunpowder. Melvin Kranzberg is with the Georgia Institute of Technology, Atlanta, and Cyril Stanley Smith is at MIT. This piece was in fact written in the early 1970s, which explains the opening references to concern about the "increasing" consumption of materials. Nevertheless, the discussion is highly relevant to today's debate. If readers wish to follow the story through to World War II, they should check out the rest of this journal article.

Introduction

A Glimpse of the Materials Field

The field of materials is immense and diverse. Historically, it began with the emergence of man himself, and materials gave name to the ages of civilization. Today the field logically encompasses the lonely prospector and the advanced instrumented search for oil; it spreads from the furious flame of the oxygen steelmaking furnace to the quiet cold electrodeposition of copper; from the massive rolling mill

From Morris Cohen (ed.), "Materials Science and Engineering: Its Evolution, Practice and Prospects," a special issue of *Materials Science and Engineering*, vol. 37, no. 1 (January 1979). Reprinted with permission.

producing steel rails to the craftsman hammering out a piece of jewelry; from the smallest chip of an electronic device to the largest building made by man; from the common paper bag to the titanium shell of a space ship; from the clearest glass to carbon black; from liquid mercury to the hardest diamond; from superconductors to insulators; from the room-temperature-cast plastics to almost-infusible refractories; from milady's stocking to the militant's bomb; from the sweating blacksmith to the cloistered, contemplating scholar who once worried about the nature of matter and now tries to calculate the difference between materials.

One of the hallmarks of modern industrialized society is our increasing extravagance in the use of materials. We use more materials than ever before, and we use them up faster. Indeed, it has been postulated that assuming current trends in world production and population growth, the materials requirements for the next decade and a half could equal all the materials used throughout history up to date.[1] This expanding use of materials is itself revolutionary and hence forms an integral part of the "materials revolution" of our times.

Not only are we consuming materials more rapidly, but we are using an increasing diversity of materials. Through most of history a few empirically selected abuse-tolerant general-purpose materials sufficed for the relatively untaxing applications that had been devised. Engineers accepted the limitation of available materials and designed in accordance with their known properties or small extensions of experience, while the producers of materials worked to balance reliability with cheapness rather than to achieve new properties. This has changed. A great range of new materials has opened up for the use of twentieth-century man: refractory metals, light alloys, plastics and synthetic fibers, for example. Some of these do better, or cheaper, what the older ones did; others have combinations of properties that enable entirely new devices to be made or quite new effects to be achieved. We now employ in industrial processes a majority of the ninety-two elements in the periodic table which are found in nature, whereas until a century ago, all but twenty, if known at all, were curiosities of the chemistry laboratory.[2] Moreover, physical structure is even more important than chemical composition. Not only are more of nature's elements being put into service, but completely new materials are being synthesized in the laboratory. Our claim to a high level of materials civilization rests on this expanded, almost extravagant utilization of a rich diversity of materials.

This extravagance is both a product of advances in materials and a challenge to its future growth. The enlarged consumption of materials means that we shall have to cope increasingly with natural-resource and supply problems—and also with energy problems, for the extraction, processing, creation, or recycling of materials requires enormous consumption of increasingly scarce energy resources. Mankind is being forced therefore to enlarge its resource base by finding ways to employ existing raw materials more efficiently, to convert previously unusable substances to useful materials, to recycle waste materials and make them reusable, and to produce wholly new materials out of substances which are available in abundance.

A Glimpse of Materials Science and Engineering

The expanded demand for materials is not confined to sophisticated space ships or electronic and nuclear devices. In most American kitchens are new heat-shock-proof glasses and ceramics and long-life electric elements to heat them; the motors in electric appliances have so-called oilless bearings which actually hold a lifetime of oil, made possible by powder metallurgy; the pocket camera uses new compositions of coated optical glass; office copy machines depend on photoconductors; toy soldiers are formed out of plastics, not lead; boats are molded out of fiberglass; the humble garbage can sounds off with a plastic thud rather than a metallic clank; we sleep on synthetic foam mattresses and polyfiber pillows, instead of cotton and wool stuffing and feathers; we are scarcely aware of how many objects of everyday life have been transformed—and in most cases, improved—by the application of materials science and engineering (MSE). It is a new profession, an inseparable mixture of both science and engineering that is becoming known as MSE for short. Moreover, as with a rich vocabulary in literature, the flexibility that is engendered by MSE greatly increases the options in substitutions of one material for another.

Quite often the development of a new material or process will have effects far beyond what the originators expected. Materials have somewhat the quality of letters in the alphabet in that they can be used to compose many things larger than themselves: amber, gold, jewelry, and iron ore inspired commerce and the discovery of many parts of the world; improvements in optical glass lay behind all the knowledge revealed by the microscope and telescope; conductors, insulators, and semiconductors were needed to construct new communication systems which today affect the thought, work, and play of everyone. Alloy steel permitted the development of the automobile; titanium the space program. The finding of a new material was essential for the growth of the laser, the social uses of which cannot yet be fully imagined. In these, as in hundreds of other cases, the materials themselves are soon taken for granted, just as are the letters of a word. To be sure, the ultimate value of a material lies in what society chooses to do with whatever is made of it, but changes in the "smaller parts" reacting responsibly to larger movements and structures make it possible to evolve new patterns of social organization.

The transitions from, say, stone to bronze and from bronze to iron were revolutionary in impact, but they were relatively slow in terms of the time scale. Changes in materials innovation and application within the last half century, however, have occurred in a time span which was revolutionary rather than evolutionary.

The materials revolution of our times is qualitative as well as quantitative. It breeds the attitude of purposeful creativity rather than modification of natural materials, and also a new approach—an innovative organization of science and technology. The combination of these elements which constitutes MSE is characterized by a new language of science and engineering, by new tools for research, by a new approach to the structure and properties of materials of all kinds, by a new interdependence of scientific research and technical development, and by a new coupling of scientific endeavor with societal needs.

As a field, MSE is young. There is still no professional organization embodying all of its aspects, and there is even some disagreement as to what constitutes the field. One of the elements which is newest about it is the notion of purposive creation. However, MSE is responsive as well as creative. Not only does it create new materials, sometimes before their possible uses are recognized, but it responds to new and different needs of our sophisticated and complex industrial society. In a sense, MSE is today's alchemy. Almost magically, it transmutes base materials, not into gold—although it can produce gold-looking substances—but into substances which are of greater use and benefit to mankind than this precious metal. MSE is directed toward the solution of problems of a scientific and technological nature bearing on the creation and development of materials for specific uses. This means that it couples scientific research with engineering applications of the end product: one must speak of materials science and engineering as an "it" rather than "them."

Not only is MSE postulated on the linkage of science and technology, it draws together different fields within science and engineering. From technology MSE brings metallurgists, ceramists, electrical engineers, chemical engineers; from science it embraces physicists, inorganic chemists, organic chemists, crystallographers, and various specialists within those major fields.

In its development MSE not only involved cooperation among different branches of science and engineering, but also collaboration among different kinds of organizations. Industrial corporations, governmental agencies, and universities have worked together to shape the outlines and operations of this new "field."

In recent years there has been a marked increase in the liaison between industrial production and industrial research, and between research in industry and that in the universities. The researcher cannot ignore problems of production, and the producer knows that he can get from the scientist suggestions for new products and sometimes help for difficulties. It should be noticed that MSE has come about by the aggregation of several different specialties that were earlier separate, not as so often happens by growth of increased diversity within a field which keeps some cohesion. This change is just as much on the industrial side as it is on the academic. Industry continually uses its old production capabilities on new materials, and the scientist finds himself forced to look at a different scale of aggregation of matter.

Most of the work on materials until the twentieth century was aimed at making the old materials available in greater quantity, of better quality, or at less cost. The new world in which materials are developed for specific purposes (usually by persons who are concerned with end use rather than with the production of the materials themselves) introduces a fundamental change, indeed. Heretofore, engineers were limited in their designs to the use of materials already "on the shelf." This limitation no longer applies, and the design of new materials is becoming a very intimate part of almost every engineering plan. MSE interacts particularly well with engineers who have some application in mind. It often reaches the general public through secondary effects, such as negatively via the pollution that results from mining, smelting, or processing operations, and positively via the taken-for-granted materials that underlie every product and service in today's complicated world. We clothe ourselves in materials of man-made fibers; we eat food from plastic containers; we drive cars made of a variety of "natural" and synthetic materials; and we watch TV on sets whose components possess special electronic properties.

The rising tide of "materials expectations" is not for the materials themselves, but for things which of necessity incorporate materials. That materials are secondary in most end applications is obvious from the name applied to the materials that remain when a machine or structure no longer serves its purpose—"junk." There is, however, at least one positive direct contact, that of waste-material processing; city waste disposal is a very challenging materials-processing problem, especially if the entire cycle from production, use and reuse of materials can be brought into proper balance.

As one cynical observer put it, "For the first two decades of its existence, materials science and engineering was engaged in producing new and better products for mankind; its major task for the next two decades is to help us get rid of the rubbish accumulated because of the successes of the past twenty years."

Importance of Materials to Man

Matter versus *Materials*

Materials are so ubiquitous and so important to man's life and welfare that we must obviously delimit the term in this survey, lest we find ourselves investigating nearly every aspect of science and technology and describing virtually every facet of human existence and social life. Unless we limit our scope, all matter in the universe will inadvertently be encompassed within the scope of our subject.

But *matter* is not the same as *material*. Mainly we are concerned with materials that are to become part of a device or structure or product made by man. The science part of MSE seeks to discover, analyze, and understand the nature of materials, to provide coherent explanations of the origin of the properties that are used, whereas the engineering aspect takes this basic knowledge and whatever else is necessary (not the least of which is experience) to develop, prepare, and apply materials for specified needs, often the most advanced objectives of the times. It is the necessarily intimate relationship between these disparate activities that, to some extent, distinguishes MSE from other fields and which makes it so fascinating for its practitioners. The benefits come not only from the production of age-old materials in greater quantities and with less cost—an aspect which has perhaps the most visible influence on the modern world—but it also involves the production of materials with totally new properties. Both of these contributions have changed the economy and social structure, and both have come about in large measure through the application of a mixture of theoretical and empirical science with entrepreneurship. And just as the development of mathematical principles of design enabled the nineteenth-century engineer to test available materials and select the best suited for his constructions, so the deeper understanding of the structural basis of materials has given the scientist a viewpoint applicable to all materials, and at every stage from their manufacture to their societal use and ultimate return to earth.

Some Influences on Man's Environment and Attitudes

The production of materials has always been accompanied by some form of pollution, but this only became a problem when industrialization and population enormously increased the scale of operation. Longfellow's poem contains no complaint about the smoke arising from the village smithy's forge; if one of today's poets would attempt to glorify the blacksmith's modern counterpart, he would undoubtedly describe the smoke belching forth from the foundry, but there would be no mention of spreading chestnut trees because all those within a half-mile radius would long ago have withered from the pollution of the surrounding neighborhood. The simple fact is that an industrial civilization represents more activity, more production and usually more pollution, even though the pollution attributable to each unit produced has been sharply decreased.

The utilization of materials, as well as their manufacture, also generates pollution. Those of us living in affluent highly industrialized countries enjoy the benefits of a "throwaway" society. The problem arises from the fact that many of the products we use are made from materials which are not strictly "throwawayable." Natural processes do not readily return all materials to the overall cycle, and in the case of certain mineral products, we can sometimes find no better way of disposing of scrap than to bury it back in the earth from which we had originally extracted it at great trouble and expense. Proposals for reuse or recycling often founder upon public apathy—but this is changing, and MSE has an important role to play.

The moral and spiritual impact of materials on both consumer and producer is both less visible and more debatable. To those reared in a Puritanical ethic of self-denial, the outpouring of materials goods would seem almost sinful, as would the waste products of a throwaway society. Such conspicuous consumption would seem almost immoral in a world where so many people are still lacking basic material essentials. A more sophisticated objection might be that the very profusion of materials presents modern man with psychological dilemmas. We are presented with so many options that we find it difficult to choose among them.[3]

It might be surprising to some that the question of the debasement of the materials producer should even be raised. Scientists have long claimed that their pursuit of an understanding of nature is innocent, and technologists have always assumed that their gifts of materials plenty to mankind would be welcomed. Hence it has come as a shock to them during the past few years that the benefits of science and technology have been questioned. Both science and technology have been subjected to criticism from highly articulate members of the literature subculture, as well as from within their own ranks, regarding their contributions to mankind's destructive activities and to the deterioration of the environment.[4]

To those engaged in materials production and fabrication, it may be disconcerting to realize that for a fair fraction of human history their activities have been viewed with suspicion and downright distaste by social thinkers and by the general public. The ancient Greek philosophers, who set the tone for many of the attitudes still prevalent throughout Western civilization, regarded those involved in the production of material goods as being less worthy than agriculturists and others who did not perform such mundane tasks. Greek mythology provided a basis for this disdain: the Greek gods were viewed as idealistic models of physical perfection; the only flawed immortal was the patron god of the metalworker, Hephaestus, whose lameness made him the butt of jokes among his Olympian colleagues. (But he got along well with Aphrodite, another producer!)

Throughout ancient society the most menial tasks, especially those of mining and metallurgy, were left to slaves. Hence the common social attitude of antiquity, persisting to this day in some intellectual circles, was to look down upon those who worked with their hands. Xenophon[5] stated the case in this fashion, "What are called the mechanical arts carry a social stigma and are rightly dishonored in our cities. For these arts damage the bodies of those who work at them or who act as overseers, by compelling them to a sedentary life and to an indoor life, and, in some cases, to spend the whole day by the fire. This physical degeneration results also in deterioration of the soul. Furthermore the workers at these trades simply have not got the time to perform the offices of friendship or citizenship. Consequently they are looked upon as bad friends and bad patriots, and in some cities, especially the warlike ones, it is not legal for a citizen to ply a mechanical trade."

The ancients appreciated material goods, but they did not think highly of those who actually produced them. In his life of Marcellus, Plutarch delivered this critical judgment: "For it does not of necessity follow that, if the work delights you with its grace, the one who wrought it is worthy of esteem." The current apprehension concerning dangers to the environment from materials production might result in materials scientists and engineers being regarded with similar suspicion today.

But there is yet a subtler way in which the triumphs of MSE

might threaten the spirit of Western man. Advances in materials have gone beyond the simple task of conquering nature and mastering the environment. MSE attempts to improve upon nature. In a sense, this represents the ancient Greek sin of *hubris,* inordinate pride, where men thought they could rival or even excel the gods—and retribution from the gods followed inevitably. This may also be "original sin," the Christian sin of pride, which caused Adam's fall. By eating the fruit of the tree of knowledge, Adam thought that he would know as much as God. Conceivably, by endeavoring to outdo nature modern man is preparing his own fall. Or perhaps his new knowledge will lead to control as well as power, and to a richer life for mankind.

Materials in the Evolution of Man and in Prehistory

Materials in Human Evolution

The very essence of a cultural development is its interrelatedness. This survey places emphasis on materials, but it should be obvious that materials *per se* are of little value unless they are shaped into a form that permits man to make or do something useful, or one that he finds delightful to touch or to contemplate. The material simply permits things to be done because of its bulk, its strength, or, in more recent times, its varied combinations of physical, chemical and mechanical properties. The internal structure of the material that gives these properties is simply one stage in the complex hierarchy of physical and conceptual structures that make up the totality of man's works and aspirations.

We do not know exactly when our present human species *homo sapiens* came into being, but we do know that materials must have played a part in the evolution of man from more primitive forms of animal primates. It was the interaction of biological material and cultural processes that differentiated man from the rest of the animal world.[6] Other animals possess great physical advantages over man: the lion is stronger, the horse is faster, the giraffe has a greater reach for food. Nevertheless, man possesses certain anatomical features which proved particularly useful in enabling him to deal with his environment.[7]

Modern physical anthropologists believe that there is a direct connection between such cultural traits as toolmaking and tool-using, and the development of man's physical characteristics, including his brain and his hand.[8] Man would not have become Man the Thinker (*homo sapiens*) had he not at the same time been Man the Maker (*homo faber*). Man made tools, but tools also made man. Perhaps man did not throw stones because he was standing up; he could have learned to stand erect the better to throw stones.

It is probable that the earliest humans used tools rather than made them, that they selected whatever natural objects were at hand for immediate use before they anticipated a possible future task and prepared the tools for it. Once this idea was formulated and man began to discover and test out things for what they could do, he found natural objects—sticks, fibers, hides—and combined materials and shapes to serve his purposes. He tried bones and horn, but the hardest and densest material at hand was stone. When he further learned how to form materials as well as select them, and to communicate his knowledge, civilization could begin; it appears there was a strong evolutionary bias toward anatomical and mental types that could do this. Although the early stages still remain the realm of hypothesis, there is general agreement that it was over two million years ago when a pre-human hominid began to use pebbles or stones as tools, though the shaping of specialized tools came slowly.

Materials in Cultural and Social Development

The recognition that there had been a cultural level to which we now give the name of Stone Age—itself a tribute to the importance of materials in man's development—did not occur until well into the nineteenth century. When, in about 1837, the Frenchman Boucher de Perthes propounded the view that some oddly shaped stones were not "freaks of nature" but were the result of directed purposeful work by human hands, he was ridiculed. Only when the vastness of geological time scales was established and it became possible to depart from a literal interpretation of biblical Genesis could credence be given to the notion that these stones were actually tools.[9]

Some of the features of today's materials engineering can already be seen in the selection of flint by our prehistoric forebears as the best material for making tools and weapons. Availability, shap-ability, and serviceability are balanced. The brittleness of flint enabled it to be chipped and flaked into specialized tools, but it was not too fragile for service in the form of scrapers, knives, awls, hand axes, and the like. The geopolitical importance of material sources also appears early. It is perhaps not surprising that we find the most advanced early technologies and societies developing where good-quality flint was available. It may even be, as Jacobs has surmised,[10] that cities arose from centers for trading flint (and perhaps also decorative or colored stones), and that the intellectual liveliness accompanying the cultural interchange of travelers then created the environment in which agriculture originated.

The pattern of human settlement from prehistoric times to our contemporary world has been determined in large measure by the availability of materials and the technological ability to work them.

Man could survive successive ice ages in the Northern Hemisphere without migrating or developing a shaggy coat like the mammoth because he had found some means of keeping himself warm—with protective covering from the skins of animals which, with his wooden and stone weapons, he could now hunt with some degree of success, but also, perhaps principally, by the control of fire, which became one of his biggest steps in controlling his environment. By the beginning of Paleolithic (early Stone Age) times—between 800,000 and 100,000 years ago—man could produce fire at will by striking lumps of flint and iron pyrites against each other to produce sparks with which tinder, straw, or other flammable materials could be ignited.

Man's control and use of fire had immense social and cultural consequences. With fire he could not only warm his body but could also cook his food, greatly increasing the range of food resources and the ease of its preservation. Claude Lévi-Strauss, the French anthropologist famous for his "structuralist" approach to culture, claims that the borderline between "nature" and "culture" lies in eating one's meat raw or eating it cooked.[11] By their role in producing and fueling fires, materials thus played a significant part in the transition from "animal-ness" to "human-ness," but more than that, fire provided a means for modifying and greatly extending the range of properties available in materials themselves. The burning of lime for making plaster for the decoration of walls and floors seems to have been the first large-scale pyrotechnical operation.[12] It was, however, preceded by the heat treatment of flint to improve its flaking characteristics.[13] It tells something about the nature of man and of discovery and the earliest of all recorded uses of fire to modify inorganic material was in the preparation of varicolored pigments by heating natural iron oxide minerals.[14]

Materials in the Beginnings of Art and Technology

Every cultural conquest, such as the use of fire, requires other cultural developments to make its use effective, and it also has unanticipated consequences in totally unforeseen areas. Containers were needed for better fires and food. The invention of pots, pans, and other kitchen utensils made it possible to boil, stew, bake, and fry foods as well as to broil them by direct contact with the fire. The cooking itself, and the search for materials to do it in, was perhaps the beginning of materials engineering. Furthermore, though the molding and fire hardening of clay figurines and fetishes had preceded the useful pot, it was the latter that, in the 8th millennium B.C., gave rise to the development of industry. Clay was the first inorganic structural material to be given completely new properties as a result of an intentional operation upon it by human beings. Though stone, wood, hides, and bone had earlier been beautifully formed into tools and utensils, their substance had remained essentially unchanged. The ability to make a hard stone from soft and moldable clay not only unfolded into useful objects, but the realization that man could change the innermost nature of natural materials must have had a profound impact upon his view of his powers; it gave him confidence to search for new materials at an ever increasing rate.

It was in the decoration of pottery that man experimented with the effects of fire upon a wide range of mineral substances. Glazing, the forerunner of glass, certainly came therefrom, and it is probable that experiments with mineral colors on pottery led to the discovery of the reduction of metals from their ores toward the end of the 5th millenium B.C. Even earlier, man's urge to art had inspired the discovery and application of many metallic minerals as pigments.[15]

From late Paleolithic times come the great cave paintings representing hunting scenes in realistic detail, executed with such mastery that, when they were discovered by chance in the caves at Altamira, Spain, in 1879 and later in Lascaux, France, many found it difficult to believe that they had been done by primitive man. These paintings provide early evidence of man's awareness of the special properties of iron ore, manganese ore, and other minerals. He sensed qualitative differences that depended on chemical and physical properties quite invisible to him, but on which much of modern industry was later to be based.

Long before this, man had sensitively used the properties of other materials in art. He had made sculpture in ivory, stone, rock, clay, and countless more-perishable materials. Though it is often said that his ability to do this came from the increased leisure time released by the efficiency of his hunting following the development of tools and weapons, it is more likely that the exercise of his explorative tendencies, his aesthetic curiosity, was one of the factors that from the very first gave him a unique evolutionary advantage among other animals. Interaction with materials at this level was both easy and rewarding, and it was probably a necessary preliminary to the selection of the more imaginative and adaptable biological mutants that were to follow. In culture as in biology, man possessed more than the rudiments of technology when he had discovered and prepared his materials for painting and had developed methods of working them with fingertip and brush, crayons, and spray. He also had used specialized tools to sculpt stone and to mold clay at about the same time he learned to finish stone abrasively, and so was freed from dependence on flakeable flint since he could then adapt commoner, harder, polycrystalline rocks such as basalt and granite for his tools.

As in the case of the use of fire by man, the next great innovation in another field of technology, agriculture, was accompanied by a diverse series of auxiliary changes. Man had to develop a whole new set of tools: the hoe to till the ground, the sickle to reap the grain, some kind of flail to thresh the grain, and the quern (mill) to grind it. These tools were made of stone and wood; they were not very efficient. Nevertheless, agriculture was able to provide man with a surer source of food than could be obtained through the older technology of hunting, and it required concomitant advances in materials. Not the least important were fired ceramics which provided the pots needed for cooking, as well as larger containers for rodent-proof storage of crops.

The introduction of agriculture meant that the supply of animal skins from hunting was diminished. Man had to find substitutes among vegetable fibers, things such as reeds, flax, or cotton, and to utilize the hair of the animals which he had learned to domesticate. Some of these fibers had been used before, especially in woven mats, fences, building components, and basketry, but mainly for clothing. So textiles developed and textiles inspired new machines: a spinning device (the spindle with its inertia-driven whorl) and a loom for weaving the threads into cloth. The patterns he worked into textiles and painted on his pots gave him practical contact with elements of geometry and with the relationships between short-range symmetry and long-range pattern which reappear in today's structure-based science.

Because the implements and weapons of our prehistoric forebears are crude and primitive in comparison with today's materials and machines, we should not be misled into downgrading the degree of skill which Stone Age man possessed. When, a few years ago, a class at the University of California was provided with a pile of flintstones and given the task of shaping simple stone implements from them, they found that even after many hours of repeated trials, they could not produce a tool that would have sufficed even for a run-of-the-mill Stone Age man.[16] But experience breeds skill, and it is a combination of tradition and invention that aided in man's progressive evolution.[17]

The great Neolithic technological revolution—with its development of agriculture and fairly large-scale settled communities—occurred some tens upon tens of thousands of years after man had already mastered his implements of stone and had achieved his intellectual and physical evolution. It set into movement a whole series of technological and cultural changes within the next two millenia which thoroughly transformed man's relations with nature and with his fellow man and, most important, his thoughts about change and his prospects of the future. Although the process might seem slow to us today, it was dynamic by the standards of the preceding ages.

Whether or not the urban society preceded or followed the agricultural revolution, it seems almost certain that the city provided conditions that accelerated man's journey along the path toward civilization; indeed, the two are almost synonymous. During the period from about 5000 to 3000 B.C., two millenia after the introduction of agriculture, a series of basic inventions appeared.[18] Man developed a high temperature kiln; he learned to smelt and employ metals and to harness animals. He invented the plow, the wheeled cart, the sailing ship, and writing. Communication and commerce based on specialized skills and localized raw materials both enabled and depended upon central government together with reinforcing religious, social and scientific concepts. The great empires in Mesopotamia and Egypt, the forerunners of our Western civilization, were based on the interaction of many institutions and the ideas and muscles of many men, but materials were necessary for them to become effective. Indeed, the characteristics of this early period are mainly an interplay between principles of human organization and the discovery of the properties of matter as they resided in a wide diversity of materials. Both tools and buildings were simple; mechanisms comparable in ingenuity to the materials used in the decorative arts of Sumer, Egypt, and Greece do not appear until much later. All, as far as we can tell, were based on experience and empiricism with little help from theory.

The Bronze and Iron Ages

Stages of Civilization Based on Materials Development

Stone was eventually supplemented by copper, and copper led to alloys, most notably bronze. Near the end of the period under discussion, bronze in turn was partially displaced by iron. So important is the change in the materials base of a civilization that the materials themselves have given rise to the names of the ages—the Stone Age, the Bronze Age, and the Iron Age. In the nineteenth century after much groundwork both literally and figuratively by geologists, paleontologists, and archeologists, these terms came to supersede both the poet's gold and silver ages and the philosopher's division of the past into periods based on religious, political, or cultural characteristics.[19] Oddly, the most critical stage, and the one that has left the best record, that of ceramics, was overlooked.

There were no sharp chronological breaking points between the three ages, nor did the switch from one material to another take place everywhere at the same time. Even, for example, in those areas where bronze tools and weapons came into use, stone tools and weapons remained on the scene for a long time. Similarly, iron did not immediately replace bronze, and indeed there were still some civilizations that passed directly from stone to iron and some that, from indifference or from lack of knowledge, never adopted either metal. As a matter of fact, the first tools and weapons of iron were probably inferior to the contemporary bronze tools whose technology had been known for over two millenia. At first, the advantage of iron over bronze was based on economics, not superior quality. Iron was laborious to smelt, but it could be made from widespread common minerals. A monarch could arm his entire army with iron swords, instead of just a few soldiers with bronze swords when the rest would have to fight with sticks and bows and arrows. With iron came a quantitative factor that had profound social, economic and political consequences for all aspects of culture.[20]

Native metals, like gold, silver, and copper, were hammered into decorative objects during the 8th millennium B.C. in an area stretching from Anatolia to the edge of Iran's central desert, during the 5th millennium B.C. in the Lake Superior region of North America, and during the mid-2nd millennium B.C. in South America. However, it was not until man learned to smelt metals and reduce them from their ores, to melt and cast them, that metallurgy proper can be said to have begun. Again, the early advantage was only an economic one, the mineral ores of copper being vastly more abundant than is the native metal, but the way was opened for alloying and the discovery of entirely unsuspected properties. Moreover, with molten metal, casting into complicated shapes became possible.

The discovery of smelting has left no records. Given the availability of adequately high temperatures in pottery kilns and the use of metal oxides for decoration, drops of reduced metal could well have been produced repeatedly before the significance was grasped. But once it was, empirical experiments with manipulation of the fire and the selection of the appropriate heavy, colored minerals would have given the desired materials with reasonable efficiency. A kiln works best with a long-flame fuel such as wood; smelting is best done with charcoal and with a blast from a blowpipe or bellows, but the time when these were first used has yet to be established. The first alloys were those of copper and arsenic, which may have been made at first unintentionally by smelting ores containing the two elements, though there is evidence for the conscious use of the metallic mineral algodonite (approximately Cu_8 As) as an intentional addition during the casting operation.[21] Whatever their origin, the alloys are superior to pure copper in castability and in hardness, without loss of the essential metallic property of malleability, and their use marks the discovery of a vastly important metallurgical principle. For a thousand years these alloys were exploited, until finally they were largely replaced by bronze, an alloy made from a heavy, readily identifiable, though scarce, mineral, and having somewhat superior properties to those of the copper arsenic alloys; there was also the added advantage that those who knew how to make it lived longer!

A lively argument is currently going on among archaeologists as to whether the original discovery of bronze took place in the region of Anatolia and the adjacent countries to the South and East or in Eastern Europe—or independently in both.[22] Whatever the evolutionary process of the development of metallurgy, there is no doubt that it had profound social, economic, and political consequences.

Though the earliest stone industry and commerce had required some organized system of production, and the division of labor was well advanced in connection with large irrigation and building projects,[23] the use of metals fostered a higher degree of specialization and diversity of skills; it also required communication and coordination to a degree previously unknown. Both trade and transportation owe much of their development to the requirements of materials technology: not only ores, requiring bulk transportation over great distances from foreign lands, but also precious objects for the luxury trade, such as amber, gem stones, gold and silver jewelry, fine decorated ceramics, and eventually glass.

The search for ways of working materials prompted man's first use of machines to guide the power of his muscles. Rotary motion had many applications that were more influential than the well-known cartwheel. Perhaps beginning with the child's spinning top, it was the basis of many devices, the most important of which were:

the drill for bead-making, stone-working and seal-cutting; the threadmaker's spindle; the quern; and above all the potter's wheel.[24] These provided the foundation for the earliest mechanized industries and were steps toward the mass-production factories of the twentieth century.

Materials development had an impact on culture in other ways than through the improvement of artifacts. This can perhaps best be seen in the development of writing. The growth of commerce and government stimulated the need for records. Little geometrically shaped pieces of fire-hardened clay seem to have been used as tallies to accompany goods in the 9th millennium B.C., and it has been postulated that sketches of them on the outside of protective clay envelopes became the first writing.[25] The materials to produce the records undoubtedly influenced the nature of the writing itself and, if modern linguistic scholars are correct, probably some details of the language structure and hence the mode of thought.

Marshall McLuhan has popularized the phrase, "the medium is the message." A painting, a poem, a print, a pot, a line of type, a ballet, a piece of carved sculpture, a hammered goldsmith's work, or a TV image conveys different sensory perceptions that form the basis of human communication,[26] and we might guess that the same process occurred at the very beginning of art and purposeful records. The Sumerians in the Tigris–Euphrates valley had abundant clay to serve as their stationery, and the sharp stylus employed with it did not allow a cursive writing to develop; did this have some impact on the ways in which they thought, spoke, and acted? The Egyptians, in contrast, could adapt the interwoven fibers of a reed growing in the Nile delta to produce a more flexible medium, papyrus, on which they could write with brushes and ink in less restricted ways. Thus the differences between the cuneiform and hieroglyphic writing were dependent on the differences in materials available, quite as much as were the mud-brick and stone architecture of their respective regions. At the time the visual arts were probably more significant than writing, for relatively few people, except professional scribes, would have been influenced by the latter. Certainly, our retrospective view of old civilizations depends on the preservation of art in material form, and the material embodiment of thought and symbol in the visual environment must have modified the experience and behavior of ancient peoples, even as it does today.

Emergence of Iron and Steel

The replacement of copper and bronze by iron began about 1200 B.C. Iron had been produced long before then because iron ores are prevalent and easily reduced at temperatures comparable to those required for smelting copper. However, the iron was probably not recognized as such, because at those temperatures it is not melted but remains as a loose sponge of particles surrounded by slag and ash, being easily crumbled or pulverized and having no obvious metallic properties. If, on the other hand, the porous mass is hammered vigorously while hot, the particles weld together, the slag is forced out, and bars of wrought iron are produced.

Though metallic iron may have been previously seen as occasional lumpy by-products from lead and copper smelting (in which iron oxides were used to make siliceous impurities in the iron more fusible), its intentional smelting is commonly attributed to the Hittites, an Anatolian people, in about 1500 B.C. The Hittite monopoly of ferrous knowledge was dispersed with the empire in about 1200 B.C., but it took almost another 500 years before iron came into general use and displaced the mature metallurgy of bronze. Each ore had its own problems with metalloid and rocky impurities.[27] Immense skill was needed to remove the oxygen in the ore by reaction with the charcoal fuel without allowing subsequent absorption of carbon to a point where the reduced metal became brittle.

Certain forms of iron—those to which the name steel was once limited—can become intensely hard when heated red hot and quenched in water. This truly marvelous transmutation of properties must have been observed quite early, but its significance would have been hard to grasp, and in any case it could not be put to use until some means of controlling the carbon content had been developed. Since the presence of carbon as the essential prerequisite was not known until the end of the eighteenth century A.D., good results were achieved only by a slowly learned empirical rule-of-thumb schedule of the entire furnace regimen. Even the process of partial softening, today called tempering, was very late in appearing (perhaps in the sixteenth century) and early "tempering" was actually hardening done in a single quenching operation, in which the steel was withdrawn from the cooling bath at precisely the right moment. It is not surprising that this was rarely successful. Carburized and quenched tools began to appear in eighth century B.C. sites, and it has been suggested[28] that the vast increase in the use of iron at that time was a result of the discovery of methods of controlling the carburization and heat-treatment processes. However, the majority of objects, then as now, were of low carbon content and not hardened or even hardenable. Yet, even without hardening, iron had no difficulty in supplanting bronze for many applications. Its abundance meant that the elite could not control it. Iron was the "democratic" metal because a rise in the living standards among larger masses of population was obtainable through its application in tools and implements.[29]

The wide distribution of iron over the earth's surface enabled it to serve for tools and agricultural implements as well as weapons of war and precious objects for the ruling households. Before 1000 B.C., there are records of iron hoes, plowshares, sickles, and knives in use in Palestine. From about 700 B.C., iron axes came into play for clearing forest land in Europe and for agricultural purposes. Iron tools together with evolving organization arrangements greatly increased the productivity of agriculture, giving a surplus which could support large numbers of specialized craftsmen whose products, in turn, could become generally available instead of being monopolized by the wealthiest ruling circles. Furthermore tools formerly made of bronze or stone—such as adzes, axes, chisels, drills, hammers, gravers, saws, gauges—could be made less expensively and more satisfactorily in iron. The new tools allowed for new methods of working materials: forging in dies, the stamping and punching of coins, and, many years later, developments such as the drawing of wire and the rolling of sheet and rod. These metalworking methods were easily harnessed to water power when it appeared and opened up ways of making more serviceable and cheaper products. Though it was not immediately exploited, the strength of metals permitted the construction of delicate machines. Iron was at first used structurally only for reinforcing joints in stone or wood, but later its strength and stability were combined with precision in creating the modern machine tool, and its large-scale fabrication also made modern architecture possible.

Materials in Classical Civilization

Development of Hydraulic Cement by the Romans—A Materials Innovation

It has been claimed that bronze made for the centralization of economic power as well as the concentration of political authority in the hands of an aristocratic few, while iron broadened the economic strength to a larger class of traders and craftsmen and so led to the decentralization of power and eventually to the formation of Athenian democracy.

Although the classical civilization of Greece rather fully exploited the possibilities offered by metals and other materials available to them from preceding ages, producing beautifully wrought ceramics, exquisite jewelry, superb sculpture, and an architecture that still represents one of the peaks of the Western cultural and aesthetic tradition, they did little to innovate in the field of materials themselves.

The same is true of the Romans who acquired a great reputation as engineers, and rightly so, but this rests largely upon the monumental scale of their engineering endeavors—the great roads, aqueducts, and public structures—rather than upon any great mechanical innovations or the discovery of new materials.

There is one exception to this generalization. The Romans did introduce a new building material: hydraulic concrete. The use of lime mortar is extremely old, probably even preceding the firing of pottery, and lime plaster was used for floor and wall covering, for minor works of art and later for the lining of water reservoirs and channels. It can be made by firing limestone at a moderate red heat; it sets hard when mixed with water and allowed slowly to react with carbon dioxide in the air. If, however, the limestone contains alumina and silica (geologically from clay) and is fired at a higher temperature, a material of the class later to be called hydraulic, or Portland, cement is formed. After grinding and mixing with water, this sets by the crystallization of hydrated silicates, even when air is excluded, and develops high strength. Similar cements based on the reactive formation of silicate crystals in an aqueous medium enabled many of the great nineteenth-century achievements in civil engineering, and they still provide the most ubiquitous artificial material in use today. The Romans were fortunate in having available large quantities of volcanic ash, pozzuolana, which, when mixed with lime, gave such a cement. They exploited it to extremely good purpose (reinforced with stone rubble or with hard bricks) in the construction of buildings, bridges and aqueducts. Massive foundations and columns were much more easily built than with the older fitted-stone construction, and unlike mortar, the cement was waterproof. By combining the new cement with the structural device of the arch, the Romans could roof-over large areas without the obstructions of columns.[30]

The Primary Role of Empiricism in Materials Advances

The case of hydraulic cement is representative of materials usage from antiquity until modern times. Namely, it was developed entirely on an empirical basis, without much in the way of any science to explain the useful properties. Most types of material in use today were discovered centuries before there was any science to aid them; in fact, the science arose from the discoveries of practice rather than the other way round. The great Greek philosophers, to be sure, had worried about the nature of matter and the three states in which it exists—solid, liquid, and gas. These, indeed, constitute three of the four famous elements of Aristotle which dominated philosophy for nearly 2000 years. His earth, water, and air represent fine physical insight, but they had to be rejected by chemists in their search for compositional elements.[31] But, in any case, it was not philosophy that guided advance. The main contribution to early understanding came from the more intelligent empirical workers who discovered new materials, new reactions, and new types of behavior among the grand diversity of substances whose properties could be reproduced but not explained except on an *ad hoc* basis. Through most of history, it has been the almost sensual experience with that complex aggregation of properties summed up in the term the "nature" of the material that has guided empirical search for new materials and modifications of old ones. The ability to go beyond such empiricism and to plan tests on the basis of an adequate theory of the composition–structure–property relationships is a twentieth-century phenomenon and had to await a quite late stage in the development of science. The science needed was a kind that was slow to emerge because of the extreme complexity of the problems involved.

Unlike astronomy, there was little place for accurate measurements of geometry in materials, and those who sought to find rules were perpetually frustrated. The curious experimenter, however, by mixing, heating, and working materials in a myriad of ways did uncover virtually all of the materials with properties that were significant to him, namely, strength, malleability, corrosion resistance, color, texture, and fusibility. Science began to be helpful much later when chemical analysis—an outgrowth of the metallurgists' methods of testing the fineness of precious metal objects or bullion, or the metal content in ores of all kinds before going to full-scale operation—advanced to the point where it showed that there was only a limited number of chemical elements and that ostensibly similar materials, differing in their nature, contained different impurities. Then it was discovered that chemical substances of identical composition could differ in their internal structure, and finally structure became relatable to properties in a definite way; only then was it possible to modify the structure purposefully to achieve a desired effect.[32]

Some Proposed Connections between Materials and Roman History

There have been many interpretations of the decline and fall of the great Roman Empire. The early Christian apologists claimed that Rome fell because it was wicked and immoral; in the eighteenth century Gibbon blamed the fall of Rome upon Christianity itself. Since that time, the "fall" has been attributed to numerous factors: political, economic, military, cultural, and the like. It is not surprising, with the recent interest in the history of technology, that technological interpretations of Rome's decline have begun to appear, and some of these center on Rome's use of materials. A few years ago Gilfillan[33] claimed that the decline of Rome was due to a decline in the birth rate of the Roman patrician class as a result of dysgenic lead poisoning. Although all Romans got a goodly intake from their lead-lined water system, the elite drank more than its share of wine from lead vessels, and this was thought to reduce the fertility of the leaders! Lately a geochemist has claimed that Rome's troubles derived from the economic effects of the enforced decline in silver production which began about 200 A.D. because the mines had become so deep that they could no longer be cleared of water with the technical means available.[34]

Medieval Materials

Materials Processing and Machine Power

Throughout the first millennium after Christ, about the only places where ancient techniques of making and working materials underwent improvement were outside Europe—in the Arab World, Iran, India, and the Far East. Textiles, ceramics, articles in silver and bronze and iron of excellent quality appeared. That portentous new material, paper, originated in China and began its Western diffusion. Though the armorers of the Western world were steadily improving their products, the Crusaders of the twelfth century had no steel which could match that of the Saracen sword. (The Japanese sword surpassed the Islamic one by an even greater margin than the latter did the European.) However, not for several centuries did these Oriental superiorities in materials processing have any effect on the materials science or technology of contemporary Western Christendom.

For all this, the first significant literature on materials is European—the *Treatise on Divers Arts* written about 1123, by a Benedictine monk under the pseudonym Theophilus.[35] He was a practical metalworker, and he described in full practical detail all the arts necessary for the embellishment of the church, such as the making of chalices, stained-glass windows, bells, organs, painted panels, and illuminated manuscripts.

Theophilus was no materials engineer in the modern sense, but he was a craftsman, probably the historical goldsmith Roger of Helmarshausen, some of whose work has survived. His knowledge of matter was the directly sensed, intuitive understanding that comes from constantly handling a wide variety of substances under different conditions. His *Treatise* is essentially a factual "how-to" book, containing many exhortations to watch carefully for subtle changes in the materials being processed but with no trace of theoretical explanation. Theory does not appear in treatises intended to help the practical worker in materials until 600 years later—well into the eighteenth century.

Although the nature of materials themselves did not change greatly in Western Europe during the Middle Ages, a number of mechanical inventions facilitated both their production and their shaping.[36] The first widespread application of power in processing materials was in grinding grain. This practice considerably increased when windpower supplemented the older waterpower, with the technique, as so much else, diffusing from the East. Textiles at first benefited only by the use of waterpower in the fulling process, but the mechanically simpler and more laborious metallurgical processes changed substantially. In ironworking, waterpower was applied successively to bellows, to hammers and eventually (fifteenth century) to slitting, rolling, and wire drawing.

A series of mechanical innovations and improvements also led to advances in the manufacturing and processing of other materials. Plant ash to make glass was replaced by more-or-less pure soda, and the furnaces to melt it in became larger. Textile looms improved, especially with the introduction (from China) of the draw loom. Even more important was the development, near the close of the thirteenth century, of the spinning wheel in place of the ancient handspun whorl, virtually unchanged since prehistoric times.

Power not only enabled the scale of operation to be increased in iron-working, but the product was more uniform because of the extensive working that was possible. In addition the use of power changed the basic chemistry of the process. Although a large furnace is not needed in order to produce molten cast iron, it is much more easily made in a tall shaft furnace driven by powerful bellows than in a low hearth. Cast iron first appeared in Europe in the fourteenth century, following a sequence of developments which is unclear but which certainly involved power-driven bellows, larger furnaces, and perhaps hints from the East. To begin with, cast iron was used only as an intermediate stage in the making of steel or wrought iron, and it was developed for its efficiency in separating iron from the ore by production of both metal and slag in liquid form. However, cast iron that contains enough carbon to be fusible is brittle, and it took Europeans some time to realize its utility, although it had long been used in the Far East.

Ferrous Metallurgy

By the beginning of the fifteenth century, cast iron containing about 3 percent carbon and commonly about 1 percent silicon and which melts at a temperature of about 1200°C in comparison with 1540°C for pure iron had found three distinct uses: as a bath in which to immerse wrought iron in order to convert it into steel; as a material to be cast in molds to produce objects like pots, fire irons, and fireback more cheaply; and, most important of all, as the raw material for the next stage of iron manufacture.

The age-old process of directly smelting the ore with charcoal and flux in an open hearth or low shaft furnace yielded a product of low-carbon material in the form of an unmelted spongy mass, which was forged to expel slag, to consolidate it, and shape it. It was inefficient because of the large amount of iron that remained in the slag, and the iron was defective because of the slag remaining in it. The wrought iron produced from cast iron by the new finery process was made by oxidizing the carbon and silicon in cast iron instead of by the direct reduction of the iron oxide ore. The two-stage indirect process gradually displaced the direct method in all technologically advanced countries. Its main justification was economic efficiency, for the resulting product was still wrought iron or steel, finished below its melting point and containing many internal inclusions of iron silicate slag. In the late eighteenth century the small hearth was replaced by a reverberatory puddling furnace which gave much larger output, but neither the chemistry nor the product was significantly different from that of the early finery.[37]

It was only with the possibility of obtaining temperatures high enough to melt low-carbon iron—essentially the time of Bessemer and Siemens in the 1860s—that slag-free ductile iron became commercially possible. The very meaning of the word "steel" was changed in the process, for the word, previously restricted to quench-hardenable medium- and high-carbon steel for tools, was appropriated by salesmen of the new product because of its implication of superiority.

Contributions of Empiricism and Theory

Earlier developments in iron and steel metallurgy had occurred with no assistance from theory, which, such as it was, was far behind practice. The Aristotelean theory of matter, essentially unchallenged in medieval times, recognized the solid, liquid, and gaseous states of matter in three of the four elements—earth, water, air, and fire. The theory encompassed the various properties of materials but was wrong in attributing their origin to the combination of qualities rather than things. Medieval alchemists in their search for a relation between the qualities of matter and the principles of the universe elaborated this theory considerably. One of their goals—transmutation—was to change the association of qualities in natural bodies. In the days before the chemical elements had been identified, this was a perfectly sensible aim. What more proof of the validity of transmutation does one need than the change in quality of steel reproducibly accomplished by fire and water? Or the transmutation of ash and sand into a brilliant glass gem, and mud into a glorious Attic vase or Sung celadon pot? Or the conversion of copper into golden brass? Of course, today we know that it is impossible to duplicate simultaneously all the properties of gold in the absence of atomic nuclei having a positive charge 79. One way to secure a desired property is still to select the chemical entities involved, but much can also be done by changing the structure of substances. Such modern alchemy is more solid-state physics than it is chemistry, but it could not have appeared until after chemists had unraveled the nature and number of the elements.

Urged on by the manifestly great changes of properties accompanying chemical operations, the alchemists worked on the same things that concerned the practical metallurgist, potter, and dyer of their day, but the two groups interacted not at all. In retrospect one can see that the alchemist's concern with properties was not far from the motivation of the present-day materials scientist and engineer. They were right in believing that the property changes accompanying transmutation were manifestations of the primary principles of the universe, but they missed the significance of the underlying structure. Moreover they overvalued a theory that was too ambitious, and so their literature is now of more value to students of psychology, mysticism, and art than it is as a direct forerunner of modern materials science. Yet the alchemists—especially the Arabic practitioners of the art—discovered some important substances; they developed chemical apparatus and processes that are basic to science today, and they represent an important tradition in theoretically motivated laboratory operations, even if they failed to correct their theory by the results of well-planned critical experiments.[38] This approach proved sterile during the Middle Ages, though the workshop tradition represented by Theophilus led to many advances. The collaboration between the two approaches, which is the very basis and principal characteristic of today's emerging MSE, simply did not occur.

Impacts of Printing and Gunpowder

Two major technological developments helped precipitate the changes that signaled the close of the Middle Ages and the beginning of modern times: printing and gunpowder. Both of these had earlier roots in Chinese technology and both were intimately related to materials.

In the case of printing,[39] all the necessary separate elements were in general use in Western Europe by the middle of the fifteenth century: paper, presses, ink, and, if not moveable type, at least woodblock printing of designs on textiles and pictures and text on paper, and separate punches to impress letters and words on coins and other metalwork. But they had not been put together in Europe. Papyrus and parchment had been known in ancient times. Paper made of vegetable fiber had been invented in China a thousand years earlier and had been introduced into Spain by the Arabs during the twelfth century. Simple presses were already in use for making wine and oil, while linseed-oil-based ink (another essential element in the printing process) had been developed by artists a short time previously.

The idea for the most important element needed for mass production of verbal communication—reusable individual type—probably came to Europe from the Orient, although the history is obscure. By the eleventh century Chinese printers were working with baked ceramic type mounted on a backing plate with an adhesive and removable for reuse. By the fourteenth century in Korea, even cast bronze type was known.

Shortly after 1440 in Europe, everything came together in an environment so receptive that the development was almost explosive. Though there may have been experiments in the Lowlands, the successful combination of all the factors occurred in Mainz in Germany, where Johann Gutenberg began experiments in the casting of metal type during the 1440s. By 1445 he and his associates were able to produce a magnificent book, the *Gutenberg Bible,* still one of the finest examples of European printing. It consists of 643 leaves, about 40 cm × 29 cm in size, printed on both sides with gothic type in two columns. Some chapter headings were printed in red, others inserted by hand. Part of the edition was printed on paper, part on vellum, the traditional material for permanence or prestige. Unlike the earlier Oriental type, Gutenberg's was cast in a metal mold having a replaceable matrix with a stamped impression of the letter, arranged so that the body of the type was exactly rectangular and would lock firmly together line-by-line within the form for each page. Both the metal and the mold were adapted from earlier pewterer's practice. Thus a new technique for mass production and communication was established, ushering in a potential instrument for mass education. Modern times were beginning.

The political and economic environment had been strongly influenced somewhat earlier by the introduction of gunpowder in Western Europe. Explosive mixtures for holiday firecrackers had been used for centuries in China; it was only in the "civilized" West that gunpowder was first employed to enable man to kill his fellowman. Here too, it is uncertain whether the introduction of gunpowder in the West was a result of independent discovery or diffusion from the Far East. At any rate, the application and development was different and prompt. As early as 1325 primitive cannon were built in the West for throwing darts, arrows, and heavy stone balls, in competition with the mechanical artillery (the ballista) familiar since the days of the Romans, which were displaced completely by the middle of the sixteenth century. By 1450 the musket had appeared and began to render the crossbow and longbow obsolete.

By 1500 bombards, mortars, and explosive mines caused the medieval elements of warfare—the fortified castle and the individual armored knight—to lose their military importance and contributed to the decline of the feudal nobility.[40] (Another technical device, the stirrup, had aided their rise.[41] Accordingly, the changes in the technology of warfare aided in the process of administrative and territorial consolidation which was to give birth to the national state and transform the map of Europe.

Even the layout of cities changed as a result of the new methods of warfare: the round towers and high straight walls no longer afforded good defense in the age of cannon; they were replaced by geometrically planned walls and arranged so that every face could be enfiladed.

Military needs sparked a great development in the scale of the material-producing industries during the Renaissance, but agriculture, construction and the generally rising standard of living also contributed and benefited. The new supply of silver coming from Spanish operations in the New World and, no less, from the development of the liquation process for recovering silver from German copper upset the monetary balance of Europe. Silver, pewter, brass, wood, and the greatly increased production of glazed ceramic vied with each other for domestic attention, and glass democratically appeared in more windows and on more tables. . . .

In the meantime the separate components of ferrous and nonferrous metallurgy, ceramics, dyeing, fiber technology, organic polymers, and structural engineering pursued their own separate lines of development, and the basic sciences of chemistry and physics slowly generated an understanding that would help explain practice, enrich and extend it. Together, this all served to provide the facts and viewpoints that would eventually knit into the new grouping of man's knowledge and activity known as materials science and engineering.

Notes

1. The most popular—and most terrifying—of the projections prophesying dire results of the current trends in materials use in relation to present rates of population growth is to be found in the report of the Club of Rome's Project on the Predicament of Mankind. See D. H. Meadows, D. L. Meadows, J. Randers, and W. W. W. Behrens, III, *The Limits to Growth* (Universe Books, New York, 1972). The Club of Rome has since modified some of the doomsday conclusions of *The Limits to Growth* study; see G. Pestel and M. Mesarovic, *Mankind at the Turning Point* (New American Library, New York, 1976).

2. Sir George P. Thomson, ". . . a New Materials Age," *General Electric Forum XI,* Vol. 1 (Summer 1965), p. 5.

3. This is one of the major theses of A. Toffler, *Future Shock* (Random House, New York, 1970).

4. There is a formidable literature of anti-science and anti-technology. Not only are there the attacks of the counterculture (represented by the writings of

T. Roszak, P. Goodman, and H. Marcuse), but more thoughtful observers, such as L. Mumford, have attacked the spirit and practice of science and technology in the modern world. Among scientists, the work of B. Commoner (*The Closing Circle,* Knopf, New York, 1971) stands out in this regard. However, a more constructive evaluation of technology is emerging in such works as: R. Pirsig, *Zen and the Art of Motorcycle Maintenance* (Bantam, Des Plaines, Ill., 1976), A. Pacey, *The Maze of Ingenuity* (Holmes and Meier, New York, 1975); S. C. Florman, *The Existential Pleasures of Engineering* (St. Martin's, New York, 1976).

5. *Oeconomicus,* Book IV; see also Sir Desmond Lee, Science, Technology and Philosophy in the Greco-Roman World, *Greece and Rome,* 20 (1973), 70, 71.

6. Th. Dobzhansky, *Mankind Evolving: The Evolution of the Human Species* (Yale University Press, New Haven, 1962). A popular account is to be found in J. E. Pfeiffer, *The Emergence of Man* (Harper and Row, New York, 1972).

7. See V. G. Childe, *Man Makes Himself* (Mentor, New York, 1951), Chaps. 1, 2; *What Happened in History* (Penguin, New York, 1946), ch. 1.

8. See S. L. Washburn, Speculations on the Interrelations of the History of Tools and Biological Evolution. In J. N. Spuhler, (ed.), *The Evolution of Man's Capacity for Culture* (Wayne University Press, Detroit, 1959) pp. 21–31. A. Brues, The Spearman and the Archer—An Essay on Selection in Body Build, *Am. Anthropol.,* 61 (1959) 457–469. See also *The Dawn of Civilization* (McGraw-Hill, New York, 1961).

9. R. F. Heizer, The Background of Thomson's Three-Age System, *Technol. Cult.,* 3 (Summer 1962), 259–266.

10. J. Jacobs, *The Economy of Cities* (Random House, New York, 1969).

11. C. Lévi-Strauss, *The Raw and the Cooked: Introduction to a Science of Mythology* (Harper and Row, New York, 1969).

12. W. H. Gourdin and W. D. Kingery, The Beginnings of Pyrotechnology: Neolithic and Egyptian Lime Plaster, *J. Field Archaeology,* 2 (1975), 133–150.

13. J. F. Epstein, Flint Technology and the Heating of Stone. In D. Schmandt-Besserat (ed.), *Early Technologies* (Proc. Symposium at University of Texas, October 1976).

14. D. Schmandt-Besserat, Ochre in Prehistory. In J. D. Muhly and T. A. Wertime (eds.), *The Coming of the Age of Iron,* in press.

15. C. S. Smith, Art, Technology and Science: Notes on Their Historical Interaction, *Technol. Cult.,* 11 (1970), 493–549; also published in D. H. D. Roller (ed.), *Perspectives in the History of Science and Technology* (University of Oklahoma Press, Norman, 1971), pp. 129–165.

16. W. Sullivan, Anthropologists Urged to Study Existing Stone-Age Cultures, *The New York Times,* April 5, 1965. (Description of conference paper by J. D. Clark, University of California, Berkeley.)

17. K. P. Oakley, *Man the Tool-Maker,* 6th ed. (University of Chicago, Chicago, 1972), p. 81 ff.

18. V. G. Childe, *What Happened in History* (Penguin, New York, 1946).

19. R. F. Heizer, The Background of Thomson's Three-Age System, *Technol. Cult.*, 3 (1962), 259–266.

20. L. Aitchison, *A History of Metals* (Macdonald and Evans, London, 1960; Interscience Publishers, New York, 1960). This provides the background for much of the discussion of metals. See also T. A. Wertime, Man's First Encounters with Metallurgy, *Science*, 146 (3649) (December 4, 1964), 1257–1267; R. F. Tylecote, *A History of Metallurgy* (Metals Society, London, 1976), and W. D. Muhly and T. A. Wertime (eds.), *The Coming of the Age of Iron*, in press.

21. E. Eaton and H. McKerrel, Near Eastern Alloying and Some Textual Evidence for the Early Use of Arsenical Copper, *World Archeology*, 8 (1976), 168–191.

22. C. Renfrew, The Autonomy of the East European Copper Age, *Proc. Prehistoric Soc.*, 35 (1969), 12–47.

23. K. A. Wittvogel, *Oriental Despotism* (Yale University Press, New Haven, 1957).

24. V. G. Childe, Rotary Motion, In C. Singer et al. (eds.), *A History of Technology*, Vol 1 (Oxford University Press, New York and London, 1954), Chap. 1.

25. D. Schmandt-Besserat, An Archaic Recording System and the Origin of Writing, *Syro–Mesopotamian Studies*, 1 (1977) 31–70.

26. M. McLuhan, *Understanding Media: The Extensions of Man* (McGraw-Hill, New York, 1964).

27. R. J. Forbes, Extracting, smelting and alloying. In *C. Singer et al.* (eds.), *A History of Technology*, Vol. 1, ch. 21 (see also note 24).

28. R. Maddin et al., How the Iron Age Began, *Sci. Am.*, 237 (1977), 122–131; J. D. Muhly and T. A. Wertime (eds.), *The Coming of the Age of Iron*, in press.

29. Samuel Lilley, *Men, Machines and History* (International Publishers, New York, 1966), pp. 9–12.

30. N. Davey, *A History of Building Materials* (Phoenix House, London, 1961). See also I. L. Znachko-Iavorskii, New Methods for the Study and Contemporary Aspects of the History of Cementing Materials, *Technol. Cult.*, 18 (1977), 25–42.

31. C. S. Smith, Matter *versus* Materials: A Historical View, *Science*, 162 (1968), 637–644.

32. C. S. Smith, *A History of Metallography* (University of Chicago Press, 1960); Metallurgy as a Human Experience, *Metall. Trans.*, 6A (1975), 603; reprinted as a separate booklet by Am. Soc. Metals, Metals Park, Ohio, 1977, and Am. Inst. Mech. Eng., New York, 1977.

33. S. C. Gilfillan, Roman Culture and Dysgenic Lead Poisoning. *Mankind Q.*, 3 (January–March 1965), 3–20; see also S. C. Gilfillan, The Inventive Lag in Classical Mediterranean Society, *Technol. Cult.*, 3 (1962), 85–87.

34. C. C. Patterson, Silver Stocks and Their Half-lives in Ancient and Medieval Times, *Economic History Rev.*, 1972.

35. There have been two recent English translations of Theophilus, C. R. Dodwell (Thomas Nelson and Sons, London, 1961), and C. S. Smith and J. G. Hawthorne (University of Chicago Press, Chicago, 1963). The former contains an excellent edited version of the Latin text; the notes in the latter place more stress on technical aspects. For a further discussion, see L. White, Jr., Theophilus Redivivus, *Technol. Cult.*, 5 (1964), 224–233.

36. L. White, Jr., *Medieval Technology and Social Change* (Clarendon Press, Oxford, 1962).

37. H. R. Schubert, *History of the British Iron and Steel Industry*—to A.D. 1775 (Routledge and Kegan Paul, London, 1957).

38. R. P. Multhauf, The Origins of Chemistry (Oldbourne, London, 1966); this provides an authoritative and lucid discussion of the development of chemical theory and practice and their relations to materials.

39. The standard work on the early history of printing is T. F. Carter, *The Invention of Printing and Its Spread Westward,* revised by L. Carrington Goodrich, 2d ed. (Ronald Press, New York, 1955). See also H. Carter, *A View of Early Typography* (Oxford University Press, Oxford, 1969).

40. A. R. Hall, Science, Technology and Warfare, 1400–1700. In M. D. Wright and L. J. Paszk (eds.), *Science, Technology and Warfare* (Office of Air Force History and United States Air Force Academy, GPO, Washington, D.C., 1970), pp. 3–24.

41. L. White, Jr., *Medieval Technology and Social Change,* (Oxford University Press, Oxford, 1962). See also L. White, Jr., *Ballistics in the Seventeenth Century* (Cambridge University Press, Cambridge, 1952).

New Materials Technology in Japan

Gene Gregory

5

As long ago as 1981 the Japanese launched a major R&D effort in new materials, which they saw as an emerging megatechnology on a par with microelectronics and biotechnology. Today Japan already dominates world production of carbon-fiber-reinforced plastics and holds a commanding position in high-performance engineering plastics, polymer membrane materials, and amorphous alloys. And as Gene Gregory, professor of International Business at Sophia University, Tokyo, explains in detail in this authoritative article, Japan is poised to take the lead in perhaps the single most important area of new materials technology—fine ceramics. This will further reduce Japan's need to import basic materials and will help confirm Japan as the world's greatest economic power in the last decade of the twentieth century. The United States, it seems, has squandered its chances.

The technology of materials has entered a new era in the 1980s. By the year 2000 intensive research and development begun during this decade will have produced a broad spectrum of new materials that will replace ubiquitous structural metals such as steel and aluminum, conductive metals such as copper, and molded materials such as thermoplastics, in a widening range of uses. Not only will existing products be improved—stronger, lighter, smaller, energy-saving, labor-saving, and lower in cost—but a steady flow of

From *The International Journal of Materials and Product Technology,* vol. 2, no. 1, 1987.

entirely new products will bring about significant changes in industrial structures and patterns of global comparative advantage.

Strategic choices will change in the process. In the longer term new materials will be designed to meet specific needs, exploiting the expanding knowledge about materials behavior. And since Japanese industry is more vulnerable than others to potential materials supply curtailments and more sensitive to materials economies because of its almost total reliance on imports, for the remainder of this century the development of new materials will be among the highest R&D priorities.

At present, hundreds of Japanese firms are engaged in advanced materials research and an even larger number have focused their development efforts on new products employing these materials. Steel and nonferrous metals producers, chemicals companies, electrical and electronics equipment manufacturers, machine builders, motor manufacturers, textile companies, as well as glass, cement, and traditional ceramic producers—all have targeted new materials as a prime mover of future growth.

By the year 2000, if current estimates are correct, the market value of new materials will reach 5,400 billion yen, a tenfold increase over the 1981 level of 500 billion yen. In addition this expansion is expected to lead to the development of new products employing the new materials, creating a market for a further 4,800 billion yen in conventional materials. All told, the pervasive effects of these new

Table 5.1
Market sizes for new materials by type (in 1,000 billion yen)

		2000 (the 1981 price)		
Function	New materials 1981	New materials	Competing conventional materials	Materials total
High-functional high polymer materials	0.2	1.5	0.5	2.0
Fine ceramics	0.2	1.9	1.9	3.8
New metals	0.1	1.5	2.3	3.8
Composite materials	—	0.4	—	0.4
Total	0.5	5.4	4.8	10.2

Source: *Japan Steel Bulletin*.

materials, including 41,600 billion yen in new products and an inducement value of equipment and intermediates of 11,200 billion yen, will total as much as 52,800 billion yen.

The current thrust in materials research will thus be paying big dividends by 1990 and by 2000 will have succeeded in juggling the basic elements of the periodic table to develop new materials and products in prodigious profusion. Since many of these new materials, fine ceramics in particular, will rely less heavily on imported resources, energy as well as basic minerals or petroleum, their development has become one of the three major cornerstones of Japanese industrial strategy, along with microelectronics and biotechnology.

The search for materials with superior qualities and economies was formalized, at the government level, in 1981 when MITI launched a program to coordinate basic research on six types of materials:

- High-performance ceramics, which will be able to operate at temperatures above 1200°C in internal combustion and jet engines.
- High-performance engineering plastics which would have properties equivalent or superior to those of aluminum.
- Composites, including carbon-fiber reinforced plastics and ceramic-fiber reinforced metals which will be essential to the burgeoning Japanese aerospace industry.
- Electrically conductive polymeric materials for use as improved plastic superconductive substitutes for metal cables.
- Advanced alloys with aligned crystals, in mono or amorphous composition, for use as semiconductors, photovoltaic elements, and such high-stress applications as in turbine blades.
- Synthetic membranes for osmotic techniques to separate out solids from liquids and gases in process industries and pollution control.

MITI's budget for this decade-long effort is of modest proportions, including research in its own laboratories, which is a relatively modest 53 billion yen. Each of these projects is a cooperative effort, however, involving private sector corporate research which will ultimately raise total expenditures many times over that amount.

The basic goals of these projects, amplified throughout the R&D of still other firms not directly involved, as well as the work

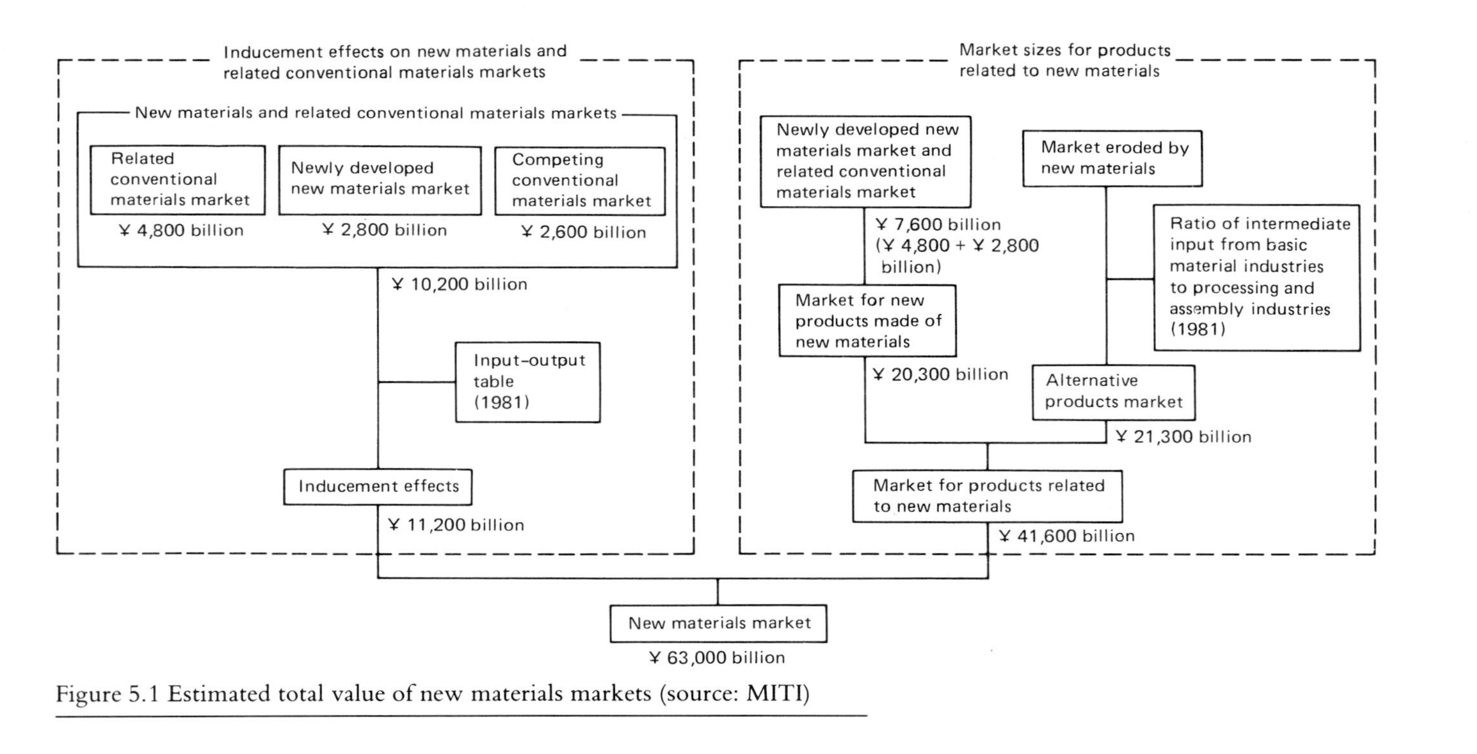

Figure 5.1 Estimated total value of new materials markets (source: MITI)

of a large number of university laboratories, are materials that are (1) light but strong, (2) functional at high temperatures, (3) of superior electrical and magnetic properties, (4) made from readily available natural resources, and (5) energy efficient to make and use.

Although Japanese materials research has not benefited from the impetus of large military and aerospace development programs, this does not mean that the level of Japanese new materials technology lags behind that of Western industries. Carbon fiber is a case in point. Invention of U.S. rayon-based carbon fiber and Japanese PAN (polyacrylonitrile)-based fiber occurred more or less simultaneously in 1954. Since then, full-scale development of composite materials using carbon fibers has progressed in the United States, fueled by NASA and U.S. Air Force requirements, while Japanese makers concentrated on commercial applications.

Although Union Carbide took the lead in 1959 with the commercialization of rayon-based carbon fiber, leading Japanese producers—Toray, Toho Rayon, and Mitsubishi Rayon—gained market share rapidly with PAN-based fiber in the 1970s to account for 70–80 percent of worldwide production. Toray's PAN-based fiber, which is approximately six times as strong and three times as hard as steel but substantially lighter than aluminum, attracted special attention in 1981 when it was selected for the cargo bay door of the space shuttle. And as much as 95 percent of the Lear Avia's airframe is made of Toray's carbon-fiber reinforced plastic. Developed originally by Osaka Industrial Testing Institute, PAN-based fiber is now challenged by pitch-based carbon fiber, a product of research at Gumma University and Kureha Chemical Company. Regarded as the next generation of carbon fibers, pitch-based fibers have the advantage of low material costs, high calcination yields, and resultant low product prices. Since pitch-based fiber is superior in rigidity and can be produced at less than 20 percent of the cost of PAN-based fiber, it is a promising material for automobile production and other large-scale uses which make it one of the principal new materials of the future.

With Japanese technology in the lead, a significant new global constellation has emerged which is expected to serve as a model for international industrial cooperation in other sectors of the new materials industry. Joint ventures have been established abroad, in the United States and Europe, by all Japanese carbon-fiber manufactur-

Table 5.2
Examples of Japanese manufacturers developing new materials

Company name	Annual sales (in 100 millon yen)	New materials as percent of total sales (estimate)	Typical new materials
Toray Industries	5,307	20	Si_3N_4, SiC, engineering plastics, RO, optical fiber, CF
Toyobo Co.	2,639	na	Engineering plastics, RO, electro-conductive plastics
Asahi Chemical Industry Co.	5,957	10	S_2N_4, RO, ion-exchange plastics, optical fiber, CF
Sumitomo Chemical Co.	6,739	1	Engineering plastics, superperformance membranes, ion-exchange plastics, alumina fiber
Sumitomo Electric Industries	4,289	1	Si_2N_4, optical fiber GaAs, electro-conductive plastics
Kyoto Ceramic Co.	1,006	100	Alumina ceramics, Si_2N_4, SiC, bioceramics
Toho Rayon Co.	710	6	PAN carbon fiber
Sumitomo Metal Mining Co.	2,858	10	GaAs, electronic materials, hydrogen-occluding alloys
Nippon Steel Corporation	31,126	—	Amorphous metals, CF, alumina fiber, steel fiber

Source: Long-Term Credit Bank.
Note: Annual sales are for the year ending March 1981, except Sumitomo Chemical Co. whose sales are for the year ending December 1980.

Table 5.3
Japanese carbon-fiber producers and their output

Manufacturer	Trademark	Annual	Outlook
Toray Industries	Torayca	1,260	Annual sales projected at 8.5 billion yen in year ending March 1983
Toho Rayon	Besfight	1,020	Annual sales projected at 6.0 billion yen in year ending March 1983
Nippon Carbon	Carbolon	420	
Mitsubishi Rayon	Pyrofil	120	
Sunika-Hercules	Magnamite	—	Finished product imported from Hercules, Inc. of America; raw material fiber supplied to same
Kureha Chemical Industry		240	
Mitsubishi Chemical Industries		—	Will construct 1,000 ton/year production facility in 1984

Sources: The Nikkei Sangyo Shimbun and Yano Research Institute.

ers, rather than use their technological lead to form wholly owned subsidiaries in those markets. As a result, in the foreseeable future the world's carbon-fiber production will be concentrated in four major groups, with vast R&D and production capabilities: Toray–Union Carbide; Toho Rayon–Celanese; Sumitomo Chemical–Hercules; and Mitsubishi Rayon–Courtaulds. All are fully integrated from raw yarn to finished products. Toray has also formed a joint venture with Elf-Aquitaine, the French national petroleum company, to produce PAN-based fiber in France.

These global structures in carbon-fiber production are significant not only because they manifest worldwide recognition of Japanese superiority in fiber technology, but because Japanese manufacturers have adopted clear strategies to develop international links with a view to avert trade friction in advanced technologies and to establish a foundation for further technological cooperation in the future.

Demand for engineering plastics is expected to increase throughout the remainder of the 1980s, with a market in 1990 al-

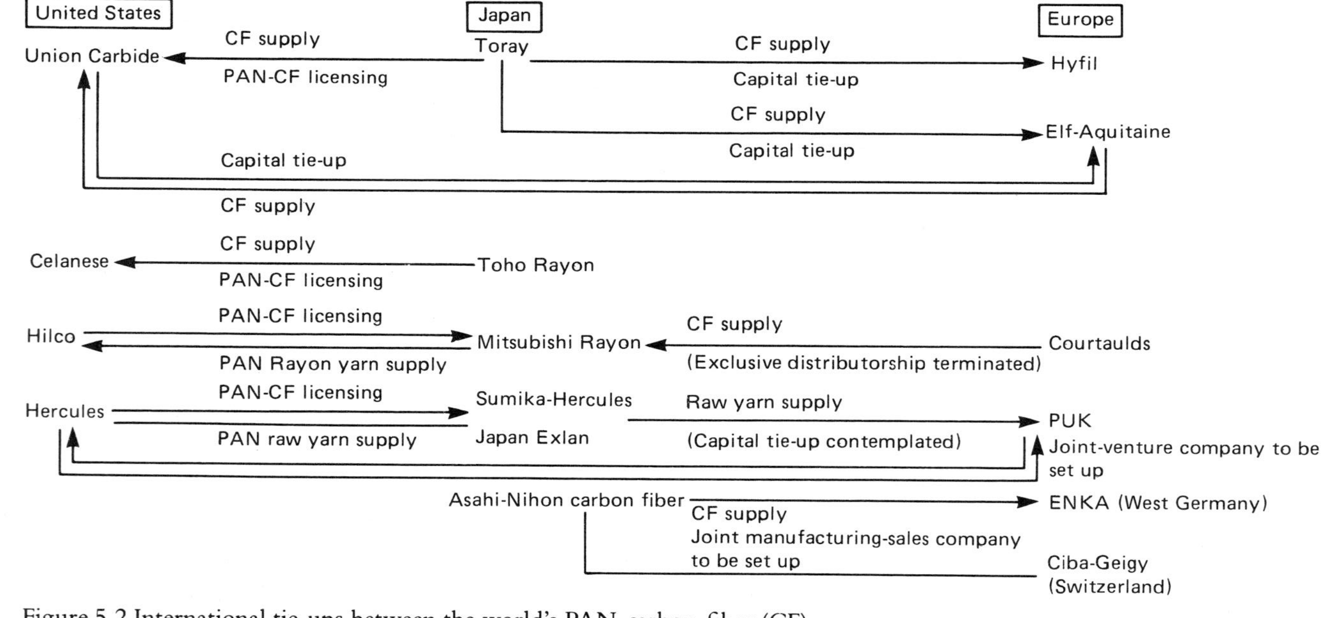

Figure 5.2 International tie-ups between the world's PAN-carbon-fiber (CF) manufacturers (source: Long Term Credit Bank)

most three times the size of 1983. But the number of producers is large, competition keen, and accordingly profitability is not expected to be as great in this sector as for other new materials. High-performance polymer membrane materials are already in use in numerous industrial applications, and ion-membrane technology is now the most advanced in the world; conductive polymers, however, are still in the R&D stage. But work presently underway at Toray Teijin, Asahi Kasei, Sumitomo Chemicals, and Sumitomo Electric is expected to result in significant new developments which, given the potentially broad base of applications in the electrical and electronics sectors, are expected to have a far-reaching industrial impact.

New metals technologies hold great promise. In the development of amorphous alloys alone, more than 100 companies are racing to bring new products to market. Sony succeeded in putting an amorphous alloy to practical use for the first time in the world in April 1980 with its cartridge for record players. Hitachi Metals has developed the production technology for an amorphous alloy just 25 microns thick and 100 mm wide, the world's highest precision in thickness and width, for use in magnetic cores. On the basis of this development, and the strength of the Hitachi group, the company expects to gain at least 10 percent of the world market for amorphous metals.

Another significant application of amorphous metals is in optical magnetic memories. Although this technology is still being developed by the KDD (Kokusai Denshin Denwa) Research Center and the NHK Basic Broadcasting Technology Research Center, when completed it could double information capacities in the future.

Solar cells are another important application of amorphous metal. Although mono or polycrystalline silicon have been considered to be the appropriate material for solar cells, the Japanese electronics industry has focused much of its R&D effort on the development of solar cells using amorphous silicon. Sanyo Electric, Fuji Electric, and Mitsubishi Electric are locked in fierce competition to develop new solar cells which will reduce the cost of energy generation from 650 yen to 16 yen/kWh. Wagering that this development will eventually break through the cost barrier to make solar energy an economically feasible proposition, leading electronics,

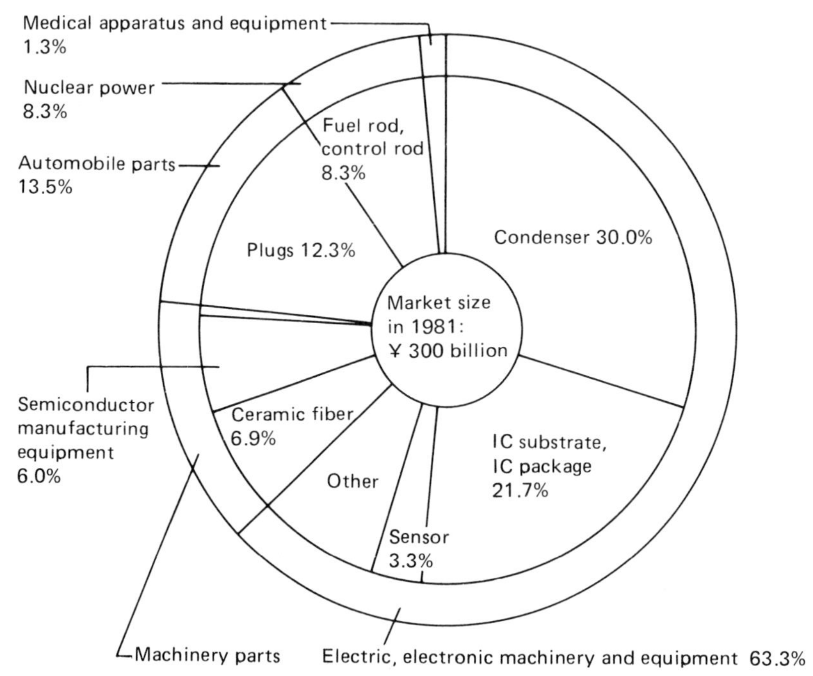

Figure 5.3 Composite ratio of new ceramics by use (source: MITI)

electrical, and chemical companies are investing heavily in production of amorphous silicon solar cells.

As important as many of these developments are, however, by far the greatest impact on industry and consumers will come from fine ceramics—the product of ubiquitous clay.

In fact, among the technological changes during the 1980s, the current wave of innovation in fine ceramics will be among the most far-reaching. Once again, a technological revolution spurred by basic research for the U.S. space program and advanced electronics seems to be having its greatest immediate impact on Japanese industry.

In the fastest growth fields of fine ceramic applications, IC packaging, and ferrites, Japanese makers have established a commanding lead in world markets. In IC packaging, Kyocera alone reportedly supplies approximately 75 percent of world demand and other Japanese firms account for most of the remainder. In the last half of the 1970s, leading Japanese makers of IC packages, piezoelectric ceramics, and ceramic capacitors were doubling output every two to three years. Patent applications on new developments in fine ceramic technology have been climbing steadily in the 1980s as an increasing number of companies sought new growth opportunities in a basic material that is destined to replace metals, and eventually plastics, in an ever-widening field of uses.

Statistics on actual production of fine ceramic products in Japan vary widely, since many are used directly by makers in the end-product manufacture (such as spark plugs) and product classifications are still not standardized. Conservative estimates by MITI, however, placed fine ceramics output at approximately 300 billion yen annually at the beginning of the 1980s, while the Yano Research Institute reported total fine ceramics production of 410 billion yen for 1981, based on its independent survey of some 85 percent of the firms in the industry. With growth in the industry in excess of 18 percent annually, production rose to 726.4 billion yen in 1984, according to a survey of the Fine Ceramics Association.

If performance during the first half of the 1980s can be taken as a prologue of things to come, these projections will very likely be over-reached. By 1981 output had already surpassed early, admittedly conservative, projections of the Long Term Credit Bank for 1985, and the Bank's forecast of 900 billion yen in fine ceramics

Table 5.4
Fine ceramics market in Japan

	Product	Market size (billion yen)
Electronics ceramics	IC packages, insulation substrates	117
	Ceramic capacitors	72
	Piezoelectric ceramics	65
	Thermistors/varistors	27
	Gas/humidity sensors	1
	Ferrites	85
	Translucent ceramics	4
	Heat generators	2
Industrial ceramics	Cutting tools	12
	Wear-resistant materials	7
	Heat-resistant materials	9
	Catalyst carrier for emission control	8
Other uses	Materials for atomic power generators	1
Total		410

Source: 1983, Yano Research Institute.

production in 1990 was exceeded by the middle of the decade. More recently, the bank revised its forecast to 1,035 billion yen by 1988, and 1,241 to 1,410 billion in 1990. Other authorities, such as the research division of Toshiba Corporation and the National Economic Research Association, predict an even more rapid growth rate with total output exceeding 2,000 billion yen by 1990. But even at the more modest growth rates projected by the LTCB, fine ceramic output during the remainder of this decade will exceed that of high-fliers such as word processors, facsimile equipment, industrial robots, carbon fiber, and fiber optics.

The truth of the matter is, of course, that no one can estimate with any degree of exactitude the amplitude of change this revolution in basic materials will bring in the remainder of the 1980s and the 1990s. In electronics alone, new ceramic applications are being announced at a quickening pace. In the first two years of this decade, Matsushita Electric, Toshiba Corp., and TDK Electronics each applied for over 100 patents on fine ceramic inventions; and established leaders in the field, such as Murata Mfg. and Toshiba Ceramics, continue to diversify into new fine ceramics production.

Demand for electro-optical ceramics is expanding in consumer electronics and in computer and communications equipment pro-

Figure 5.4 Forecast of fine ceramics market size (source: *Fine Ceramics Basic Problems Study Council Report,* May 1984)

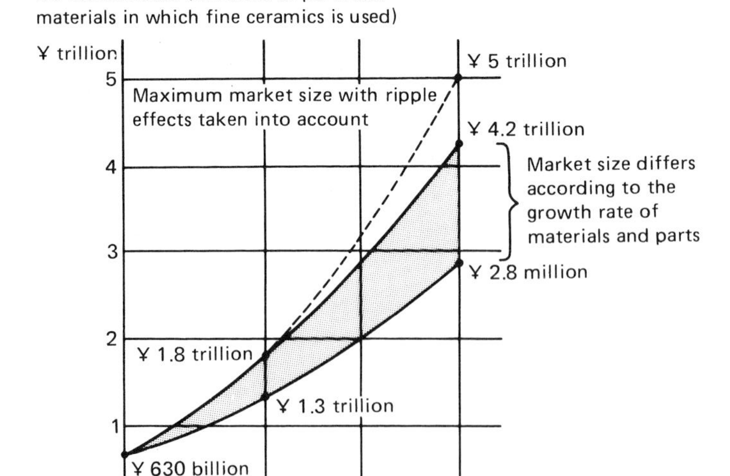

duction by geometric proportions: magnetic iron oxide for facsimile equipment and VTR tapes, gadolinium-gallium-garnet (GGG) crystals for magnetic bubble substrates for use in the next computer generation, and various materials used in VLSI production.

But, as important as these developments in semiconductor and other electronics applications are for the immediate and medium-term growth of fine ceramics output, the takeoff is yet to come when technology presently being perfected makes possible the replacement of iron and steel in automotive engines and an increasing number of machinery parts.

The basic economics of fine ceramics is sufficiently compelling to propel a technological revolution that will alter the structure of the materials industry and the geographical pattern of materials supply. Ceramic materials are generally more abundant and evenly distributed throughout the earth's crust than conventional metals used in modern industry, and many of the alumina and silicon ceramic materials are plentiful in Japan. Their refinement and downstream fabrication processes tend to require relatively less energy than metals, need not be located in congested industrial areas, and are more pollution-free.

On the demand side, the qualities of many ceramics are superior in many industrial applications, resulting in economies in materials and energy as well as products (spacecraft protective shields, for instance) that cannot be made of conventional metals. Down-to-earth applications in automotive engines will not only mean lighter and more fuel-efficient motor vehicles, they will also bring about changes in suppliers of engines and their production at different locations.

These changes are not far-off speculative developments. Their pace is being quickened by the increasingly fierce competition for otherwise mature automotive markets. Detroit automakers have begun to introduce ceramic parts and carbon fiber in new models and expect to have ceramic turbine engines that will be mass-produced in the early 1990s. For the Japanese auto industry it has therefore become vitally important to establish technological capacity in this field and, if at all possible, to gain a competitive advantage by the end of the 1980s.

The major thrust of research on ceramic engines comes from twelve motor manufacturers and producers of components. Some of them began R&D in this field more than fifteen years ago and have research teams of more than a hundred people working to develop this critical technology.

In recent years Japanese ceramic materials makers have also made significant progress in ceramic engine design, with prototypes produced by both Kyocera and NGK Spark Plug. New ceramics that withstand turbine mouth temperatures of up to 1,500°C are being developed under MITI's "Moonlight" program for comprehensive energy conservation technologies which, once perfected, will speed the introduction of ceramic engines. The Toshiba Corporation has already developed a new silicon nitride ceramic with bending strength of 150 kg per square millimeter at 1,200°C, the world's highest level.

Joint R&D by Japanese motor manufacturers with ceramic firms gives them a head start in ceramic engines because the two combine all the essential expertise of basic materials development with engine design and production. Although the lead has been taken by Isuzu, both Toyota and Nissan have important R&D programs in this field, as does Honda. Initially, production will almost certainly be undertaken by the manufacturers of diesel engines, not only because the engine is the critical part of a vehicle but also because the leading makers have the R&D organizations capable of continuing a sustained effort and their competitive future depends, at least in part, on their success in new materials development. Materials producers and fabricators will become important automotive parts manufacturers—in joint ventures for development and production of head plates, cylinder liners, piston heads, piston pins, turbocharger rotors, and other engine parts—thus challenging the market positions of established metal parts makers.

Isuzu and Hino already have perfected diesel engines with ceramic pistons, valves, and other parts that eliminate the need for water cooling and reduce fuel consumption by one-half, with 30 percent higher output than conventional engines. Two of Japan's largest turbocharger makers, Mitsubishi Heavy Industries and Ishikawajima Heavy Industries, are now supplying domestic and overseas motor manufacturers with ceramic turbines of molded silicon nitride, which are 40 percent lighter than conventional types and achieve significant improvements in acceleration.

Meanwhile the MITI-sponsored fifteen-company cooperative

research project for the development of a gas turbine engine completed its work last year. Production of gas turbine parts such as rotors, stators, burners, nosecones, heat exchangers, and others is now approaching the commercial stage—completing the first step toward the ultimate all-ceramic gas turbine engine.

Major advantages of these new engines which make them especially attractive to motor manufacturers are:

1 Sixty percent less thermal conductivity than present engines.
2 Elimination of radiator systems.
3 No need for lubricants.
4 Increased fuel efficiency by more complete combustion at higher temperatures.

Together with the introduction of carbon-fiber reinforced plastics, new ceramics will therefore hasten replacement of conventional metals used in a widening range of applications. The amount of steel used in a 1990 motor car alone will, in all probability, be at least 20 to 30 percent less than those of 1975, and iron consumption per car is expected to fall by as much as two-thirds.

Other important applications for structural ceramics are also in advanced stages of development. Significant progress has been made in the design of special materials for tritrium breeder reactors used in nuclear fusion, cement for geothermal wells, and other heat-resistant applications likely to spur the development of new energy sources. Although these markets are likely to be important, the major objective is clearly perfection of mass production applications for automotive engines.

Before fine ceramics are used as structural materials, however, several critical problems remain to be solved. Heat-resistant properties must be improved for use in engines operating at high temperatures. And, more difficult yet, are remedies for brittleness, which is the common property of conventional ceramics. This problem can be overcome, however, by using nonoxidized raw materials, such as silicon nitride and silicon carbine. Other possibilities are being actively explored in the use of binders for fastening ceramic particles in the sintering process.

Finding solutions to these problems in the laboratory is one thing; their application in manufacturing is another. Mass production of ceramic products has traditionally been characterized by problems of poor yield and disparities in strength. Unlike metals, whose properties can be controlled rather precisely in present refining methods, even ceramics produced at one time in the same kiln can have strength differentials as high as 30 to 40 percent. To complicate matters, cutting and working ceramics is difficult because the material is extremely hard, and detection of faults is much more difficult than in metals.

Since in mass production it is critically important to avoid using high-cost labor in overcoming these problems, competition to develop new ceramic materials requires increasing concentration of resources on the perfection of the best and lowest-cost production techniques. Current manufacturing methods allow only three to four cycles of the sintering process per day, in furnaces of limited size, and entail costs of 20,000 yen per kg, which is far too expensive for use in automotive engines that are currently produced of steel costing 50 yen per kg. Before fine ceramics for these mass-produced applications become competitive, costs will have to be reduced to 5,000 yen per kg or less.

Some cost reduction has been attained by eliminating pressurization and sintering at normal temperatures. Improved techniques have achieved product quality comparable with that formerly attainable only with expensive pressurized firing processes. Asahi Glass Co., for one, has already succeeded in commercially producing ceramics by normal pressure sintering, with strengths approaching 90 kg per square millimeter, more than sufficient for most machine parts. Its products are currently being used on conveyor rollers in steel rolling mills, and are expected to find wider applications as production costs are reduced.

New structural, or engineering, ceramics have recently been developed by other Japanese makers using special production methods with high-purity powders. These methods are performed in three stages, often by different manufacturers in the production chain:

- Ceramic powder production. Various oxides and nonoxide compounds of aluminum, silicon, zirconium, boron, and other metallic elements are synthetically made to specification for specific types of products and applications.
- Mixing and sintering under various pressures and chemical gas

atmospheres. By selecting various combinations of basic materials and production conditions, many different high-performance ceramics can be made for use as structural materials.

- Shaping, cutting, and polishing in final products is followed by testing to ensure strength and lack of defects. Nondestructive testing methods are now under development as part of the multifrontal effort to perfect the mass production of fine ceramics.

Among the various engineering ceramics, nonoxide compounds such as silicon nitride, silicon carbine, SIALON (silicon-aluminium-oxide-nitride), boron nitride, and zirconia are emerging as important industrial materials. Nozzles, pumps, valves, and other machine parts used at high temperatures and under high pressure are now being made from these superceramic materials. The industry expects the engineering market for fine ceramics produced by these processes to double or triple annually for the next few years.

Sharply mounting demand for new engineering ceramics is further propelled by three other promising fields for development: sensors, bioceramics and "cermet."

Ceramic sensors are used in continuous casting systems in steelmaking, for household gas detection and humidity controls in household appliances, medical electronics equipment, and measuring instruments. Ceramic bones and teeth are being marketed by Kyocera, Mitsubishi Mining & Cement, and Hoya Corp. and are expected to find a vast market considering the rising age level of the Japanese and world populations generally. And a family of hybrid products, cermet, which combine ceramics and metals, is finding increasing use in machine tools, brake linings, and fiber-reinforced metals such as Nippon Carbon's "Nicaron" (silicon carbide fiber).

In response to this vast and growing array of opportunities for superceramic applications, a small army of firms has been intensively developing new materials. Spearheading the advance are traditional ceramic manufacturers NGK Spark Plug, NGK Insulator, Narumi, and Toshiba Ceramics, along with the new breed of highfliers in electronic ceramics led by Kyocera, Murata, and TDK Electronics. But moving in behind these forerunners in growing numbers are glass-makers, cement and chemical companies, metal refiners, synthetic fiber makers, and all of the major integrated electrical machinery and communications equipment manufacturers.

By mid-1983 more than 150 firms had joined the Japan Fine Ceramics Association, and two to three times that many were contemplating some form of new fine ceramics development. MITI officials estimate that by 1990 the industry will include as many as 500 to 600 firms. But this is only a fraction of the total industry; in the Seto district of Aichi Prefecture alone, there are between 1,200 and 1,300 ceramic companies, 70 percent of which are involved in fine ceramics production.

A megatechnology like microelectronics and biotechnology, fine ceramics is serving to induce a broad restructuring of Japanese industry.

Structurally depressed industries whose products are no longer competitive in domestic or world markets, finding new growth potential in fine ceramics, are shifting both R&D and production resources to take advantage of these opportunities. At the same time, many highly integrated high-technology firms are adding fine ceramics to their diversified range of new growth activities. And, just as new entrepreneurial upstarts such as Sony, Casio, and Sord found their places in the firmament of the electronics industry, providing vital explosive energy at critical development stages, innovative newcomers are finding new opportunities for venture business in fine ceramics.

The traditional ceramics industry itself is typical of the depressed industries that are finding new life in this technological revolution. One of Japan's oldest industries, with its modern structures founded on a millenium-old experience of ceramic production, in the mid-1970s all but a few of its 2,113 member firms had seemingly run out of steam. Output of ceramic wares in 1974, the year after the first oil shock, totaled just over two million tons, of which the production share of small and medium-scale enterprises amounted to 70.1 percent.

In its modern form the industry has been dual-structured: kitchenware, ornamental ceramics, construction tiles, and diverse industrial ceramics are produced by 1,411 minor enterprises; while ceramic sanitary ware, electrical ceramics, and most industrial ceramics, which have scale economies in production, are manufactured by a few relatively large-scale enterprises. Paralleling this dual structure, electronic ceramics are produced by both small- and large-scale enterprises. Beginning small with ceramic capacitor production in the early days of the transistor radio development, TDK Electronics, Kyocera, and Murata have emerged as world leaders in electronics ceramics with global manufacturing and sales organizations. Despite the preeminence of these firms in ceramic capacitors, piezoelectric ceramics, IC packaging and ferrites, there are several dozen other small manufacturers sharing in the markets for these products.

These makers of electronic ceramics are among the pacesetters in the transformation of the ceramics industry to new high value-added production. In the first half of the 1970s, total output of the ceramics industry grew only 12 percent, at an annual rate of 2.3 percent. Fine ceramics (especially electronics ceramics) production increased 88 percent, however, or at an annual rate of 13.5 percent. Even more remarkably, while exports of ceramic products as a whole had levelled off and were declining, exports of fine ceramics rose by 7.72 times in that early stage of development.

By 1980 production of ceramic IC packaging and insulation substrates alone had grown to 107 billion yen, equal to almost a third of the total 341 billion yen ceramics output of the entire industry in 1974. Piezoelectric ceramic sales in that same year amounted to another 65 billion yen, ceramic capacitor output had risen to 76.6 billion yen, while ferrite production was more than 53.8 billion yen. Taken together, production of these four categories of electronic ceramics alone had risen at the beginning of the 1980s to almost 90 percent of the ceramic industry's total output just six years earlier. When other production is added, fine ceramics output in 1980 had already surpassed the production of all ceramics in 1974.

Recent growth rates in fine ceramics have been impressive even by the standards of such high-growth industries as advanced semiconductors. In the last two years of the 1970s, ceramic IC package and insulation substrate output of the top fourteen makers rose by a respectable 85 percent, led by increases of 130 and 110 percent by NGK Spark Plug and Toshiba Ceramics in their race to trim the market share of front-running Kyocera. But even Kyocera, with well over half the domestic market, boosted its production by 80 percent in the two-year period. Only slightly less remarkable were the average 76 percent growth of piezoelectric ceramics and the 48 percent rise in ceramic capacitor output for the same brief time span.

While a number of major ceramic companies have shifted their resources to electronic products, many more are moving into the rapidly growing engineering ceramics field, which is closer to the experience of many firms in the industry in producing industrial ceramic wares. And simultaneously, since the growth potential in engineering ceramics is especially high, electronic ceramic makers such as Kyocera themselves are also diversifying into this field.

The rush into superceramics production is not confined to the traditional ceramics industry, however. Nonorganic chemical firms, led by Showa Denko and Denki Kagaku Kogyo, have developed a wide range of basic ceramic powders. Nippon Steel and Mitsubishi Metal, and other leading metals companies, have also entered the field in response to the growing threat of fine ceramics as a metals substitute. Likewise, furnace materials makers Kurosaki Refractories and Shinagawa Refractories have taken advantage of their experience with heat-resistant ceramics to develop new ceramics in this field. And new ceramics for use in consumer products have been developed by automotive industry leaders such as Toyota and Nippon Denso, as well as by appliance makers Matsushita Electric, Toshiba Corp., Hitachi, and NEC.

Of the 150-odd companies that formed the Japan Fine Ceramics Association in 1982, only 35 were originally ceramic manufacturers; 29 were chemical companies, 25 were from the electronic industry, 22 from the steel and nonferrous metals industries, 18 from the machinery industry, 7 from the automobile industry and another 7 were heavy engineering companies.

Other effects of the fine ceramics revolution promise to be equally as far-reaching as the changes in the structure of manufacturing and in corporate product strategies. To the extent that ceramics replace metals and plastics, the locus of basic materials production will move to the sources of raw materials or sites, ensuring the best access to markets.

Heavy metals and chemical production in Japan has traditionally been sited at deep-water ports for most efficient handling of raw material imports and finished goods export. If the ports were not blessed by nature with the necessary depth and plant locations on shore, massive investments were made to create those facilities through dredging and filling. Ceramic production, on the other hand, has historically been concentrated close to the source of raw materials, which is principally centered around Nagoya, in Gifu and Aichi Prefectures.

In 1965 three-quarters of Japanese ceramic tableware was produced in those two prefectures, and much of the integrated output of sanitary ware, electrical insulators, and automotive ceramics was located in the same region. Kyoto, another traditional center of ceramic production, became a principal center of electronics ceramics manufacture.

Since ceramic materials and production know-how are found in at least 116 different locales throughout the Japanese archipelago, the fine ceramics industry has tended to develop outside the congested areas of the Kanto-Kansai conurbia. Indeed, centrifugal forces of land and labor costs increase the pressures on ceramic makers to locate in hitherto industrially remote regions of Japan. Responding to these pressures, Kyocera recently located its largest production facility and research laboratories at Kagoshima, where both land and labor were available at more advantageous conditions than in Kyoto. Significantly, the proximity of the plant site to a newly constructed international airport that enables expeditious handling of export air shipments was an important added attraction.

As most fine ceramics used in the electronics industry, as well as many engineering ceramic products, are high in value, and low in volume and weight, they are most economically transported by air. Hence, the emerging geographical pattern of fine ceramics production, given the general availability of materials and the high value to weight ratio of the final product, is centered around international airports that ensure optimal access to domestic and world markets.

But, as is already evident, the pull of the market in some instances will force the internationalization of production by Japanese fine ceramics makers. There are compelling advantages, in addition to avoidance of protectionist barriers, that induce location of production facilities close to the final consumer. Kyocera has responded accordingly with plants in the United States and Europe, ceramic capacitor makers are producing in major markets abroad, and other firms seeking to maximize their growth rates in this new field of high technology are considering overseas investments or cooperative undertakings with foreign firms in Japan.

One of the results, and not at all the least significant, of the new ceramics revolution will be to reduce the relative importance of those heavy basic materials industries which were formerly optimally organized along national lines. By increasing the propensity for global industrial structures in the new materials industry, and at the same time reducing the reliance on scarce natural resources, some of the pressures that have heretofore given rise to tension in the world economy may be mitigated.

Beyond the Era of Materials

Eric D. Larson, Marc H. Ross, and
Robert H. Williams

6

Expanding on a point made by Gregory, this important paper puts the materials revolution into global and historical perspective. The authors identify a "cycle" of demand for materials and argue that economic growth is no longer accompanied by increased consumption of basic materials. Indeed, we are moving from an Age of Materials to an Age of Information, this fundamental and perhaps irreversible shift being brought about by materials substitution, design changes, saturated markets, and a shift to high-tech goods with a low materials content. Larson and Williams are with the Center for Energy and Environmental Studies at Princeton University, while Ross holds key positions at the University of Michigan and the Argonne National Laboratory. This article first appeared in Scientific American, *June 1986. For a critique, see "Letters" in the October 1986 issue of the same journal.*

The significance of materials for human society is reflected in the fact that several epochs are named for materials that were exploited intensively during their span: the Stone Age, the Bronze Age and the Iron Age. The Industrial Revolution was based in large measure on radical improvements in the methods for modifying basic materials such as cotton, wool, iron, and later steel. Continuing improvements in production techniques have since made a growing number of material products available to more markets. Indeed,

From *Scientific American,* vol. 254, no. 6 (June 1986).

since the Industrial Revolution, a hallmark of economic growth has been an increase in the consumption of materials.

In recent years there appears to have been a fundamental change in this pattern of growth. In North America, Western Europe, and Japan economic expansion continues, but the demand for many basic materials has leveled off. It appears that the industrial countries have reached a turning point. They are now leaving the Era of Materials, which spanned the two centuries following the advent of the Industrial Revolution, and are moving into a new era in which the level of materials use will no longer be an important indicator of economic progress. The new era may turn out to be the Era of Information, but it is probably too soon to name it with confidence.

It is not too soon, however, to examine the causes of the change. Our work suggests four causes. Substitution of one material for another has slowed the growth of demand for particular materials. So have design changes in products that increase the efficiency of materials use. Perhaps more important, the markets that expanded rapidly during the Era of Materials are by and large saturated. And new markets tend to involve products that have a relatively low materials content.

The passage of the Era of Materials has some significant implications for the further economic development of the industrial countries. Those implications constitute a set of lessons that are only now beginning to be learned. The first lesson is that, contrary to much that has been said and written, imports had only a minor role in the weakening of the U.S. basic-materials industries through the 1970s. Instead, that weakness was due largely to a leveling off of demand. Since about 1980, however, imports have begun to undermine U.S. strength in materials processing. It is probable that in the decades ahead more of those operations will be carried out overseas, where costs are lower. The affected U.S. industries will fight to retain the later stages of manufacturing and will increase their role in the processing of recycled materials.

Although such trends appear to be disturbing, they by no means imply the death of manufacturing in the United States and other industrial countries. There are many opportunities for innovation in high-technology products characterized by a low materials content per dollar of value added, such as high-strength and corrosion-resistant steels and specialty chemicals for the fast-growing

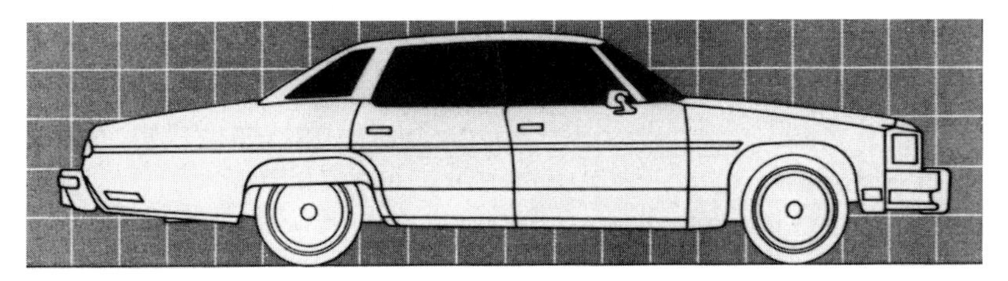

Figure 6.1 The 1976 *Caprice Classic* has a curb weight of 2,007 kilograms (4,424 pounds). During the mid-1970s rising energy prices provided a strong impetus for reducing the size of U.S. automobiles. For the *Caprice* the major "downsizing" took place between 1976 and 1977: the 1977 model weighed 300 kilograms (661 pounds) less than that of the year before. By means of changes in design and substitution of lighter, stronger materials, weight was saved throughout the car. For example, 54 kilograms were saved in the grille, sheet metal, and electrical work, 45 in the suspension and brakes, 30 in the frame, and 25 in the wheels, tires, and steering gear.

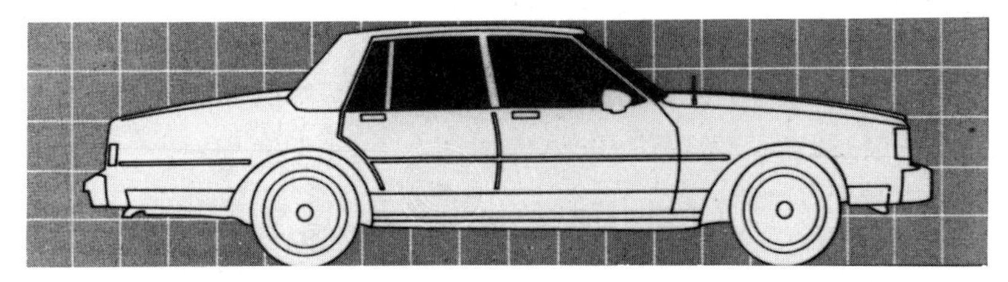

Figure 6.2 The 1986 Chevrolet *Caprice Classic* has a curb weight of 1,617 kilograms (3,564 pounds), or 390 kilograms less than the 1976 model. The disparity in weight between the two *Caprices* reflects a general trend among U.S. automobiles. Between 1975 and 1985 the average weight of U.S. cars fell from 1,727 to 1,450 kilograms. Part of the reduction was due to decreases in the iron and steel content along with increases in the fraction of aluminum and plastic. More efficient use of materials is among the causes of a leveling off of demand for basic materials in the industrial countries. The leveling off represents a significant turning point: since the beginning of the Industrial Revolution economic growth in industrial countries has been accompanied by increased demand for basic materials.

pharmaceutical, electronics, pesticide, herbicide, and biotechnology markets. Nevertheless, the end of the Era of Materials can be followed by a new era of growth in manufacturing only if the watershed through which industrial societies are passing is widely understood and appropriate adjustments are made.

Since in our view the primary factors that are affecting the changing role of materials involve consumption rather than production, we shall begin by describing the forces that underlie the demand for materials. We measure the consumption of materials in physical quantities (kilograms) rather than economic ones (dollars) to give a sharper picture of the changing role of materials. We represent the physical quantities in two main ways. If the physical quantity is divided by the population, the result is the consumption of a material per capita at a particular time. If the quantity is divided by the gross national product (GNP) in constant dollars, the result is an index of the significance of a material in the economy as a whole. By examining these two measures over time, it is possible to identify significant trends in the demand for a particular material. When the trends for various materials are compared, it is found that for many materials the historical pattern of demand follows a similar course.

When a material is introduced, consumption rates (measured in kilograms per capita) are typically low, and there are vast potential markets. In the early part of the cycle, consumption of the new material generally increases much faster than the economy as a whole. The rapid growth is reflected by a rise in the index of kilograms per dollar of GNP. Such rapid growth encourages improvements in processing technology that increase productivity and the quality of the product as they lower its price. These changes further stimulate the growth of demand.

In the second phase of the cycle more sophisticated products are made, the material begins to be a less significant component of the production process and the ratio of value added in manufacturing to the quantity of material increases. Additional innovations make it possible to use the material more efficiently. In this phase the demand for the material measured in kilograms per dollar of GNP peaks and begins to decline, even though consumption in kilograms per capita may still be increasing. Then, in the final phase of the cycle, the markets for the material in bulk become saturated; new markets are largely for specialty products, which have little effect on the total quantity of consumption. The reversal of growth is so complete that even per capita consumption levels off and may begin to decline.

The history of steel consumption in the United States provides an excellent example of the demand cycle for a material. During the late nineteenth century steel made a significant contribution to the construction of cities, heavy industry and railroads. The rapid growth in demand created a favorable climate for innovations such as the Bessemer process, which reduced costs and encouraged continued expansion. As a result of these trends the demand for steel grew faster than the GNP until about 1920. In the 1920s, however, the market for heavy steels of the type employed in building infrastructure stopped growing rapidly. As a result the consumption of steel per unit of GNP began to decline from the historic peak that had been reached in about 1920.

Although the demand for steel had begun to decrease in relation to the economy as a whole, per capita consumption continued to increase: steel was being produced for consumer goods. For example, steel sheets were needed for automobiles and household appliances. Such markets for consumer products were far from saturated in 1920. In the 1950s, however, the growth of per capita demand began to decelerate even as more consumer goods were produced. When consumer markets approached saturation in the 1970s and no new bulk markets developed, per capita demand leveled off. It is currently falling, and the significance of steel consumption in the national economy as measured by kilograms per unit of GNP is roughly what it was one hundred years ago, having decreased to 40 percent of the peak level of 1920.

Most other basic materials seem to be following a similar trend. We carried out case studies for six materials in addition to steel. Among them cement and paper are commodities that have been used extensively since the early days of industrialization. In contrast, aluminum, ammonia, chlorine, and ethylene are materials that came into widespread use only in the twentieth century. In every case we found the cyclical pattern of demand that is predicted by classical theories of industrialization. The duration of the cycle was much

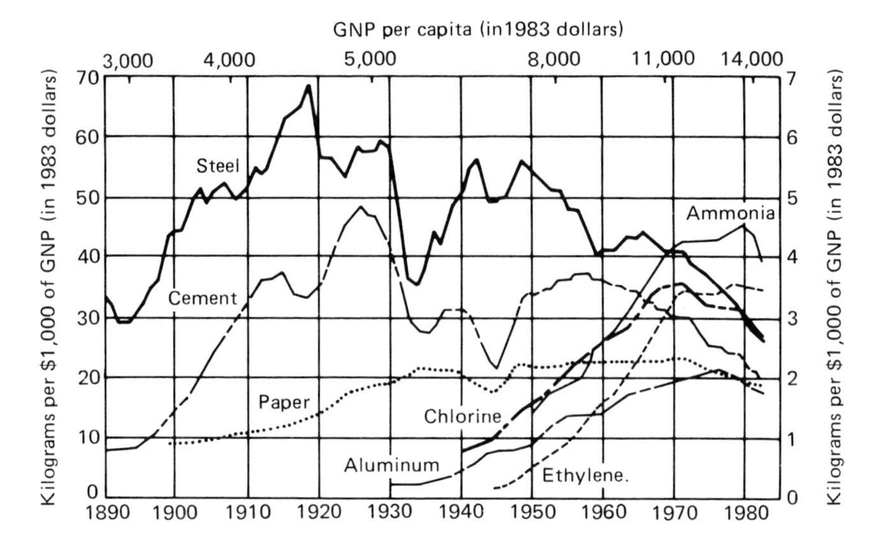

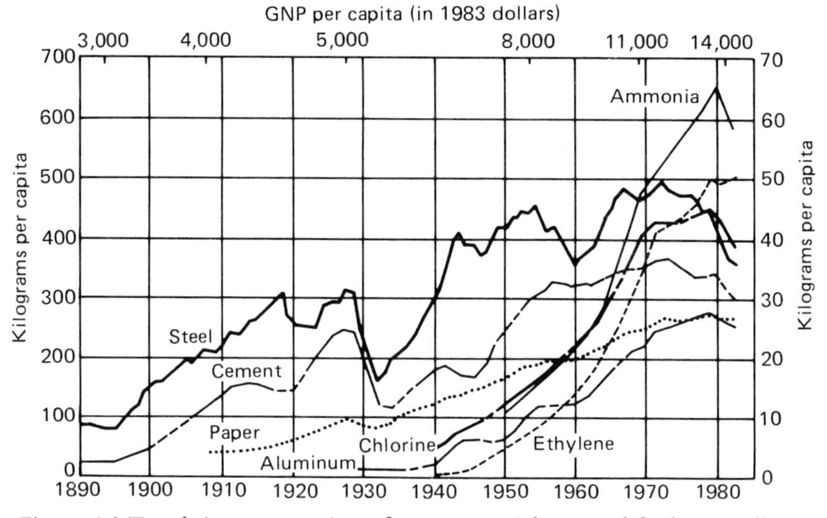

Figure 6.3 Trends in consumption of seven materials exemplify the overall fate of basic materials in the United States. Among the seven are two "traditional" materials in addition to steel: cement and paper. The others are "modern" materials: aluminum, ammonia, chlorine, and ethylene. Consumption of the traditional materials peaked long before that of the modern ones, but use of all seven is now declining relative to GNP (*upper panel*). For most of these materials per capita consumption is also declining (*lower panel*). The scale at the top of each panel shows GNP per capita. As it suggests, the diminished role of basic materials is related to increasing affluence.

longer for the traditional materials than it was for the modern ones, but the outcome was the same. For all seven materials, consumption per dollar of GNP had begun to decline by the 1970s, and consumption per capita has now essentially leveled off. Similar trends are seen in the industrial countries of Western Europe, including France, Germany, and the United Kingdom.

Is the leveling off of demand for materials merely a pause in a historical pattern of growth, or does it mark a fundamental structural change in the economy? Because the departures from long-term trends are recent, statistics alone cannot provide the answer. When the statistics are considered along with the underlying causes of the leveling off, however, they suggest that a profound shift is under way. As we noted earlier, four factors appear to be responsible: substitution of materials, more efficient use of materials, saturation of markets, and shifting consumer preferences. Whereas the first two have been operating almost since the beginning of the Industrial Revolution, the second two are relatively recent phenomena that are associated with the final stages of the demand cycle.

The substitution of modern materials has been slowing the growth of demand for traditional materials such as steel and paper. In the automotive market, which accounts for between 15 and 20 percent of the demand for steel in the United States, lighter materials are replacing steel. The tendency toward substitution was reinforced in the 1970s by the promulgation of minimum standards for fuel economy. To increase fuel economy manufacturers began to build smaller, lighter cars that contained less steel. The average weight of cars made in the United States decreased from more than 1,700 kilograms in 1975 to less than 1,500 in 1985. Over the same span the fraction of the average vehicle made up of iron and steel fell from 81 to 69 percent: the fraction made up of aluminum and plastic rose from 6 to 11 percent. Many such substitutions are taking place throughout U.S. industry.

Since early in the nineteenth century, improvements in the strength and durability of materials have also made it possible to decrease the amount of material in products without vitiating any of their functions. Indeed, the function of the product may be improved at the same time, as is shown by the history of the locomotive. In 1810, when boilers were made of cast iron or sheet iron, the

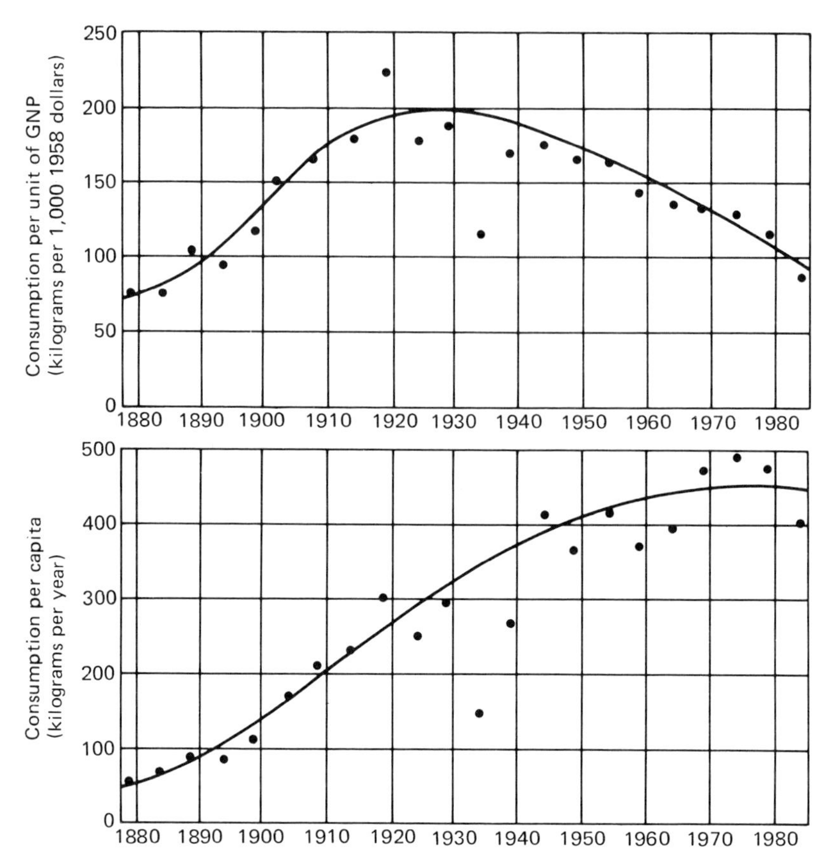

Figure 6.4 Steel construction in the United States illustrates the classic cycle of changes in demand for a basic material. The upper panel shows the consumption of steel in kilograms per $1,000 of gross national product (GNP). The lower panel shows consumption in kilograms per capita. In the early part of the cycle demand increases rapidly according to both measures. At some time consumption per unit of GNP, which indicates the relative importance of the material in the economy, reaches a peak and begins to decline. Steel consumption reached that point in about 1920. Per capita consumption continues to grow after that, but in the last stages of the cycle per capita consumption levels off and may even decline.

ratio of weight to power in a typical locomotive was about 1,000 kilograms per horsepower. Some 50 years later steel boilers were introduced, making possible significant reductions in weight: by 1900 the ratio was about one-tenth of what it had been in 1810. As electric locomotives were widely introduced around 1950, the ratio reached 25 kilograms per horsepower. By 1980 it was 14.

The dramatic reduction in the weight-to-power ratio of locomotives was due to a combination of many improvements in materials and design. Similar refinements can be seen in many modern products. Plastic films available today are stronger, yet thinner, than those sold a decade ago; the new films represent a general trend toward a lower materials content in plastic products. The radial tire, which is much more durable than the bias-ply tire, is an example of an improved design that leads to demand reduction (in this case for rubber). The aluminum can has also undergone an evolution in design: its average weight decreased by about 20 percent between 1970 and 1984, owing largely to improvements in production technology. Thinner walls of the can, together with design changes that reduced the size of the lid (in particular, "necking" of the ends of the can), account for the reductions.

At least two factors drive the increasingly efficient use of materials. One is the rise in the cost of producing the materials themselves, much of it due to higher prices for energy. Since 1972 the average price of electricity and coal (corrected for inflation) doubled for industrial customers in the United States; the price of natural gas quadrupled. Many materials-processing industries are quite energy-intensive: they consume large amounts of energy per unit of output. About three-fourths of the cost of producing ammonia is accounted for by natural gas (which for making ammonia is both a feedstock and an energy source). About one-third of the cost of making aluminum is accounted for by electricity. Such industries are naturally quite sensitive to increases in the cost of energy. The increases thus provide a powerful stimulus for making more efficient use not only of energy but also of materials.

The second factor that stimulates efficiency is competition from substitute materials. Such competition particularly affects the manufacturers of traditional materials, whose products must compete against modern materials with more desirable properties. Partly to

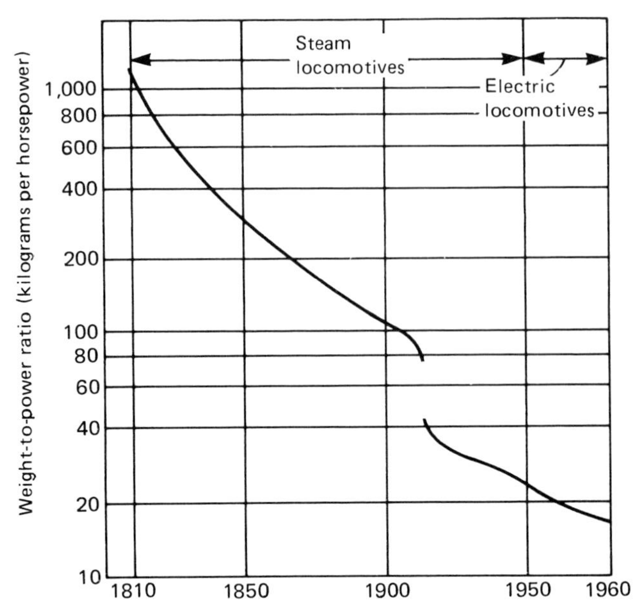

Figure 6.5 Weight-to-power ratio of locomotives underwent a decrease of nearly seventyfold between 1810 and 1980; the decrease reflects many improvements in design and materials. In the mid-nineteenth century iron boilers were replaced by boilers made of steel, a change that made possible lighter equipment and higher internal pressures. By 1900 the ratio had decreased by a factor of ten, and it continued falling during two world wars, reaching a level of about 25 kilograms per horsepower as electric locomotives were introduced around 1950. (The gap between 1910 and 1920 results from the disruption of data collection during World War I.) Similar (albeit less dramatic) improvements have been made in many industrial products. Substitution of materials and design changes that lead to more efficient use of materials are two of the factors responsible for the leveling off of demand for basic materials.

compete against the makers of aluminum, plastics, and fiberglass, the steel industry has put increasing emphasis on the production of specialty steels with properties such as high strength or resistance to corrosion. The automotive industry has responded by utilizing more specialty steels in its products. In 1975 about 5 percent of all the steel in an average car was high-strength or stainless; by 1985 the fraction was 14 percent, and most observers expect it to rise above 20 percent by early in the next decade. Such changes are accompanied by a reduced demand for steel, since each kilogram of high-strength steel typically replaces 1.3 kilograms of ordinary carbon steel or cast iron.

Although materials substitution and improved efficiency of materials use have become more significant in recent years, they are probably not as important in reducing demand as the relatively recent phenomena of the saturation of traditional markets and the shift in consumer markets to new, less materials-intensive products. Consider steel and cement. Much of the underlying demand for steel and cement in bulk in the United States in the nineteenth and early twentieth centuries came from the building of infrastructure, including highways, railroads, public transportation, commercial buildings, and housing. The era of infrastructure building is now all but over. The rail network has been shrinking for many years, and highway construction slowed in the 1960s with the completion of most of the interstate system. Demand for new housing will undoubtedly decline as the young-adult cohort of the population ceases to expand. The only area where demand for steel and cement continues to be strong is the market for commercial buildings, which is not a large one for materials in bulk.

Other basic-materials industries are facing similar, if less epochal, shifts in the pattern of demand. The most rapidly growing market for paper during the 1950s and 1960s was in corrugated boxes to serve as shipping containers. By about 1970 that market had encountered its limit because all except large commodities and bulk goods were being packaged in corrugated boxes. A parallel saturation of the market for inorganic chemicals in bulk is illustrated by recent trends in the utilization of nitrogen fertilizer, which accounts for about 80 percent of all the ammonia produced in the United States. From 1955 to 1980 the amount of fertilizer increased more than sixfold. Future growth will be slow, however, both because most cropland is currently receiving high levels of fertilizer and because today's high prices for fertilizer provide a powerful incentive to use fertilizer more effectively. Saturation of the market for organic chemicals is illustrated by the case of synthetic fibers, which have captured most of the apparel, automotive, and home-furnishing markets; there are no apparent opportunities for further expansion.

Markets for heavy consumer goods are also reaching the limits of growth, partly owing to the affluence of the industrial countries. Increases in income in the past few decades have made it possible for most households to buy a wide range of basic consumer goods. In the United States almost all adults who might drive have cars. Furthermore, people are now keeping their vehicles longer. From the late 1950s until 1975 the average age of passenger cars held fairly steady at between 5.5 and six years. After that it rose quickly, partly as the result of improved durability, reaching 7.4 years by 1983. Most observers of the automobile industry expect new-car sales to remain at the levels of the mid-1970s for the foreseeable future. Similar trends apply to such heavy household appliances as stoves, refrigerators, washers, and dryers, because most households own a full set. Rather than supplying growing markets, the function of production in these areas is to replace consumer goods as they wear out.

Saturation of traditional markets is reinforced by another and relatively new phenomenon: consumer preferences that change with rising income. The affluent tend not to spend additional income on more of the same—yet another car, for example. Instead, marginal income is often spent on items such as a videocassette recorder, a personal computer and the accompanying software, membership of a health club, better health care or a service that provides stock-market information. Although such goods and services are disparate, they are characterized by a low materials content per consumer dollar. There are now no significant new markets for consumer goods having a high content of materials per dollar.

Our analysis of the four factors responsible for the declining consumption of basic materials suggests that these recent trends in the use of materials will not be reversed: they signal a historic shift in the economies of the industrial countries. It is clear that the transformation has already had a profound effect on the structure of U.S. industry. Indeed, it appears that declining demand rather than the threat of imports was responsible for the stagnation in many basic-materials industries in the United States during the past decade. In the 1970s the overall production of basic materials followed a declining trend very similar to that of overall consumption. The fact that the production trend followed the consumption trend so closely implies that although trade was important for certain materials, its overall role did not change much.

Since 1980, however, the pattern has in fact begun to change. For many basic industries the first stages of materials processing have begun to move to countries where production costs are lower.

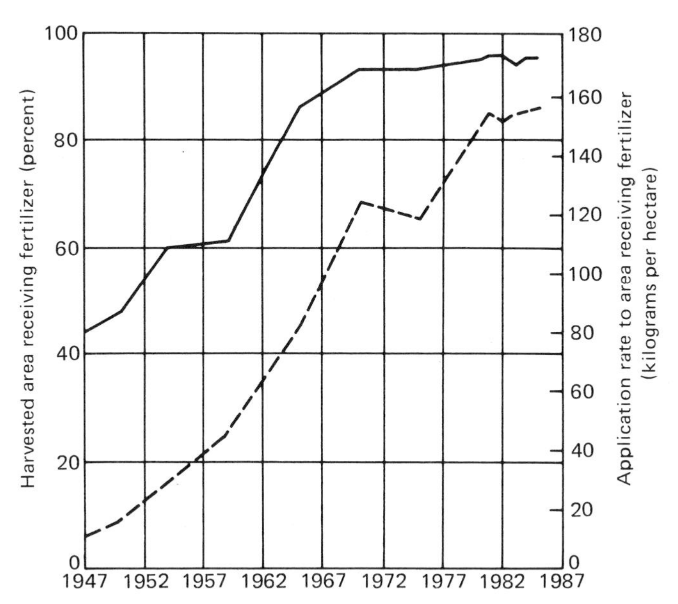

Figure 6.6 Saturation of markets is among the factors that have contributed to a leveling off of demand for basic materials. The data in the illustration show why the market for nitrogenous fertilizer (which accounts for about 80 percent of all ammonia produced in the United States) is largely saturated. The solid curve indicates the proportion of U.S. farmland on which corn is being grown that receives nitrogenous fertilizers. The dashed curve indicates the amount of fertilizer applied to each hectare of such land. (A hectare is about two and a half acres.) Two factors suggest that demand for fertilizer will not grow much more in the years to come. Almost all corn land already receives fertilizer. Moreover the benefits gained from applying more fertilizer diminish rapidly above the level reached in the past few years.

This trend will undoubtedly continue, with the result that production in the United States will decelerate even more quickly than demand. The response being made on the part of U.S. manufacturers is to cede primary processing operations to foreign plants while attempting to secure a continuing role in the later fabricating and finishing steps, in part through the development of new products. U.S. manufacturers will also expand production based on recycled materials. Both strategies are promising. The United States can draw on its diverse technological resource base to help bring new products to market. Moreover the fraction of materials needs that can be met by recycling is in principle much greater when demand is not growing than it is in the early phase of a rapid growth of demand.

A close look at the situation of steel, aluminum, ethylene, and paper producers suggests how the outlook can vary from industry to industry under the new market conditions. The steel industry has been affected by a leveling off of demand longer than any of the other three, and the stagnation of demand has had a stifling effect. Since 1950 only two all-new integrated steel mills have been built in the United States, the last one in the 1960s. (Integrated mills carry out the entire production process, beginning with the processing of iron oxides.) Building new integrated mills is not economic because the capital charges exceed the reduction in operating costs achieved by building the new mill. The introduction of new components into existing mills has also proceeded slowly in the United States—far more slowly than in, say, Japan. New steel plants are being built in countries where production costs are not as high as they are in the United States. The disparity stems from lower labor costs and in some cases from government intervention, which can reduce the cost of capital or even provide direct subsidies. Also many new foreign mills are situated at major ports and so have the advantage of low ore-transport costs.

Whereas integrated steelmakers in the Unites States may be affected by a terminal illness, the secondary steel industry, made up largely of minimills (whose raw material is scrap), is in robust health (see "Steel Minimills," by Jack Robert Miller, *Scientific American,* May 1984). A key part of the successful strategy of the secondary industry has been to choose products that can be manufactured with low capital costs and labor requirements per unit of output. In the late 1970s and early 1980s the minimill sector was growing by about 10 percent per year, and in 1983 it accounted for 18 percent of all steel made in the United States. Until now the minimills have by and large been limited to products that can be made from steel containing significant impurities (particularly copper). Technological advances may overcome that limitation, either by removing the impurities or by finding ways to improve properties of the steel (such as ductility) despite the impurities. Such advances would make it possible, for instance, to produce sheet steel for new cars from automobile scrap. If that were done, secondary producers might take a substantial part of the remaining market from integrated producers.

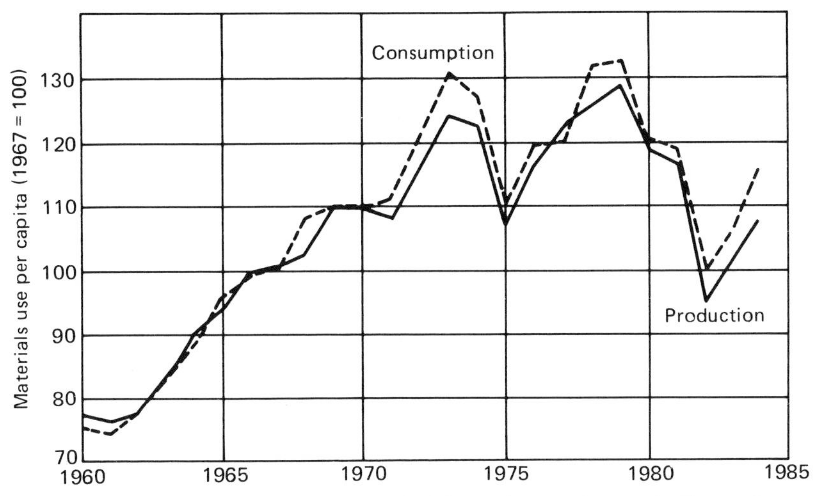

Figure 6.7 Production of basic materials followed a trend very similar to the trend for consumption in the United States during the 1960s and 1970s. The solid curve shows production per capita, the dashed curve consumption per capita. The close resemblance between the curves before 1980 suggests that until then imports had little effect on the overall level of production. If imports had been a factor, production would have fallen relative to consumption. Hence it appears that before 1980 the leveling off of demand was the main cause of slow growth in the industries that process basic materials. In the current decade the level of production has fallen with respect to consumption, which suggests that since 1980 both imports and the leveling off of demand are constraining domestic production.

Aluminum is another industry in which primary capacity is diminishing while the secondary sector flourishes. As we noted earlier, a large fraction of the cost of making primary aluminum goes to pay for electricity. Indeed, each increase of one cent per kilowatt-hour in the price of electricity leads to about a 10 percent increase in the cost of producing aluminum from ore. The cheapest electric power available in the United States is found in the Pacific Northwest, where the price is about 2.7 cents per kilowatt-hour to the aluminum smelters; power from new coal or nuclear plants costs between 5 and 6 cents. Unexploited hydropower sites and natural gas and coal at remote sites are natural resources that make possible low-cost electricity generation. In Australia, Brazil, and Canada new power plants that utilize such possibilities and that have been built or are under construction can provide electricity at less than two cents per kilowatt-hour. Hence expansion of primary capacity in the United States would not be economically competitive.

Making aluminum from scrap, however, consumes only from 5 to 10 percent as much energy as making it from ore. As a result the secondary sector is still capable of rapid growth: between 1970 and 1983 its share of total output increased from 20 to 33 percent. The recent expansion, which is based largely on the recycling of aluminum cans, is likely to continue.

The manufacturers of ethylene are in a somewhat different position than are the manufacturers of steel or aluminum, because the United States is still a net exporter of the chemical: in 1984 exports accounted for about 8 percent of total production. Yet the factors that are causing primary metals production to leave the United States are also affecting the makers of ethylene. Ethylene, a product of petroleum or of petroleum gas, is an important component of many intermediate chemicals (those that are further modified to yield the final product). From ethylene come polyethylene and other polymers, along with ethylene glycol and ethyl alcohol. Large new plants to make those substances as well as ethylene itself are under construction or have recently been completed abroad in areas having certain specific advantages. In the Far East the advantage is proximity to markets; in the Middle East, Canada, and Mexico it is proximity to raw materials, such as ethane, that are hard to transport. Completion of these facilities will substantially reduce the U.S. share of world ethylene production. In the medium term exports of basic ethylene derivatives are likely to decline sharply; in the long term the realization that it is cheaper to make ethylene abroad will inhibit construction of new plants in the United States. What is true for ethylene and its derivatives also applies to other basic organic chemicals.

Among the basic materials paper is the one that provides the greatest solace to U.S. industry. For at least the time being the paper industry is assured a niche in the marketplace of the Information Era, owing in large part to rapid growth in the demand for office paper. (The "paperless office" remains a scheme of the future.) Although overall consumption of paper is declining in relation to GNP, it is not declining on a per capita basis. Moreover the likelihood is that the international demand for pulp and paper will increase sharply in the next few years. The United States has a unique combination of forest resources, production facilities, and technical skills, which may make it possible to export more paper goods to supply the needs of other countries. In most cases those needs are for two commodities: linerboard, the high-strength facing from which corrugated shipping containers are made, and high-quality pulp, which goes into a variety of products. Many countries that have their own paper mills still must import these materials, and so aggressive marketing could lead to considerable growth in U.S. exports.

The combination of mature domestic markets and increased foreign competition facing the basic-materials processing industries of the United States and other industrial countries has far-reaching implications for economic planning. Those who make economic policy in the industrial countries must recognize that reforms such as subsidizing the steel industry would not restore rapid growth, because they would have no effect on the underlying stagnation in demand. The materials industries cannot be sustained by protecting antiquated technology. Substantial innovation will be necessary to bring these industries into conformity with the present reality. In the past the rapid growth of demand was a spur to technological innovation, but that stimulus is gone. Perhaps external stimuli, such as tax and research-and-development policies that encourage innovation, will be needed if the materials industries are to adapt to changed circumstances.

The recent changes certainly call for increased effort on the part of economic planners. Yet in some respects they make planning easier. Specifically, the shift away from primary materials processing could stabilize or even reduce overall energy requirements for industry. Since the processing of basic materials consumes much more energy per dollar of output than fabrication and finishing activities do, even a small shift away from processing can have a profound effect on the energy consumption of industry (which in 1984 accounted for about two-fifths of all energy consumed in the United States). Our analysis suggests that aggregate materials production will remain roughly constant in the United States between 1984 and 2000 (when measured in kilograms of output weighted by the energy consumed in manufacturing each product). Since we expect industry to improve the efficiency of its energy use at a rate of from 1 to 2 percent per year during that period, the result may well be a decrease in industrial energy consumption, perhaps of as much as 20 percent, so that considerably less capital would be needed to provide energy for industry than is suggested by conventional estimates.

A decrease in energy requirements would also slow the upward trend in the world oil price that will inevitably occur in the decades ahead as the market economies of the world again become heavily dependent on oil from the Persian Gulf region, where the bulk of the earth's remaining oil resources lie. The economic benefit hence would be complemented by a reduction in the world tensions that arise from competition for increasingly scarce resources. In addition

Figure 6.8 Recycled materials may provide a way for manufacturers in industrial countries to lessen the effect of imported materials. The solid curve represents the fraction of all steel products used in the United States that were made elsewhere. The dashed curve shows the corresponding fraction of minimill products. (Minimills are steel mills that employ scrap as raw material.) Although imports are increasing in the steel industry as a whole, they are being driven out in areas where minimill products are concentrated. The authors contend that manufacturers in the industrial countries will have to cede much of the processing of primary materials to countries with advantages in labor or raw materials costs. Such losses may be partially offset by growth in industries based on recycled materials, along with expansion of industries that entail extensive fabrication and finishing of diversified products.

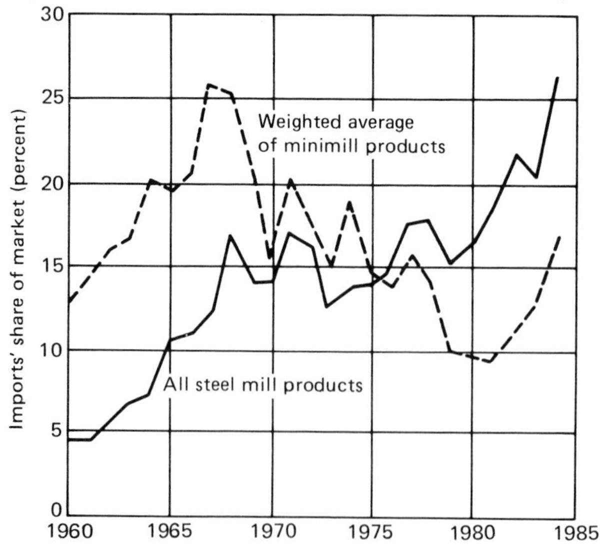

declining demand for materials might bring with it better prospects for resolving environmental problems that tend to grow worse as the level of materials consumption increases.

The end of the Era of Materials is therefore by no means to be regarded only with distress. Like any other profound historical transformation, it brings with it benefits as well as heavy costs for those with an investment in the passing era. What is of crucial importance is to recognize that a fundamental and perhaps irreversible change is taking place. The industrial countries are witnessing the emergence of an information-centered society in which economic growth is dominated by high-technology products that have a relatively low materials content. In this society basic materials will continue to be used, and at very high rates compared with the rates in other societies. But the critical economic fact is that their use will no longer be growing. In the years to come economic success and failure will be determined by the ability to adapt to this reality.

Materials and the Economy

III

Advanced Materials and the Economy

Joel P. Clark and Merton C. Flemings

7

Progress in materials science and engineering stimulates growth in many sectors of the economy. Conversely, new materials and processes are invented to meet the needs of new and existing industries. In this keynote article Joel P. Clark and Merton C. Flemings of the Department of Materials Science and Engineering at MIT argue not only that materials science sets limits to the rate of economic growth but that advanced materials have the potential to solve basic problems such as the finiteness of natural resources. Complex decisions about what materials or production processes to use can now be made by computers. However, in using advanced materials, we should be aware of possible environmental and health hazards.

A fundamental reversal in the relationship between human beings and materials is taking place. Its economic consequences are likely to be profound. Historically humans have adapted such natural materials as stone, wood, clay, vegetable fiber, and animal tissue to economic uses. The smelting of metals and the production of glass represented a refinement in this relationship. Yet it is only recently that advances in the theoretical understanding of the structure of physical and biological matter, in experimental technique, and in processing technology have made it possible to start with a need and then develop a material to meet it, atom by atom. In addition to meeting needs, materials science and engineering creates opportu-

From *Scientific American,* vol. 255, no. 4 (October 1986).

nities and provides society with new ways to address such problems as the scarcity of resources, the maintenance of economic growth, and the formation of capital. Productivity and the structure of the labor force are also profoundly affected by advances in the field. For industrial, financial, and governmental leaders the definition and implementation of strategies that exploit opportunities created by materials science is a central challenge of the last quarter of the century.

Advances in materials science and engineering have impact quickly throughout the economy. On the average, every person in the United States requires the securing and processing of some 20,000 pounds of nonrenewable, nonfuel mineral resources each year. Industries engaged in the direct production of primary materials employ approximately 1.5 million wage and salaried personnel, or about 1.5 percent of the labor force. On each person employed in the primary materials industries depend the jobs of from two to three workers in other sectors.

The value of shipments of advanced materials is about $70 billion, or approximately 14 percent of total materials shipments. The production of such materials occupies about 10 percent of the total labor force of the materials industries. As in the case of employment, the indirect effect of the presence of these materials on the rest of the economy is highly significant. The reason is that advanced materials are not an end product; they are assembled into components critical to the successful performance and operation of such large, complex systems as aircraft and aerospace vehicles, electronic devices, and automobiles. Advanced materials are essential to the future growth of these and other industries. In fact, progress in materials science sets ultimate limits on the rate at which key sectors of the economy can grow.

If, for example, the computer industry is to maintain the rate at which its products' information-handling power has increased, ways must be found to circumvent fundamental physical barriers. Beyond a theoretical limit of 100 million elements per memory chip, technologies must be resorted to that depend on quantum effects involving only a few particles, or even on biological models. The burgeoning demand for communications capacity forces the development of ultrapure glasses and other materials for generating and processing beams of light.

In aircraft and launching vehicles, structural-weight savings can be directly translated into increases in payload (whether of passengers, scientific instruments, or warheads) or into fuel economy. To achieve the speeds envisioned for the Orient Express, a Mach 8 aircraft that could fly from New York to Tokyo in three hours, materials must be developed that can withstand temperatures of up to 3,200°F. Because high operating temperatures correspond to fuel efficiency, turbine blades, and other engine components require progress in the development of composites and advanced alloys. In the automobile industry economies of operation continue to compel recourse to high-strength plastics for body construction, as well as the development of ceramic engine components. New materials originally developed for various fields are found to be of value for repairing the effects of injury on the human body.

Almost every item of military hardware is manufactured from the group of materials that includes ferrous and nonferrous alloys, ceramics, plastics, and composite materials. Moreover industries vital to the military's need for equipment, transport, and support are

Figure 7.1 The competition between processes can influence materials use just as competition between materials themselves can. The charts break down the cost of producing automotive connecting rods by conventional forging and by a new process, powder-metallurgy forging. The mold needed for powder metallurgy costs more than a conventional forging tool, but the new method requires less labor and yields a more precisely shaped rod; less material is lost in finishing. Hence the new process is the more economical one.

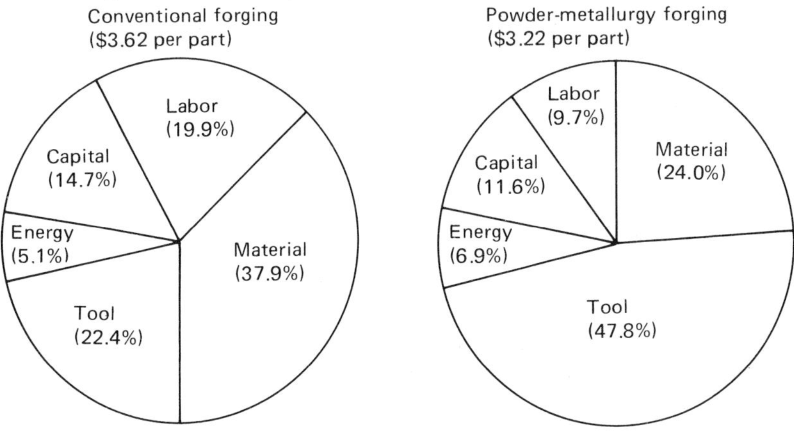

finding it increasingly important to exploit the properties of advanced materials to enhance the performance of the systems they supply.

Materials science stands as a critical element in the resolution of such fundamental economic issues as the finiteness of resources, the scarcity of strategic materials, the maintenance of economic growth and productivity, the creation of capital, and competitiveness in the market. Copper nicely illustrates the first point. Continued demand for this elemental metal has been reflected in the fact that very low-grade ores are still being mined, a supply that is supplemented by recovery and recycling. Yet copper's place as a conductor is now challenged by other metals such as aluminum; electrically active polymers are also being developed. In communications applications copper is already obsolete, its place taken by optical fibers.

The ultimate prospect of an energy famine, currently pushed to the back burner of public awareness by the petroleum glut, will find its resolution in technologies based on nuclear fission, nuclear fusion, and solar energy, all of them dependent on advanced materials.

Figure 7.2 The cost of rapidly solidified metal powders could drop if plant output increased. The curves show the decline expected in the cost of rapidly solidified nickel and aluminum powders. Rising demand for the powders might come from the aircraft industry: rapidly solidified aluminum might be used in airframes, rapidly solidified nickel in jet engines.

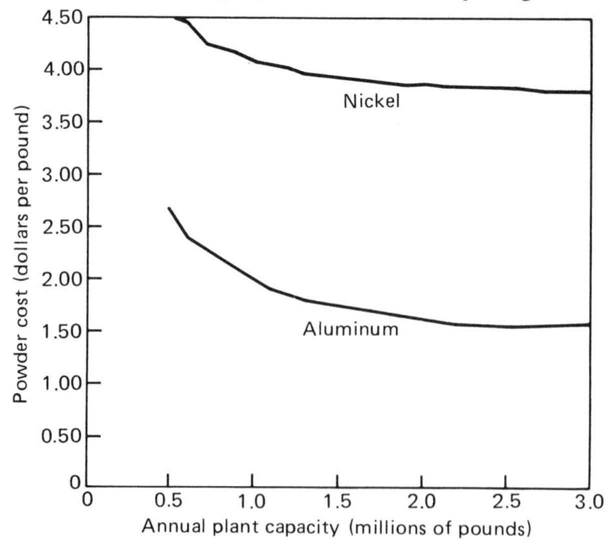

Transmission losses will be sharply reduced or even eliminated by superconducting alloys and by the products of powder metallurgy.

Advanced materials offer comparable leverage in the area of such strategic minerals as chromium, manganese, cobalt, and the platinum-group metals, which must now be imported. Materials scientists and engineers are now developing new metallic alloys and processing methods for them, as well as ceramics, polymers, and composites that require little or no reliance on imported materials and often yield more efficient or cost-effective products. Advanced ceramics are particularly significant in this context. To be sure, the production of structural ceramics involves sophisticated chemical processing methods, and their cost and reliability are uncertain. Yet the solution of such problems would considerably reduce dependence on cobalt and tungsten in cutting and wear-resistant applications and on chromium, cobalt, manganese and platinum-group metals in automotive components.

Nowhere is the role of advanced materials more critical than in coping successfully with the interrelated issues of competitiveness in the marketplace, productivity and the efficient use of capital. A most significant sector in which this interrelation is being played out is the automobile industry. Automobile manufacturers are finding their market increasingly international and competitive. Both price and the ability to offer a wide range of distinct models are important—market requirements that demand economy and flexibility in the manufacturing process. A number of models must be produced from a few assembly lines in small volumes at a cost that allows competitive pricing.

To meet the challenge manufacturers have begun to adopt the *kanban* ("just in time") inventory strategy, highly automated assembly lines and computer-assisted design. Advanced materials play a critical role because they help the manufacturer to address the issue of cost and to achieve flexible manufacturing, and they result in improved performance.

For example, a set of steel-stamping dies for a fender costs about $2 million, whereas a set of plastic-forming tools for the same component costs about $250,000. For a five-year run of 2.5 million fenders the steel die is the more economical because several plastic-forming tool assemblies would be necessary to make the same number of parts. Yet at the low production volumes favored by con-

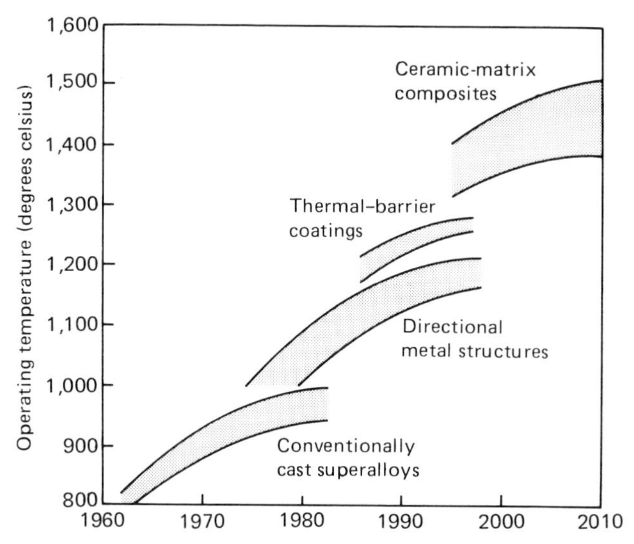

Figure 7.3 A rise in the operating temperature of jet engines (measured at the turbine inlet) will be made possible by new turbine-blade materials. The lowest band on the graph indicates the temperature increase that has been achieved so far through improvements in nickel-based superalloys, the standard turbine material. In coming decades alloy turbine blades made of metals strengthened by directional crystal structures, blades protected by coatings of ceramics or special alloys, and ultimately ceramic-matrix composites will allow an increase in turbine-inlet temperatures. At higher temperatures an engine can operate more efficiently, burn lower-quality fuels, and generate greater thrust.

temporary market conditions and strategies, the economics of capital costs make plastic the more attractive material.

Like bumpers, body panels can easily be formed from polymeric materials. Each of the "big three" U.S. automobile makers is testing the effectiveness and acceptability of plastics for both applications. Ford offers a plastic bumper on certain models of its *Escort* line and General Motors has two lines with plastic body panels—the *Corvette* and the Pontiac *Fiero*. The implications of this shift toward plastic automobiles are sizable. The density of steel is about 7.8 grams per cubic centimeter, whereas the average plastic weighs only about 1.2 grams per cubic centimeter. Switching to the lighter material yields improved mileage. There is another benefit. Because plastics do not rust and are more resistant to minor damage, they extend the life of the average automobile.

Since 1974 the weight of the average U.S. automobile has decreased by 15 percent, or approximately 550 pounds. The bulk of the loss has been at the expense of carbon steel and cast iron. It is paradoxical, then, that although a further weight loss of 27 percent is projected for automobiles manufactured in 1992, steel, as a percentage of vehicle weight, is expected to remain at about the 1974 level. The resolution of the paradox is the use of advanced high-strength, low-alloy steels, which can be fabricated into lighter products than carbon steels can. Concomitantly the use of aluminum has increased by about 50 percent since 1975, to 135 pounds per automobile today, and it is expected to be only slightly higher in 1992.

The substitution of high-strength, low-alloy steel for carbon steel has yielded benefits for both the consumer and the manufacturer. More than a gallon of gasoline will be saved over the lifetime of a car for each pound of carbon steel replaced by the new steels. Consequently the 300 pounds of high-strength, low-alloy steel used in 1986 will result in a saving to the consumer of about 300 gallons of gasoline over the lifetime of the average car. The use of aluminum saves about 0.6 gallon per pound of the metal, about 180 gallons over the car's lifetime.

The advantage of high-performance materials to the consumer is clear. What is perhaps surprising is that, in this case at least, the higher-performance materials also increase the producer's revenue. The high-strength steels now sell at premiums of about 20 percent above the price of carbon grades, and aluminum sells at a premium of 150 percent—premiums that more than compensate for the reduced tonnage. Plastics, contrary to popular belief, can also be expensive, high-value-added materials. Teflon costs $21 per pound. Vespel, a high-temperature plastic made by Du Pont, is priced at about $2,700 per pound. These are significant trends in a materials industry that has matured to the point where the demand for structural materials such as steel has peaked and is largely generated by replacement needs.

Decisions about the advantages of choosing one material rather than another can be framed and made quite precisely in quantitative terms. At the materials systems laboratory of the Massachusetts Institute of Technology one of us (Clark) has applied two such methodological approaches to problems associated with the materials industries. One technique, termed multiattribute utility analysis, enables the analyst to estimate the value of the characteristics of a material in a specific application. The second technique, process-cost modeling, employs computer models to simulate the cost of producing components using various materials and processing technologies. The models enable the analyst to define the cost implications of a set of assumptions and analyze the sensitivity of cost to changes in such variables as process yields and production volumes.

An automobile engineer could use the first technique to select a bumper material and manufacturing process from a series of choices, all of which meet an engineering performance criterion such as the Federal Motor Vehicle Safety Standards for bumper impact. The choices might be standard forged steel, aluminum or a polymer composite. The analysis enables the engineer to measure the utility of the bumpers according to installed cost versus weight. This analysis is useful because it yields a ranking of the alternative systems.

Often the competition is less between materials than it is between processes. Most automotive connecting rods, which link pistons to the crankshaft, are forged. In order to reduce cost in the final forming process, certain sections are significantly more massive than they need to be. There is, however, a penalty: considerable material is lost during forging and machining. A relatively new process avoids the difficulty. Called powder-metallurgy forging, it begins

Figure 7.4 The growth in demand for structural ceramics over the next decade will be concentrated in the areas listed in the chart. The listed applications capitalize on the resistance of ceramics to heat and wear; in addition ceramic turbocharger rotors profit from the light weight of the material, which enables the rotor to reach operating speed more quickly, and mechanical seals benefit from the resistance of ceramics to chemical attack. The requirements for ceramic powder indicated for 1995 assume that each market area will have grown to a size intermediate between the lower and higher 1995 projections.

	Demand for ceramics (millions of dollars)		Powder requirements in 1995 (millions of pounds)
	1985	1995	
Cutting tools	55	150-200	1.5
Turbocharger rotors	—	40-60	1.0
Mechanical seals	80	150-250	3.0
Automotive valve guides	—	100-200	5.0
Total	135	440-710	10.5

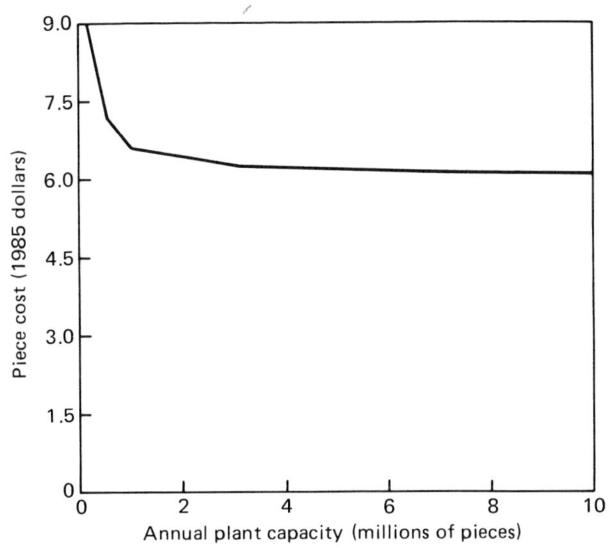

Figure 7.5 Economies of scale markedly reduce the cost of injection-molded silicon nitride turbocharger rotors because of the investment required for production. The curve shows the relation of rotor cost to production volume in a hypothetical plant. The yield of the manufacturing process (the ratio of a product's content of a material to the input of the material into the manufacturing process) is taken to be 80 percent (much higher than now); the silicon nitride powder is assumed to cost $10 a pound (much less than now).

with a powdered metal. The powder is loaded into a preshaped form or cast and typically subjected to extremely high temperature and pressure. Because the cast shape is close to that of the finished product, wastage of material is minimized; the process also reduces the need for labor. Yet mold costs are high, which is a problem that should be amenable to computer-assisted design and manufacture.

The competition between forming processes is nowhere more intense than it is in the aircraft-engine market. Here cost is less important than the issue of which process yields a component that most closely meets the specifications. For example, the operating temperature of a jet engine is a key factor in its efficiency. The temperature has increased dramatically over the past three decades, primarily as a result of materials developments. Turbine blades were originally forged. Now they are cast in many cases with special "directionally solidified" structures that result in improved life and high-temperature strength. In the future some kinds of blades will be made using powder-metallurgy and rapid-solidification processes. By promoting the rapid loss of heat from the casting, rapid solidification enhances the tendency of powder processes to give rise to highly homogeneous structures with improved properties.

As a consequence of developments such as these, operating temperatures have been increasing at the rate of about 12°F per year, a significant figure in view of the fact that each 150° increase in the temperature can result in a 20 percent increase in engine thrust and a significant improvement in fuel economy. Whatever the importance of operating specifications, the cost of producing rapidly solidified materials is regarded as being very high. Nevertheless, it is estimated that such materials could be sold at almost the same price as that fetched by conventional powders if an annual production volume of from two to three million pounds in a single plant can be achieved.

Rapid solidification appears to have provided a solution to a major problem in another economic sector. Each year 400 billion kilowatt-hours of energy are lost worldwide as electrical power is delivered to its users. A significant part of the loss is sustained in distribution transformers, devices that reduce the high voltage of the transmission line to the low voltage required in most household and industrial uses. The strong magnetic fields in which the transformer operates set up eddy currents of electricity: random, useless meanderings of current that heat the transformer core. Certain amorphous metal alloys have the wanted magnetic properties for a transformer core but are rather poor conductors, and so they are not good conveyers of eddy currents. Therefore they resist heating and their electrical loss is correspondingly low.

An amorphous metal alloy is one in which the individual atoms assume a more nearly random pattern rather than a crystalline one. Such a material is best made by rapid-solidification techniques. Because the metal cools rapidly, the atoms are caught before the forces among them have time to array them in crystalline fashion. Since 1976 funding from the General Electric Company, Allied-Signal, Inc., and the Electric Power Research Institute has been going into various research and development programs designed to create manufacturing processes for the material and demonstrate its usefulness.

Transformer cores made of amorphous metal have been introduced into the marketplace and are expected to capture a significant

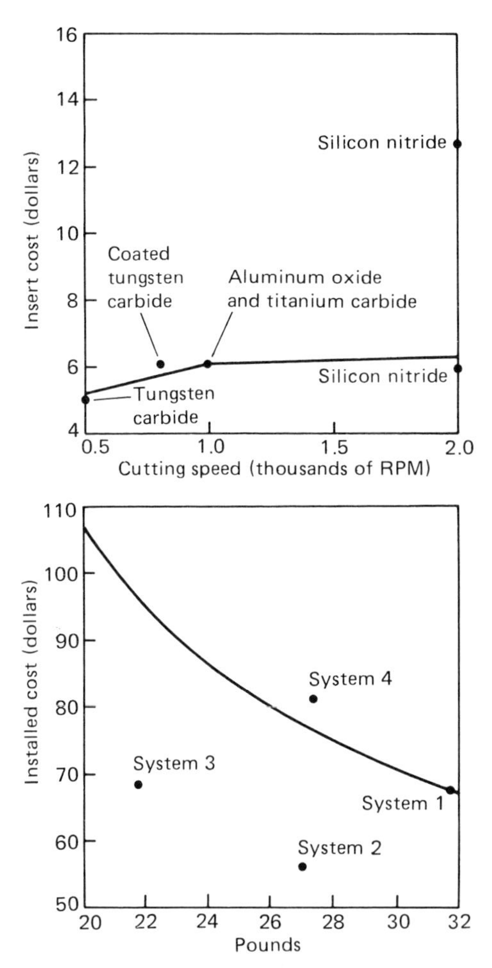

Figure 7.6 Iso-utility lines indicate the various combinations of a component's cost and its performance that would be attractive to manufacturers of plastic bumpers (*left*) and to users of ceramic cutting-tool inserts (*right*). The lines are the product of multiattribute utility analysis, a technique applied by one of the authors (Clark), in which data from interviews with the users of a material are transformed into a mathematical function. In these cases, if the cost and performance of an alternative material system place it below the iso-utility line, it will be more attractive than the current system; if it falls above the line, it will not be competitive. The graph for bumpers plots price against performance (weight) to compare the "utility" of two bumpers now in use (*systems 1 and 2*), both consisting of polyurethane on a steel backup beam but differing in their energy-absorbing systems, with a similar bumper having a third energy-absorbing mechanism (*system 3*) and a system made of a different polymer (*system 4*). For cutting-tool inserts, performance is equated with permissible cutting speed in revolutions per minute. Although silicon nitrate makes possible very high cutting speeds, the ceramic will not be competitive until the price per insert falls by at least $6 (lower right-hand label in the second panel).

share within the next few years. A startling 75 percent reduction in average core loss can be obtained through its use. The importance of the reduction is demonstrated by the fact that conversion to distribution transformers with amorphous metal cores would save an estimated 40 billion kilowatt-hours per year in the United States and 100 billion kilowatt-hours worldwide.

Both economic promise and the necessity for technological progress invite a close look at three areas of materials science: ceramics, composites and semiconductors. Advanced structural ceramics are light and capable of withstanding intense wear and heat. These qualities allow them to be fabricated into components that enhance fuel efficiency, increase productivity in industrial processes and replace materials that are scarce or strategic. Investigators in the field are attempting to overcome two major defects in ceramics: brittleness and high processing costs. Major engineering and economic advantages justify the effort. If ceramics can be successfully used in jet engines, the maximum operating temperature could be increased from about 1,830°F to as much as 2,700°. Consequently fuel efficiency could be dramatically improved and lower-quality fuels could even be used.

Such opportunities create the potential for rapid growth and profitable investment in markets for advanced structural ceramics in the next five to twenty years. This growth may fall short of the multibillion-dollar estimates forecast in some recent reports. Our own more guardedly optimistic outlook is that the market will approach $1 billion in ten years. Major applications have been predicted in cutting-tool inserts, turbocharger rotors, mechanical seals, and automotive valve guides. Growth rates for areas of current use (cutting tools and seals) will average about 12 percent on an annual basis.

Several technical obstacles will have to be overcome before these markets can develop. Large-scale production must achieve consistency and reliability. Today world production of silicon nitride powder is roughly 250,000 pounds. If the midpoint of the aforesaid 1995 consumption projection for cutting tools and turbocharger rotors holds, the powder requirements for these two applications alone would be approximately 2.5 million pounds, or almost 10 times the current level. A chicken-or-egg problem stands in the way of this target. Producers of raw materials have been traditionally reluctant to invest in production capacity until a market is developed; such large users as the automotive industries have been traditionally slow to adopt new materials until alternative sources and assured supplies exist.

For all four major applications predicted, costs must be reduced if the projected growth is to be realized. Multiattribute utility analysis can be used to quantify the process. For example, it shows that although users of cutting tools will even pay a small premium for the benefits of the silicon nitride insert (up to about $6 per piece, compared with $5 for tungsten carbide inserts), significant cost reductions will still be necessary in order to offer such a price. One set of assumptions that would allow insert producers to sell their products in this price range includes silicon nitride powder prices of from $15 to $20 per pound, an overall processing yield of 85 percent, and production levels of at least 250,000 parts per year.

Similar assumptions regarding production rates and powder costs are required if prices for silicon nitride turbocharger rotors are to fall into the range of $14, a necessity if such rotors are to compete successfully with rotors made of superalloy materials. An analysis of costs leads to the conclusion that process yields will need to increase to 70 percent or more (a significant improvement over the current levels of about 25 percent), powder prices must be less than $20 per pound, and production levels must be higher than 250,000 parts per year.

Another crucial barrier to the use of ceramics in virtually all structural applications is poor reliability. Ceramics tend to fail catastrophically when cracks that originate from small defects propagate. For instance, the critical flaw size for most ceramics is less than 100 micrometers in diameter, compared with several millimeters for most metallic alloys. There are several methods currently under development to improve the reliability of ceramic components.

Composites are the fastest growing advanced materials. The term composite material refers to a matrix that is reinforced with fibers, whiskers, or dispersions of another material. The aim of producing a composite is to have a material whose performance characteristics combine the strengths of each constituent material.

In the past twenty years a number of composite systems have been devised, a familiar example of which is fiberglass-reinforced plastics (FRP). A new class of materials, generically known as advanced composites, is now being developed. (The term advanced composites was coined to distinguish materials incorporating the new high-performance fibers from the rather "low-tech" FRP.) Of this group polymer-matrix composites are furthest along in commercialization; there is also much interest in ceramic-matrix composites and metal-matrix composites.

The superior mechanical and physical properties offered by composites compared with homogeneous materials can include a better stiffness-to-density ratio, a better strength-to-density ratio, improved fatigue resistance, greater fracture toughness, and good thermal-shock resistance (in the case of ceramic-matrix composites).

Interest in the development of advanced composite materials is guaranteed in the aerospace industry. An improvement in frame weight can reduce direct operating costs, improve payload capacity, and increase speed and range. Benefits to both the commercial and the military sectors are therefore great. For several years the aircraft energy efficiency program of the National Aeronautics and Space Administration, in conjunction with the major aircraft manufacturers, has been engaged in a major effort to develop composites for incorporation in airframe structures. Currently only subassemblies made of composite materials have been employed. They include the rudder on the DC-10, the elevator on the Boeing 727, and the aileron on the Lockheed L-1011. All the control surfaces on the Boeing 757 and 767 aircraft will be made of graphite-epoxy composites. The total weight saving of 845 pounds of material will reduce fuel consumption by 2 percent, saving approximately 100,000 gallons of fuel per aircraft per year. In addition composite components are made up of fewer parts than conventional components, further reducing costs.

For instance, it has been estimated that the substitution of a graphite-epoxy composite material for aluminum alloys in a large aircraft fuselage (one that is more than 12 feet in diameter) will reduce the total production costs by 30 percent. The structure would be 30 percent lighter and would exhibit improved fatigue and corrosion resistance. A number of issues, however, still need to be addressed before the various composite materials are accepted on a widespread basis. These include impact resistance, cabin noise, and, most important, Federal Aviation Agency approval.

Semiconductors are undoubtedly the most critical family of materials from an economic point of view. The electronics industry in the United States alone had an annual aggregate sales value of $172 billion in 1984. Future economic advances depend to a large extent on advances in materials and processes.

The relentless miniaturization of integrated circuits and the irresistible pressures for higher process reliability and lower cost are driving all but the strongest producers from the market. Today the merchant electronics industry in the United States (which excludes the major computer and communications companies) is being buffeted by foreign competition, and its very future is in doubt. The merchant industry accounts for about $80 billion in sales. Its loss would be a serious blow to the industrial strength of the country; its future depends on being able to produce advanced materials reliably and economically.

Further progress in this field calls for investigators to plunge into the most profound questions of quantum theory and solid-state physics. Dimensions become so small in the newer devices that surface effects alter electronic behavior in new and often unexpected ways. Topographical dimensions are already being controlled at submicrometer scales and boundary thicknesses within a few tens of angstrom units, a span comparable to that of the wavelength of light. Devices are envisioned consisting of small aggregates of atoms. Reliability in the manufacturing of complex structures must be very high in each of many process steps, and all of this will have to be accomplished in the most demanding competitive climate.

The implications of increased consumption of advanced materials are not all beneficial. As such materials become widely used, they may well exacerbate critical economic and environmental problems. The economies of developing countries, many of which are already experiencing extreme pressure from the decline of oil prices, could experience further reduction or even reversal of growth rates as advanced materials replace natural ones. Copper, nickel, cobalt, chromium, and manganese are particularly vulnerable. Since countries such as Gabon, South Africa, Zaire, Zambia, Zimbabwe, Brazil, Chile, Peru, and Papua New Guinea derive a substantial fraction of their foreign exchange from the export of commodity metals, decreased consumption could lead to problems of global economic and political stability.

Advanced materials may do little to help industrial society cope with the health hazards and environmental effects of manufacturing. Effluents produced by the manufacture of ceramic powders or organic resins from energy feedstocks are not qualitatively different from those emitted in the mining and manufacture of ferrous and nonferrous metals.

Hazards to health in the workplace may also accompany the production of some advanced materials. The manufacture of ceramic and composite materials may entail the dispersion of particles. Although the carcinogenic effects of these materials on human beings have not been documented, the smallest fibers used in the production of high-performance ceramics have been shown in controlled laboratory studies to give rise to tumors in the lung lining of animals. No regulations have as yet been established to limit exposure to ceramic fibers in the workplace.

The fact that composites may be chemically active in the production stages increases their threat to health and safety. Despite exhaust systems, spray booths, and air masks, some people become sensitized to the vapors and cannot tolerate the environment of a reinforced-plastics plant. Polymers and polymeric composites present special disposal problems. On the other hand, thermoplastic-matrix composites may even be recyclable.

The developments in materials science and engineering described in this article are sometimes the progenitors and sometimes the followers of new technologies. Whether it is creating opportunities or meeting needs, the field of materials science and engineering is in the midst of a widely acknowledged revolution, and one that is certain to be a key factor in shaping the global economy.

Ceramics Take on Tough Tasks

Gordon Graff

8

In the first of four articles on different examples of new materials, technology writer Gordon Graff provides a lively overview of the new ceramics. These strong, lightweight materials are boosting efficiency in a range of mechanical and electronic applications, such as ceramic diesel engines and chip carriers. But it's not all plain sailing, and much needs to be done to overcome the disadvantages of ceramic parts, especially their brittleness and their propensity to sudden failure, as Graff makes clear.

A quiet materials revolution is sweeping industry. The excitement centers on a group of solids stronger than steel, almost as hard as diamonds, and impervious to furnacelike temperatures. They are the fast-emerging technical ceramics, now bidding for a larger role in everything from auto engines and batteries to integrated circuits and cutting tools. Indeed, perhaps more than any new materials since plastics, they seem destined to make an indelible imprint on the way people live.

Already, in fact, superefficient ceramic-based auto engines capable of achieving 42 miles per gallon are going through their paces at the research labs of several auto companies. Meanwhile, an experimental fuel cell with a solid-ceramic electrolyte promises to make today's turbine-power generators obsolete. And advanced ceramic packaging is in the works to house the delicate circuitry of fifth-generation computers.

With all these new outlets on the horizon, technical ceramics are poised for a major boom. According to MIT ceramics researchers George B. Kenney and H. Kent Bowen, global sales of technical ceramics will soar over the next decade from today's $4.25 billion a year to as high as $20 billion.

Although the United States is believed to have the technological edge in advanced ceramics, and Britain has also been a traditional leader, a strong challenge is coming from Japan, which already commands about half the worldwide technical ceramics market. In 1981, for example, the country's Ministry of International Trade and Industry launched a ten-year $120 million program with industry and academia to develop fine ceramics for industrial applications.

A major focus of this work is processing technology, an area where experts say the United States lags. And some Japanese manufacturers have ambitious plans, including the world's first ceramic-based turbocharger auto, which may hit the market as early as this spring. Not surprisingly, U.S. firms are worried. "We've got to stay on top of these things," warns a scientist at a major U.S. ceramics company. "If we don't, the Japanese will kill us."

Stressing Performance

The allure of technical ceramics—not to be confused with such humble cousins as bricks, teacups, and bathroom sinks—stems mostly from improved performance. Technical ceramics are defined, in fact, as solids that deliver superior mechanical, thermal, or electrical performance (relative to materials such as metals or plastics) in high-stress environments. Ceramic-based gas turbine auto engines, for instance, are expected to boost fuel efficiency by more than 30 percent by operating at 2,300°F or higher—temperatures that would turn ordinary metal parts into useless putty. And fuel cells with solid ceramic electrolytes would end reliance on corrosive and unstable liquid electrolytes.

Other factors driving the ceramics boom are the dwindling reserves or uncertain supplies of certain "strategic" materials—chromium, cobalt, titanium, and manganese—which have long been essential for heavy-duty applications. Compared to metals, "ceramic materials are a relatively widely distributed, inexpensive resource," says Bowen, director of MIT's Materials Processing Center. In fact, some leading elements in ceramics—silicon, carbon, nitrogen, and oxygen—are among the most abundant on earth.

For all these advantages, however, ceramics have seen little use as load-bearing machine and engine components. "The design engineer in the past hasn't liked to use ceramics, because they're too brittle," observes Arthur F. McLean, manager of the Ford Motor Co. ceramics research dept. Indeed, this brittleness can cause tiny, inconspicuous flaws in a ceramic material to suddenly grow with lightning speed, leading to catastrophic failure of the part.

But advances in processing and design engineering are improving the reliability of ceramics. New processing methods such as injection molding, slip casting, and hot isostatic pressing are making more uniform parts with fewer of the flaws that lead to failure. Advanced analytical methods employing X-rays and ultrasonics are now able to spot dangerous voids before they cause trouble. And a computer-aided design technique called finite element analysis is helping engineers gauge a part's probability of failure.

Engine Work in High Gear

All these evaluation techniques are being put to work in ceramic gas turbine engines, where reliability will be essential. Two major U.S. demonstration programs, each costing in the neighborhood of $60 million, are underway at corporate labs, both supported largely by the Department of Energy (DOE) and administered by NASA Lewis Research Center, in Cleveland. The two projects, which complement each other, began in October 1979 and are slated to run through November 1985.

One effort, under way at GM's Detroit Diesel Allison and Pontiac div.'s, is focusing on an engine known as the AGT-100. It is a lightweight 100-horsepower two-shaft motor with a regenerative system that reuses exhaust gases with the aid of a rotary ceramic heat exchanger. Ultra-heat-resistant silicon carbide is used in the engine's high-temperature areas, including the power turbine, gasifier, and combustion liners. (Silicon nitride serves as a backup material

New Processes Make Tougher Ceramics

Ceramics makers agree that the key to stronger and more reliable ceramics lies in new processing techniques that result in fewer of the flaws and voids that lead to failure. Several promising new processes cut down on flaws by evening out the distribution and size of particles in finished parts.

Right now the first step in making technical ceramics is to blend different mineral powders—alumina, titania, sand, feldspar, or whatever else is needed, in a recipe—in different proportions. (The blending has traditionally been more of an art than a science, but it is becoming more systematic.) The powders are then typically either pressed together mechanically or mixed with a solvent in a slurry and poured into a mold (slip casting) from which the solvent is then drained.

The resulting "green" parts are then usually sintered (fired) in a furnace—a process that induces partial melting and fusing of the powder particles to create a hard solid ceramic part. In some cases the forming and sintering may be done simultaneously, as in hot pressing where the powders are compacted between two pistons and heated. Or the powders may be melted and poured into a mold. Whatever process is used, the final step usually involves machining the part to the desired size and shape.

As useful as these techniques have been, they have not been particularly suitable to very complex shapes, such as a turbine rotor. As a result some of the older methods are being modified or replaced by entirely new processes.

For example, researchers at Ford Motor are replacing the traditional plaster mold used in slip casting with a wax mold. (Wax can accommodate more complex shapes than plaster.) After the slurry is poured into the mold, the solvent drains out the bottom into a plaster enclosure around the wax. Finally, another solvent is used to dissolve away the wax, leaving behind the part, which goes to the furnace for sintering.

Another promising process is injection molding, in which the powders are mixed with a pliable plastic resin and the mixture is injected into a steel mold of the desired shape. Gentle heat is applied to melt and drive off the polymer. The resulting part is then fired. Injection molding, like slip casting, can create complex parts to near-net shape—that is, they require little expensive machining.

One variation of hot pressing now coming to the fore is hot isostatic pressing (HIP). Whereas ordinary hot pressing is limited to simple shapes, HIP is adaptable to the more complex parts now in demand in industry. An unfired part is encased in an envelope (often made of heat-resistant glass) and placed in a pressurized furnace, where it is subjected to intense heat (around 3,600°F) and pressure (about 30,000 psi) from an inert gas such as argon. The result is an extremely dense finished part. While HIP has been used to make machine tools, it has not been widely employed for larger parts, because of the long cycle times required and potential dangers of the high-pressure furnace.

Not as far along, but extremely important, is sol-gel processing. In this method, organic chemical derivatives of a ceramic's key elements are mixed with a water-ethanol solvent. A reaction between ingredients and solvent forms a uniform colloidal "sol," which is poured into a mold. Gentle heating drives off most of the solvent and organic groups to form a semisolid gel. Firing of this gel creates the finished part.

Sol-gel's chief advantages are uniformity of particle size and ability to control particle size and distribution by varying the chemistry of the individual steps. While the technique has been confined so far mostly to coatings, it is now being adapted to solid shapes. Sol-gel "is going to be a very important feature of ceramic technology in the future," says General Electric investigator Richard Charles.

Meanwhile researchers looking for ways to make ceramics stronger are turning to special composites. Two approaches are being taken. One is to combine ceramic fibers with a polymeric or metallic matrix—silicon carbide in aluminum, for instance. The other is to mix two different ceramics—say, 90 percent alumina and 10 percent zirconia. Often the difference in strength between composites and single-phase ceramics can be striking. In the alumina-zirconia blend cited here, for example, the fracture-toughness rating (a standard index in the industry) is eight. By contrast, plain alumina has a rating of four. Composites are under study for such uses as ablative heat shields for spacecraft, rocket nozzles, and structural members of aircraft.

Corning researchers have developed an extremely tough composite based on silicon carbide embedded in an unidentified matrix. According to Pellegrino Papa, an investigator at the firm, the fracture toughness of the material, 25–26, makes it a good candidate for use in the turbines of advanced automobile engines.

Although composites perform well enough in bench-scale tests, design engineers still have much to learn about their properties. "The theory of composites is less well developed than that of the monolithic ceramics," notes ceramics project manager William E. Goette, of the NASA Lewis Research Center. "People are starting to look at them" for heavy duty applications, he adds, "but not much has been done."

in some of these applications.) Another ceramic, aluminum silicate, is used in the regenerator disks. And some of the flow-path walls in the engine are fabricated from lithium aluminum silicate.

The AGT-100 engine has a turbine inlet temperature of 2,350°F; that of earlier, superalloy-metal gas turbine engines is 1,800°F. When the ceramic engine is incorporated into a 3,000-pound passenger car, says H. E. Helms, chief project engineer of the AGT-100 program, the higher temperatures should permit a 30 percent increase in fuel efficiency—42.5 mpg, total—over standard spark-ignition engines and superalloy gas turbines. (Road tests have been eliminated from the current program due to DOE budget cuts.)

Right now, says Helms, GM engineers are "working the bugs out" of the engine on test stands. Assuming these tests are successful, he says, at least five to seven years more work would be needed before the engine would be ready for production.

The other gas turbine engine effort is being mounted jointly by Garrett Turbine Engine Co. and Ford. The 100-horsepower AGT-101 employs various forms of silicon carbide and silicon nitride in the turbine assembly, combustor liner, regenerator, and other components. The flow-path housing contains lithium aluminum silicate. Like its GM counterpart, the Garrett/Ford engine will have to withstand high turbine inlet temperatures—about 2,500°F. Here, too, the high temperatures will afford high fuel efficiency—42.8 mpg in a 3,000-pound passenger vehicle. (As in the other program, road tests are no longer scheduled.)

Because the AGT-101 is a single-shaft unit, the rotational stresses on the turbine rotors—running at up to 100,000 rpm—will be considerably higher than on the double-shafted AGT-100. Ford's McLean says that a special "sintered reaction-bonded" silicon nitride looks like a good bet for the highly stressed rotor.

The two engines have their pluses and minuses, points out William E. Goette, manager of the gas turbine and diesel projects at NASA Lewis. The Garrett/Ford unit, he notes, "is a simpler engine that requires a more complex transmission," whereas the GM model is "a more complex engine that uses a simpler transmission." Both, he concedes, will require a lot more work before they hit the auto assembly line. Goette says similar engines may someday power aircraft as well as automobiles.

However, the path to commercialization in either autos or planes could be a rocky one, claim outside critics. Although the engine demonstration projects have been successful, says MIT's Bowen, "not enough money has been spent on developing manufacturing technology." By contrast, he says, the Japanese have made ceramics production processes a national priority.

In the U.S. several government-backed industrial research programs on power-generating ceramic gas turbines have ground to a halt in the past two years. According to Westinghouse ceramics investigator George Wiener, the company abandoned its power generator project because it was "too long range and too expensive." (The company is now looking at ceramic-based fuel cells for advanced power generation.) And General Electric has likewise scrapped its ceramic power-generating turbine program, though a spokesman affirms that "we're following developments in the area very closely." The Japanese, however, are still bullish on power-generating ceramic turbines, with Kyoto Ceramics reported to be particularly active.

Diesel Progress

Although ceramic gas turbines for cars or generators are still a decade or more away, the ceramic diesel is likely to make it to the auto showroom a lot sooner—probably before 1990. With the diesel the goal is to retain the roughly one-third of an engine's heat output that is normally lost to the water cooling system. This heat then becomes available for useful work.

Heat loss is trimmed by insulating key engine components with ceramics, which allows elimination of the watercooling system entirely. Among the diesel components that have been ceramic-insulated on a trial basis are pistons, liners, cylinder heads, valve seats, and valve guides.

The most successful insulating material to date has been a special "partially stabilized" zirconia (ZrO_2); but companies such as GM, Ford, and Japan's Hitachi and Kyoto Ceramics are also looking at silicon carbide and silicon nitride liners and coatings. Because the ceramic diesel (sometimes referred to as the "adiabatic" engine) will have no water cooling, it will reach higher temperatures than ordi-

A Ceramic's Properties Are Tied to Its Structure

Ceramics are loosely defined as crystalline solids composed of metallic and nonmetallic elements. Certain inorganic materials without metal atoms—silicon carbide and silicon nitride, for instance—are also regarded as ceramics. The technical ceramics now leaping to prominence are mostly polycrystalline, containing conglomerations of microcrystals. Common ingredients in engineering ceramics are oxides of aluminum, magnesium, zirconium, and titanium. Other popular compositions are carbides, nitrides, and borides of silicon.

The extreme strength, chemical inertness, and heat- and abrasion-resistance of the technical ceramics stems largely from their structures. The atoms in ceramic crystals are typically locked in rigid lattices. The powerful bonding in these crystals is covalent, meaning that electrons are shared by dissimilar atoms and are not free to participate in chemical reactions. The application of heat and pressure in processing causes these already strong microcrystals to interlock and fuse, forming an extremely tough network. However, the rigidity of the crystal lattices in ceramics means that bonds cannot be stretched: they tend to break all at once, leading to ceramics' well-known problem of brittleness.

The tightly held electrons also account for ceramics' low electrical conductivity at ordinary temperatures. By contrast, the mobile bonding electrons in metals is what makes them good conductors.

The strength of a given ceramic is strongly dependent on the size of its crystal grains. A material with large particles or with a random distribution of particle sizes is generally weaker than a ceramic with fine-grained or uniform-size particles. This is because large or random-sized grains fit together with relatively large spaces in between; smaller or more uniform grains can fit together more densely, with fewer spaces in between.

Ceramics' strength is different from metals'. Most ceramics are noticeably inferior to metals in tensile strength (resistance to lengthwise stress)—at least at room temperature. For example, silicon carbide has a room-temperature tensile strength of about 15,000 psi, while superalloys often have values exceeding 100,000 psi. At 3,000°F, however, the tensile strength of many metals plummets to near zero, while ceramics often retain appreciable tensile strength.

Ceramics excel in compressive strength (resistance to crushing), even at room temperature, with values often exceeding those of metals. The compressive strength of technical ceramics is generally five to ten times their tensile strength.

In their bending strength (resistance to flexing), metals and ceramics are often competitive at room temperature. But here too metals weaken rapidly at temperatures above 1,500°F; ceramics retain a good degree of bending strength at much higher temperatures.

Ceramics are also stiffer than metals. While the modulus of elasticity (stress-to-strain ratio) of technical ceramics is generally 30 to 60 million psi, that of steel hovers around 29 million psi.

Ceramics' reputation for heat resistance is well justified. The maximum service temperatures of alumina (Al_2O_3) and silicon carbide are 3,540°F and 3000°F. But even the highly heat-resistant nickel alloys are seldom serviceable above 1,500°F.

Finally, some ceramics have exceptional hardness. Cubic boron nitride has a hardness rating of 5000 kg/mm^2—second only to that of diamond (8,000 kg/mm^2). Silicon carbide, another superhard material, has a rating of 2,800 kg/mm^2; in comparison, hardened tool steel has a value of only 740 kg/mm^2.—Mark E. Alsop

nary engines. As a result, new heat-tolerant engine lubricants are being developed. GM's Helms, for instance, says new formulations remain stable at temperatures of 450°F, whereas conventional oils break down at 280°F.

The most advanced ceramic diesel program in the U.S. is a joint project of Cummins Engine and the U.S. Army Tank Command. A prototype zirconia-coated diesel engine recently powered a five-ton military truck in over 5,000 miles of driving. The engine achieved a 30 to 50 percent increase in fuel efficiency relative to uncoated engines. Further tests on modified versions of the engine are due this year. The Army's long-range goal is to deploy, by 1992, a fleet of tanks powered by ceramic-diesel engines 40 percent smaller and lighter than conventional engines but just as powerful.

Still, the first to bring ceramic-powered diesel vehicles to the civilian market will probably be the Japanese. Hitachi has already built a complete passenger car powered by a silicon carbide-insulated diesel and has run it through fifty hours of road tests. And Kyoto Ceramics has demonstrated a vehicle driven by a silicon nitride-insulated four-stroke diesel engine. Other firms active in ceramic diesels are NGK Spark Plug and the Toshiba group. Although the Japanese firms are close-mouthed about their commercialization plans, U.S. sources believe that the first ceramic diesel passenger cars will go on sale by the end of the 1980s.

Although ceramic-based gas turbines and diesels are not quite ready for the consumer market, turbocharger engines with ceramic rotors are set for debut on the passenger car scene soon (see *High Technology,* May 1983, p. 22). The Japanese will reportedly introduce a version during the first quarter of 1984. Ceramic rotors should solve the nagging problem of sluggish acceleration that has plagued turbo cars with normal nickel alloy rotors. The reason is the lower rotational inertia of ceramic rotors, which permits them to speed up faster. Ford's McLean, in fact, reports that test vehicles with reaction-bonded and sintered (fired) silicon nitride turbocharger rotors have shown a 30 percent improvement in acceleration.

In the Factory

Technical ceramics are already making a big splash in heavy industrial equipment. In particular, sintered silicon carbide, available commercially for the past four years, is rapidly finding its way into such parts as pumps, valves, nozzles, bearings, seals, and rings. Silicon carbide's stellar properties—resistance to wear, corrosion, and chemical attack—are proving valuable in chemical-processing and oil- and gas-recovery equipment, where exposure to high temperatures and pressures, abrasive substances, and strong acids is commonplace.

A tough competitor with silicon carbide for the same applications is silicon nitride, which has lower thermal expansion and lower elastic modulus (stress-to-strain ratio)—characteristics that make it more head- and shock-resistant. GE, for one, hopes to offer nitride parts to complement its carbide line. But Richard Charles, of the firm's Schenectady, New York, R&D center, says the nitride is still in the laboratory stage.

Ceramics are also getting a big lift from the energy conservation trend in industry, in particular from duty in heat exchangers that recover the 2,000 –3,000°F waste heat once vented to the outside by aluminum smelters and steel furnaces. A variety of materials are being pressed into service here, including silicon carbide and nitride, zirconia, lithium aluminum silicate, cordierite (magnesium aluminum silicate), and the sialons (silicon nitride and alumina mixtures). One of the leading lights in ceramic heat exchanger development is Garrett AiResearch, which has been looking at various materials and designs, under contract to DOE and the industry-backed Gas Research Institute (Chicago) and Electric Power Research Institute (Palo Alto, California).

Cutting tools loom as another major new outlet for technical ceramics. Such materials as alumina (Al_2O_3), titanium carbide, titanium nitride, silicon nitride, and the sialons are challenging the long-held position of tungsten carbide for this application.

All Charged Up

Even as some manufacturers capitalize on the hardness, toughness, and heat resistance of ceramics, others are harnessing their unique electrical properties. Some experimental batteries, for example, are using alumina as a solid electrolyte (ion conduit) in place of such liquid electrolytes as sulfuric acid or molten carbonate salts. Westinghouse is working on a fuel cell employing solid zirconia as the electrolyte instead of phosphoric acid (as in other cells). In both cases the payoff is higher operating efficiency.

A special ceramic called beta-alumina, which can conduct sodium ions, is the electrolyte in high-performance sodium-sulfur batteries now getting trials at GE, Ford, and elsewhere. As the battery is charged, sodium ions pick up electrons and form elemental sodium; meanwhile, sulfide ions lose electrons and become molten sulfur. Discharging is just the reverse.

Although the sodium-sulfur battery (also known as the beta-battery) operates only at temperatures in the 650°F range, that is still lower than the operating temperatures of several prototype electric-vehicle batteries tried in the past. In 1981 GE joined forces with Britain's Chloride Silent Power to demonstrate the beta-battery in electrically powered milk delivery trucks. The batteries were short-lived, however, due to degradation of the ceramic electrolyte. Since then the company has come up with a more stable electrolyte called beta double-prime alumina.

According to GE battery researcher Kenneth Browall, the improved electrolyte allows "much better efficiencies"—up to 90 percent in some cases. And although no new electric-vehicle road tests have been run, Browall says that "we project very long lifetimes" for the beta double-prime battery in operation. In addition, he notes that GE has been talking to utilities about using the battery in storing off-peak power (load leveling).

But ceramics may eventually play a role in generating power rather than storing it. That, at least, is the hope of Westinghouse, which is grooming its zirconia-based "solid-oxide" fuel cell for that purpose. (The work is being funded by DOE under a two-year $5.5 million contract that runs until December 1984.) The cell's electrolyte is zirconia doped with small amounts of yttria (Y_2O_3). This material has the ability to conduct oxygen ions at the elevated operating temperatures of the cell (around 1,800°F).

According to Westinghouse researcher Wiener, the cell has unusually high efficiency—a fuel-to-electricity conversion ratio of "over 50 percent"—and can use a wide array of different fuels: natural gas, medium- or low-Btu coal gas, liquefied natural gas, or oil (with some additional preprocessing). Hydrocarbon fuels would first have to be reacted with steam in a catalytic reformer to produce mixtures of carbon monoxide and hydrogen that would be fed into the cell along with air.

"We're very excited" about the cell, Wiener declares. He cautions, however, that commercialization is still iffy, improbable before the early 1990s. At this point, only individual solid-oxide cells have been built; but Wiener says that Westinghouse researchers will start work on a prototype 5-kW generator, consisting of about 300 linked cells, by the middle of 1984. Enclosed in a black box about 12 inches square and 2 feet high, it will provide enough power to run a small house. Any commercial generation system, says Wiener, would be modular, with utilities adding on cells as needed.

The material in the solid-oxide cell is similar to the zirconia used in oxygen sensors of automobile engines and of furnaces. But other ceramics are being adapted for sensing other substances. Various Japanese firms, for instance, are developing sensors for hydrocarbons and fluorocarbons. The devices are based on such ceramics as zinc oxide and spinel. And Ford is developing advanced automotive sensors based on titania (TiO_2) as well as zirconia.

In electronics, ceramics already have a strong foothold. This holds true particularly for alumina, which is widely used to package higher-priced integrated circuits to keep out dirt, air, and moisture. (Plastics or porcelain-metal composites are used for the lower-priced ICs.) But other ceramic packaging materials are appearing.

Hitachi has announced a new silicon carbide material that is especially applicable to high-power electronics, where heat dissipation problems are especially severe. In addition, researchers at Rutgers University's ceramics research center (Piscataway, New Jersey) are using an advanced ceramics fabrication technique called sol-gel processing to yield electronics packaging that could be sintered at lower temperatures than present packaging. (High sintering temperatures sometimes damage delicate components underneath.)

Meanwhile new materials are challenging the preeminence of the ceramic doped-barium titanate in microelectronics capacitors. A new type of higher-efficiency "boundary-layer" capacitor was introduced by Japanese firms several years ago. It employs a strontium titanate ceramic. No U.S. companies have marketed similar devices, but several firms are said to have them in the works.

Research Still Needed

Though there is little dispute that electronics will be a key growth area for ceramics over the next decade, experts say the progress of ceramics in engines and heavy machinery could be slowed by some of the materials' limitations. "There is still a belief in industry that the overall reliability of ceramics should be better," says Roger R. Wills, manager of an ongoing market study of technical ceramics at Battelle Columbus Laboratories.

But many believe that basic research now advancing on all fronts will quell some of the lingering doubts. "In the last five years, the level of knowledge and understanding of structural ceramics has probably doubled," says Pellegrino Papa, manager of ceramic technology at Corning Glass Works. "And my guess is that it will probably double again in less than five years." This knowledge, says Papa, will be matched by equal determination: "The industrial sector has decided it's finally going to understand what causes [ceramic] failure—and do something about it."

Ceramics Engineers Seek Improved Reliability

"You can make two identical ceramic parts the same way and one of them may have a third less strength than the other." That, as NASA researcher William E. Goette describes it, is the problem faced by design engineers trying to incorporate ceramics into high-temperature turbine rotors and other high-stress environments. Unlike metals, which usually have specific tensile strengths, ceramics have a scatter of strengths that makes their behavior in service hard to predict.

But new nondestructive testing methods are helping to take the guesswork out of ceramics design. Scientists now know that ceramic failure usually starts at a small crack or void that propagates throughout the part catastrophically. Therefore a major thrust in analytical work today is to spot these flaws in advance.

One approach is to X-ray a part to detect the flaws visually. Other techniques involve ultrasonics, in which sound waves passed through a material give clues about the number and type of flaws inside. But these techniques can spot only the larger defects—those in the 150- to 200-micron diameter range.

One technique that can spot smaller flaws is a X-ray tomography method now under development at Ford Motor. Using a combination of fluoroscopic X-ray pictures of a part at different angles and a computer that analyzes the picture elements, researchers can get a computer-generated cross-sectional picture of the part. Ford ceramics investigator Arthur F. McLean says that flaw sizes down to 50 microns are detectable this way.

Another discipline that aims to make ceramics more predictable is probabilistic design methodology. As the name implies, it is aimed at giving engineers a numerical estimate of a part's probabilty of failure. In the first step of this approach, test bars of a material are stressed to determine their range of strengths. Then, in a method called finite element analysis, a part made of the same material is broken down mathematically into many rectangular elements. The dimensions of each element are fed into a computer, which uses the previously acquired experimental data to identify the levels of stress in each segment. From this information the computer estimates the probability of failure of the entire part.

At the University of Pennsylvania, ceramics professor Richard E. Tressler is trying to determine how high temperatures affect a ceramic's probability of failure. Working with silicon carbide "infiltrated" with silicon, he has found that microscopic flaws change with temperature and stress. At certain temperatures (800 to 1,100°C) the flaws grow when a part is subjected to stress. But they blunt (shrink) at these temperatures when stress stays below a critical value; in other words, the material actually becomes stronger. Tressler is investigating how a ceramic's microcrystalline structure affects this blunting-to-growth transition, in order to design parts that might become stronger at elevated temperatures.

What's Sexier and Speedier Than Silicon?

Gene Bylinsky

9

The strategic importance of the semiconductor industry can hardly be overestimated, so the arrival of a new material for making chips will have major implications for manufacturer nations and user nations. The synthetic compound gallium arsenide is being used increasingly in certain types of chips and chip applications, as award-winning science writer Gene Bylinsky explains in this upbeat article. Critics still scoff at claims that gallium arsenide is the material for chips of the future ("always has been, always will be"), but the growth of optoelectronics is giving it a considerable boost.

Once in a while, a material comes along that's made to order to meet the needs of a new generation of technology. Silicon, an excellent conductor of electrons when properly processed, has powered the twentieth-century revolution in microelectronics by making possible data-processing chips that shuttle electrons around in billionths of a second. Because it's even more sensitive to light than photographic film, silicon has also contributed to advances in optoelectronics, the application of light to information processing—replacing bulky cathode-ray tubes in TV cameras, for example. Solar cells of silicon are commonplace today, too, as power sources in pocket calculators, atop houses, and on the unfolded arms of spacecraft probing the solar system.

Now this familiar workhorse of electronics is being nudged by an even better, more versatile material: gallium arsenide. Proclaimed for about twenty-five years as a key to the future of microelectronic technology and already in limited use in such devices as satellite dish antennas and radar detectors, gallium arsenide finally seems ready to step into the big time in data processing and to enlarge enormously its hold on optoelectronics.

Nobody expects gallium arsenide to replace silicon across the board. For one thing, it's too expensive: silicon, found in ordinary sand, is the second most abundant element in the earth's crust after oxygen, while gallium makes up less than 0.01 percent of it. (Gallium turns up most commonly as a byproduct of aluminum making; arsenic, well known as a poison, is either mined or extracted as a byproduct of lead or copper refining.) Gallium arsenide is emerging not as a substitute for silicon but as an important complement to it. This remarkable material has a battery of useful qualities:

- It moves electrons around three to six times faster than silicon.
- It emits light—something silicon can't do.
- It absorbs sunlight more efficiently than silicon, making possible better solar cells.
- It has a higher resistance to radiation than silicon—important for space satellites, which are exposed to damaging particles such as electrons from the sun.
- It can operate at much higher temperatures than silicon, reducing the cooling requirements for computers and other electronic systems.
- It uses less power than silicon.
- It can combine the processing of both light and electronic data on a single chip.

Scientists and engineers have long known about gallium arsenide's unique properties, but only in the past year or so have they been able to capitalize on them in a big way. Horrendous barriers to making gallium arsenide of sufficiently high quality stood between the designers' desire to use its extraordinary speed and their ability to make data-processing chips with it. Now many of those problems are being overcome, and gallium arsenide is leaping from the high-tech doghouse to the high-tech penthouse. Gallium arsenide integrated circuits, or chips, first became available on the open market last year, and many analysts in the industry think billion-dollar-a-year markets are in the making. Most forecasters see gallium arsenide chip sales soaring from about $30 million this year to at least $1 billion in 1990. By the end of the century, gallium arsenide could account for one-third of the semiconductor industry's business—which is expected to spiral by then from last year's $15 billion to an astounding $150 billion annually.

Despite the relatively poor quality of gallium arsenide in the past, it's been used for about a decade to make so-called discrete, or singly packaged, components such as transistors, and small circuits designed specifically for signal processing in high-frequency microwave radio, radar, satellite communications, and related areas—wherever silicon bumps into its speed limit. Silicon simply can't process very high frequencies, which makes the costly and difficult gallium arsenide alternative worthwhile. These specialized applications of gallium arsenide amount to about $150 million a year.

What was needed to take gallium arsenide into the mainstream of electronics was better integration of individual components on chips, making them dense enough so that gallium arsenide's speed advantage wouldn't be offset by the length of time it took electrons to travel long distances between individual transistors. Higher-density gallium arsenide chips are becoming a reality. They are expected to launch a quantum leap into a new, more advanced age of electronics, making possible products that will far overshadow earlier uses of gallium arsenide.

The military is the biggest consumer of gallium arsenide, and it's likely to remain so. Already, for example, gallium arsenide components aboard military aircraft can receive a hostile radar signal, distort it, and send it back so that the airplane seems to be somewhere else—all thanks to that remarkable processing speed. If the Star Wars program goes ahead, military spacecraft will almost certainly be controlled by compact gallium arsenide computers, now under development, and powered by efficient gallium arsenide solar cells. A Hughes Aircraft subsidiary has just started large-scale fabrication of such cells near Los Angeles. They're unlikely to be available soon for commercial use because of their high cost, but other nonmilitary applications of gallium arsenide in computers, communications, and instrumentation are expected to grow rapidly.

Things started to tilt gallium arsenide's way in high-speed-chip data processing last year when two new California companies—GigaBit Logic of Newbury Park and Harris Microwave Semiconductor of Milpitas—introduced the first small logic circuits made of gallium arsenide into the open market. To be sure, such big vertically integrated companies as Hewlett-Packard in the United States and Fujitsu in Japan had been making experimental circuits for their own use, but not selling them to anyone else. Now designers at dozens of companies without their own gallium arsenide supplies could buy the chips and start designing them into their own systems. GigaBit Logic has already signed up 200 customers, ranging from supercomputer maker Cray Research Inc. (*Fortune,* March 18, 1985) to builders of smaller computers, instrument producers, mainframe computer manufacturers, and telecommunications firms.

The first systems incorporating gallium arsenide chips are beginning to trickle in. Hewlett-Packard, for one, has just introduced a microwave frequency sampling instrument with such a chip; the instrument, used in building radio and radar transmitters, makes possible measurements at higher frequencies. The initial computers and other data-processing systems with gallium arsenide chips will probably start showing up in about two years. By then engineers should have supercomputer power on their desks in packages the size of today's tabletop work station terminals. Cray has set 1987 as the introduction date for its supercomputer based on gallium arsenide components.

Since speed is to electronics what money is to Wall Street, gallium arsenide should make possible not only supercomputers with capabilities undreamed of today, but also Dick Tracy-style wristwatch radio-telephones that would receive and transmit signals via orbiting satellites. Gallium arsenide chips will be absolutely essential to the success of direct-broadcast satellites that beam TV programs to small rooftop dish antennas. Again, silicon simply can't process signals fast enough. Gallium arsenide chips should also bring down the price of a cellular car telephone, and they're at the heart of a collision-warning radar system now being developed for passenger cars. Smarter and faster robots, better medical instruments, quicker processing of seismic data, navigation systems that will use satellites to pinpoint positions of ships and airplanes—all these will be made possible by gallium arsenide.

A gallium arsenide wafer for making chips costs about $200, compared with $10 for a silicon wafer. Because gallium arsenide is so expensive, at the outset computers won't be built completely out of it—only the critical portions of their central processors and memories, to take maximum advantage of the material's exceptional speed. The transition to gallium arsenide will be gradual for still another reason: the state of development of chips made from it lags behind silicon by at least a decade. The highest-density silicon chips already hold several million microminiaturized transistors and related components in a sliver not much bigger than a baby's fingernail. So far gallium arsenide designers can squeeze only thousands or at best tens of thousands of components on a chip that size. Specialists predict, however, that gallium arsenide chipmaking will improve faster than silicon did, partly because gallium arsenide chip producers can conveniently borrow a lot of manufacturing equipment from their silicon counterparts.

In optoelectronics, by contrast, gallium arsenide and related compounds quickly grabbled most of the $400-million-a-year market because silicon couldn't even begin to compete with it as a light emitter. The relatively poor quality of the starting material wasn't a problem in making solid-state lasers and light-emitting diodes. Those lasers are only about the size of a grain of salt, so even an imperfect slab of gallium arsenide contains a lot of good ones. The tiny lasers pump light pulses through hair-thin glass fibers at dizzying billions-a-second rates to transmit voices, data, and video signals in the familiar binary "yes–no" language of data processing. More recently gallium arsenide lasers have found their way into such consumer products as digital audio disk players and laser disk data-recording systems, which use the tiny lasers to inscribe and read information. Light-emitting diodes are widely used in small read-out screens in VCR time displays, computers, and a variety of instruments.

In the near future, the combination of gallium arsenide's optical and electronic properties will be exploited in novel ways. Already Honeywell and others have made prototype integrated circuits that combine tiny lasers with signal-processing elements—both made of gallium arsenide—on a single chip. Attached to glass fibers, these circuits will use ephemeral packets of light, or photons, to transmit both speech and data—a faster and easier way than pushing around bulkier electrons. With the use of fiber optics spreading rapidly in both long- and short-distance communications, the upshot will be substantially reduced costs. Eventually this kind of circuit will make computing with light a reality.

Enticed by the possibilities of this new material, companies big and small around the United States have started a spectacular building boom in gallium arsenide chipmaking facilities. Privately held GigaBit Logic, the first venture capital-backed company set up expressly to make gallium arsenide chips, will double its production capacity by next year. Fifteen miles up the California coast on Highway 101, at Vitesse Electronics Corp. in Camarillo, huge tractor-trailers are disgorging production equipment into the company's brand-new buildings; Vitesse raised $30 million from Norton Corp., a Worcester, Massachusetts, conglomerate, to build gallium arsenide chips and computers. Expansion of gallium arsenide facilities is also under way at such Los Angeles area defense contractors as Rockwell International and Hughes.

Even in Silicon Valley, the enemy stronghold, new gallium arsenide companies are beginning to appear, and five-year-old Harris Microwave Semiconductor is expanding. In Beaverton, Oregon, the instrument maker Tektronix has just elevated its fast-growing gallium arsenide chipmaking operation to the status of a subsidiary, called TriQuint Semiconductor. Ford Microelectronics Inc., wholly owned by the carmaker, completed a $33-million gallium arsenide chip mill in Colorado Springs earlier this year. Honeywell starts production in July at a new facility in Richardson, Texas. In New Jersey, small start-up companies—among them Pivot III-V, Anadigics, and Lytel—have lately sprung up near AT&T's Bell Labs, a pioneering center in gallium arsenide research, forming the nucleus of a nascent Gallium Gulch. Similarly, the boom is on in the Boston area: the electronics conglomerate M/A-Com Inc. is putting finishing touches on the country's largest gallium arsenide fabricating facility in Lowell, and Raytheon plans to expand its gallium arsenide chipmaking.

One conservative estimate puts at $120 million the amount that has been invested in gallium arsenide facilities and new company start-ups in the past six months alone. "Never before has this level of interest and momentum accompanied a new technological direction," says a report by Strategic Inc., a Cupertino, California, research firm that specializes in electronics.

Since silicon is almost certain to remain the dominant material, silicon chipmakers profess unconcern about the rush into gallium arsenide. Of the ten top silicon chip sellers, according to a recent report by the New York City-based market research firm Frost & Sullivan, only three—Texas Instruments, Motorola, and RCA—are involved in gallium arsenide digital chip development, probably because they are defense contractors. About the only silicon chipmaker to go into gallium arsenide in a big way is Harris Corp., which acquired the start-up now known as Harris Microwave Semiconductor. Silicon chipmakers are preoccupied these days with staying afloat in the worst recession yet to hit their industry, caused partly by a brutal shakeout in the personal computer business. Even innovators in one technology can become myopic about advances in another, however. "Gallium arsenide will surprise a lot of people," says Richard C. Eden, senior vice president for research and development at GigaBit Logic.

Some of the surprises will be coming from Japan. In keeping with their excellence in materials processing, the Japanese have led in improving the quality of gallium arsenide, to the point that they are now the dominant world suppliers. Japanese researchers have surged ahead—at least in the lab—in making complex gallium arsenide memory chips. Last year Nippon Telegraph & Telephone built an experimental 16K (16,384-cell) random access memory (RAM) chip, whereas U.S. researchers have so far made only a 1K (1,024-cell) chip. Random access memory chips allow immediate deposit or retrieval of data in any of their cells, much as any telephone can be dialed in a telephone network; such chips form the main memories of big computers. The Japanese appear somewhat behind the United States, however, in building out of gallium arsenide complex logic, or arithmetic-processing, chips, which are more difficult to design.

The Japanese, says Ted Wakayama, Japanese-born analyst for Strategic Inc., have made successful development of gallium arsenide a matter of national pride: they are eager to show the world they can succeed in pioneering a new technology. Approaching development of gallium arsenide as a national program, the Japanese see the material as their long-sought wedge to capture control of the world computer markets. With government financing, Japanese giants such as Fujitsu, Hitachi, and NEC are all rushing development of gallium arsenide components.

Says Takahiko Misugi, director of Fujitsu Laboratories Ltd.: "Gallium arsenide's hour has come." But if the Japanese had hoped to surprise their U.S. competitors with a sudden flood of gallium arsenide computers, the great American drive into gallium arsenide will come as a shock to the Japanese. Big U.S. systems companies such as IBM, Hewlett-Packard, ITT, and GE all have large efforts in progress. IBM Vice President John A. Armstrong, who directs the company's semiconductor research, describes work on gallium arsenide as "the largest alternative technology [program] at IBM."

What makes gallium arsenide such a marvel of a material for both electronics and optoelectronics? Chemically speaking, it's the nature of its atomic and crystalline structure and the spacing between its atoms. Gallium arsenide provides electrons a fast track—and emits light when a current of a specific frequency is passed through it—because of the unique energy levels in the orbits of electrons that surround its atomic nuclei. When an electron in gallium arsenide is made to jump from a higher energy band to a lower one, the electron emits a photon of light. Electrons can move rapidly through the material because of its highly symmetrical crystalline lattice and because of those unique energy levels. When the age of microelectronics began in the 1950s, scientists started experimenting with various compounds as the raw material for chips. Gallium arsenide turned out to be a hit, but producing it in commercially usable form took some doing.

Unlike silicon, gallium arsenide doesn't occur in nature. Silicon is a nicely homogeneous element with highly ordered crystals, and it's easily purified. Not so gallium arsenide. Both gallium and arsenic contain impurities that are difficult to eliminate and that distort the electronic properties of the material that results when they are combined. Moreover, putting gallium and arsenide together in just the right way, matching them atom for alternating atom throughout the structure, turned out to be a hellish task. For one thing, during processing at high temperature arsenic turns into a poisonous vapor. To confine it, technicians mix gallium with arsenic in a sealed flask.

The slabs of material produced that way weren't of high enough quality for chips, though they were adequate for single transistors, those tiny lasers, and light-emitting diodes. To make chips, engineers not only needed higher-quality gallium arsenide, they needed it in the form of round wafers. In the manufacture of silicon chips, such wafers are sliced from big ingots of silicon made in special furnaces. In this process, high-purity silicon is melted in a quartz crucible. A seed crystal about the size and shape of a pencil is then lowered into the molten silicon and slowly pulled and rotated to form a single larger crystal with perfectly positioned atoms.

Adapting the silicon crystal-pulling machines to the making of gallium arsenide crystals met with only partial success. At best, the gallium arsenide workers could produce crystals only one inch in diameter. The economics of making semiconductors—silicon or gallium arsenide—depends in part on the size of the wafers sliced from the crystals, because the more usable chips that can be manufactured on a single wafer, the higher the "yield," as semiconductor people call it, and the higher the profit. At a minimum, the gallium arsenide developers needed a wafer three inches in diameter. They finally got that in 1979 when Cambridge Instruments, a British company, developed crystal pullers specifically for gallium arsenide.

By that time Fred A. Blum, Richard C. Eden, and coworkers at Rockwell International were well on their way toward building up denser and denser gallium arsenide logic chips. Unable to persuade their employer to go into commercial production with the chips, Blum and Eden quit. They started GigaBit Logic in 1982 after raising $30 million from several sources, including Standard Oil of Indiana and Analog Devices Inc., a Norwood, Massachusetts, maker of electronic components. GigaBit will probably sell at least $15 million worth of gallium arsenide chips this year. The fact that IBM had dropped research in a competing technology and begun a big effort in gallium arsenide gave the whole field the equivalent of a Vatican blessing.

Because gallium arsenide is a man-made material, the first such substance to be used for electronics and optoelectronics, it can be altered to meet specific requirements by adjusting those energy band gaps—to make tiny lasers that operate at different wavelengths, for example. Even this remarkable material may soon have some competition, however. Scientists are already looking at other synthetic materials that are even faster, including one called indium phosphide.

High-Performance Plastics

Gordon Graff

10

A new generation of tough and durable plastics is being rapidly adopted by makers of cars, computers, and food packaging. New composites created by mixing and matching existing polymers like nylon and polyester—or combining them with ceramics, glass, or carbon fibers—can be made stronger, lighter, and tougher than steel. The boom in optical storage discs such as CD-ROMs is boosting demand for the polycarbonate plastic from which they are made. But plastics still have to overcome prejudice and an image of cheapness, says Gordon Graff.

Plastics are finally getting some respect. Once regarded as flimsy, throwaway substitutes for costlier materials, the new generation of high-performance plastics (also called engineering resins) are often tougher and more durable than the "real thing."

Cars with plastic bodies, for example, are lighter and more corrosion-resistant than their all-steel counterparts; chips insulated with plastic may help computers achieve far greater circuit densities than possible with today's glass and quartz materials; foods packaged in new aseptic "barrier" resins that keep out air, moisture, and germs can be popped into the microwave oven in their containers, something not possible with foods in tin cans; and semiporous plastic films are being engineered to separate virtually any type of molecule from a complex solution.

Although the new materials still face some lingering resistance from cautious designers, the spate of new outlets has generated high hopes among producers. "This is a very exciting business to be in right now," says Charles R. Eggert, manager of market development at Monsanto Chemical's plastics division in St. Louis. In fact, industry consultants Charles H. Kline and Co. (Fairfield, New Jersey) estimates that by 1991, worldwide sales of all engineering resins will be $8.3 million in 1986 dollars. Sales this year are expected to reach $5.8 billion.

The renaissance in plastics stems not so much from the development of new polymers as from the mixing and matching of different types of existing polymers—such as nylon, polyesters, polycarbonates, or polyphenylene oxide—or from their combination with reinforcing fillers such as glass, metal, or carbon. These innovations are enabling the industry to come up with a steadily expanding menu of novel properties. Some of the new materials are stronger than steel; others are electrically conductive, or stable at temperatures that would have melted older plastics.

Such properties come at a price, of course. While older commodity plastics such as polyethylene, polypropylene, and polyvinyl chloride generally sell for under $1 a pound, the new resins command up to $15 a pound—prices that could spell important new profits for producers beset by lackluster commodity sales. But the ultimate beneficiary is "the end user who doesn't want to worry about product recalls or customer problems," says Bennett Nathanson, manager of commercial research and new business development at Celanese's engineering resins division in Summit, New Jersey.

Coming: A $10 Billion Market

The advances in plastics technology are especially attractive to the auto industry, which has ambitious plans to boost the amount of exterior plastics in its cars. Older resins on the inside of the car, meanwhile, are rapidly giving way to superior new ones. With a few exceptions—the *Corvette* and the Pontiac *Fiero*, for example—most cars still have steel bodies, but Detroit is moving inexorably toward plastic body panels and bumpers. In fact, as many as 70 percent of new cars will sport plastic body panels by the year 2000, according to a recent University of Michigan study; only about 5 percent of today's new cars are so equipped. At Du Pont's polymer products department in Wilmington, Delaware, R&D director Richard G. Bennett interprets that as a future $10-billion-a-year business for resin suppliers.

The most obvious reason for the switch to plastics is weight reduction—700 pounds worth in the case of an experimental GM "concept car" in which metal exteriors were replaced by plastics. But automakers note other advantages, including corrosion resistance and the ease of styling changes. Plastics could also prove to be less labor-intensive, since a metal part assembled from many smaller pieces can be replaced with a single molded plastic component. Plastics also permit configurations that would be impossible with other materials, such as the increasingly common forward lighting systems made of polycarbonate lenses contoured with the rest of the car body. Glass lenses would be more vulnerable to shattering if incorporated into the vehicle body.

Still, Detroit has been slow in converting to all-plastic vehicles. One reason is that such conversions require "quite a bit of retooling," notes Irvin E. Posten, manager for composites in GM's engineering department. As a result only a few "dedicated" plastic car plants have been built.

Among the products of these factories is what many consider to be the prototypical plastic car—the *Fiero,* which debuted in 1984. The hood, roof, deck lid, and upper-rear quarters are made of sheet-molding compound (SMC), a rigid composite consisting of a polyester resin reinforced with glass fibers. The door panels, bumpers, and bumper cushioning are made of polyurethanes fabricated by reaction-injection molding, or RIM, in which liquid precursors of the plastic react inside the mold to form the finished part. And the rocker panel (along the lower edge of the body) and lower-rear quarter panels use a variety of thermoplastics (the general name for materials that are pumped into a mold in liquid form and solidify upon cooling).

Although the *Fiero*'s exterior is all plastic, the underlying frame is still made from steel. Automakers hope eventually to change that as well, but "structural components made of plastic will take a lot more development work than the exteriors," says Marilyn Perchard, principal engineer for plastic materials at Ford (Dearborn, Michigan). The reason, she adds, is that the frames bear critical loads and that the plastics involved would "require different processing parameters" from those in the exteriors.

Not all the bugs have been worked out of the technology for plastic exteriors either. For one thing, it is hard to find thermoplastics that will stand up to the 400°F ovens used to cure anticorrosion coatings on the rest of the exterior, which is still largely made of steel; this step is followed by the finish coat, which requires subjecting the auto body to temperatures of about 300°F.

Although the SMC used in large panels can take these temperatures, the thermoplastics found in bumpers and doors normally can't. Affixing a plastic part such as a bumper after the anticorrosion treatment but before the painting won't do, because the entire assembly line would have to be retooled. So the bumpers, which are often made of polycarbonate/polyester alloys, are usually painted separately, then attached near the end of the production line. To avoid this costly step, the auto industry is seeking plastics that can be affixed early on and then left to go through the ovens. Ford is looking at one such heat-resistant resin for the bumpers of its *Taurus*—a nylon/polyphenylene oxide (PPO) alloy made by General Electric. Perchard calls the material, trade-named GTX, "very promising." Another heat-resistant GE product, an alloy of PPO and polyesters, is being evaluated by Ford for door panels.

An additional problem is that SMC's cycle time (the time it takes to stamp out a finished part) is two to three minutes—rather slow by auto industry standards. Automakers are also hindered by SMC's high porosity, which makes painting difficult, and by the glass fibers that typically are visible on the surface.

Although thermoplastics have far better surface characteristics, they are usually not strong enough to compete with SMC for horizontal panels (hood and roof) and are therefore confined to vertical panels such as doors and fenders. But GE and PPG, the Pittsburgh-based chemical producer, recently joined forces to make a stampable grade of fiberglass-reinforced thermoplastic sheet suitable for horizontal sections. According to Herbert K. Hoedl, a general manager in GE's plastics division (Pittsfield, Massachusetts), the new sheet not only equals SMC's performance "but exceeds it in many cases." He notes further that the cycle time for the material is under one minute and that it withstands the searing oven temperatures used with steel and SMC. The product is set for commercial introduction later this year.

Auto headlamps are also receiving plenty of attention. The new streamlined polycarbonate designs that are showing up in many late models are decidedly more rugged than glass; however, they have some problems of their own. For one thing, ultraviolet light from the sun causes polycarbonate to become brittle; the plastic is also heat-sensitive, so it must be placed quite far from the light bulb fixtures.

Celanese thinks it has the answer to these limitations: a new resin called polyarylate. Formed by a reaction of the monomers terephthalic acid and isophthalic acid with the chemical bisphenol A, polyarylate is more UV- and weather-resistant than polycarbonate. It also stands up better to heat, allowing a lens to fit closer to a light bulb. Says Robert H. Jackson, director of business development at Celanese's engineering resins division, "we've had a tremendous response" to polyarylate from potential customers in the automotive, aerospace, and other sectors.

Under the hood, plastic parts have been introduced piecemeal over the years for a variety of fuel-handling, mechanical, and electrical applications. But one small New Jersey company has built an entire engine predominantly out of plastics. The engine, a V-6 turbo for racing cars, makes extensive use of a stress- and heat-resistant material trade-named Torlon by its maker, Amoco Chemicals (Chicago). The 2-liter, twin-cam, 4-cylinder engine, built by Polimotor Research (Fairlawn, New Jersey), weighs only 200 pounds, about half as much as its all-metal counterparts; it uses plastics nearly everywhere but in the immediate vicinity of the combustion chamber.

Although the engine has thus far powered only racing cars, Polimotor president Matthew Holtzberg predicts that "it may also appear in passenger cars before long." GM's Posten, however, notes that few tests have been done on the engine's performance in passenger cars, so "it may be quite a while" before it is adapted for that purpose.

Staking Out Electronics

Plastics are also expanding their role in electronics. Polymer photoresists have long been used in fabricating integrated circuits, and almost all electronic equipment is packed with plastic circuit boards. But the new engineering plastics are expected to be used as chip insulation layers, optical fiber components, and data storage discs.

The need for new insulation materials for electronic chips becomes more critical as circuits grow in density and complexity. The interconnection lines on these chips typically consist of several levels of metal conductors with insulation between them. A layer of glass or quartz is usually deposited on each level of circuitry before the next level is laid down, but glass and quartz often harden with tiny bumps on top, making it harder to lay down the new lines and increasing the possibility of short circuits between the layers. As a result electronics firms are searching for a polymer that would "planarize"—that is, form a completely flat surface when deposited. Such a material would also have to be a good insulator and be capable of withstanding high temperatures; the deposition of vaporized metal can heat chip surfaces to more than 700°F.

Closest to meeting these needs are a class of materials called polyimides, says James Economy, polymer science and technology manager at IBM's Almaden, California, research center. He is confident that polymers will be developed to meet all the requirements.

Electronics manufacturers are also interested in polymers that would improve the way chips are mounted on circuit boards. Right now, chips are usually mounted on a ceramic module, which is itself attached to the board. However, the different rates of expansion and shrinkage of the silicon and ceramic when exposed to repeated heating and cooling during processing cause stress at the point of attachment, which can lead to weakened connections and subsequent failure. Also ceramics tend to slow down the flow of electrons in the wires embedded in them, reducing circuit speed. Electronics researchers thus hope to find polymers for the modules with the same thermal expansion characteristics as silicon, thus reducing strains during processing. Such materials should also speed the flow of electrons in the embedded wires. Again, IBM researchers are studying polyimides for this application, although Economy notes that his staff hasn't yet found the ideal material.

Growth in Optical Discs

Plastics will also play a prominent role in the optical disc technology that is expected to blossom in the 1990s. Today's laser-read audio compact discs are the best-known example of optical discs, but ultrahigh-capacity optical data storage discs—known as CD-ROM (read-only memory)—are now arriving on the computer scene. A single 4.7-inch disc can hold some 550 megabytes of data, equivalent to 28 20-megabyte hard disks or 1500 360-kilobyte floppy disks. Both audio and computer versions are typically made of polycarbonate plastic coated with an aluminum layer; information is encoded on the aluminum surface in the form of tiny pits that can be read by a laser beam.

The growth of audio and computer optical discs could bring parallel growth in the polycarbonate resins that go into them. In fact, companies such as Du Pont, Dow (Midland, Michigan), GE, and Mobay (Pittsburgh) are boosting production of such resins to cater to the disc market, which Du Pont estimates will reach $4 billion by 1990.

Future developments could put a crimp in this otherwise rosy picture, however: the next logical step in the evolution of optical discs—an erasable format for audio, video, and computers—may not be able to use plastics at all. One erasable system that is being considered by developers including IBM, 3M, Eastman Kodak, Control Data, and Xerox is a technology called magneto-optic recording. These systems rely on a laser beam that writes data on a magnetic film—typically consisting of rare-earth mixtures such as terbium-cobalt-nickel—deposited on a substrate. The problem is that if the substrate is made of polycarbonate, the magnetic films rapidly degrade when exposed to air and moisture. So some method of depositing the films inside hermetically sealed glass discs is being examined as an alternative, even though this precludes flexibility and adds to the cost. Some developers such as 3M say plastic mag-

Autos and Electronics Drive the Plastics Boom

High-performance plastics (also called engineering resins)—which include such materials as high-grade nylon, polycarbonates, polyimides, acetals, and thermoplastic polyester—are finding growing use in a variety of commercial applications. The electronics industry accounted for 30 percent of the total of 839 million pounds of engineering resins sold in the United States last year, according to the Society of the Plastics Industry, a New York-based trade group. Transportation (primarily autos, but also aerospace) made up 17 percent of the total, followed by building and construction (9 percent), industrial machinery (4 percent), and packaging (2 percent). About 24 percent of U.S. engineering resins are exported.

A boom in plastics use in autos and electronics in the next few years should push the production of engineering resins from a worldwide volume of 3.5 billion pounds this year (worth about $5.8 billion) to around 5 billion pounds (worth $8.3 billion) by 1991, according to Peter Gavrushenko, project manager for Charles H. Kline (Fairfield, New Jersey), a market research firm.

This expansion will be a boon to engineering plastics producers. In the U.S., companies with a broad product line include Du Pont (Wilmington, Delaware), Celanese (New York), Dow Chemical (Midland, Michigan), Monsanto (St. Louis), and General Electric (Pittsfield, Massachusetts). Important overseas manufacturers with U.S. facilities include ICI, Hoechst, and Mobay.

Among automakers, the chief attractions of plastics are lighter weight and less corrosion than metals, ease of fabrication, and amenability to styling changes. Ford, General Motors, and Chrysler all have ambitious programs to replace metals, glass, and other traditional materials used in their vehicles, with plastics. Fenders, doors, and wheel covers will be new applications for some plastics. But the largest single use of plastics will be in car body panels, rising from 10 to 20 million pounds this year to 300 million pounds by 1995, according to Roger K. Young, a manager for engineering thermoplastics at Dow Chemical.

Electronics will be another key outlet, particularly in circuit boards, chip insulation and packaging, and new products such as optical discs. The extreme heat and electrical resistance of some of the engineering resins is a prime factor in their growth, particularly as circuits continue to grow denser. The new plastics will edge out such traditional materials as glass and quartz insulation, as well as some older plastics such as epoxies used in circuit boards. The net result is that last year's worldwide production of 25 million pounds of heat-resistant engineering resins for electronics will grow at a healthy 10 percent a year over the next decade, according to Carl H. Eckert, director of the materials group at Kline.

Several new markets are opening up for producers of engineering plastics. For example, new food packaging could consist of shelf-stable "barrier" resins. These sturdy plastic laminates, which keep out air, moisture, and germs, are already used to package ice cream, catsup, and other foods. And "all you have to do is look at a supermarket shelf and you see all sorts of new opportunities for barrier resins," says Thomas H. Lyon, market development manager for specialty plastics at Dow Chemical. He envisions them replacing cans and bottles in pre-prepared single-portion servings that can be heated in their own containers. Such packages could replace some $16 billion worth of cans and bottles annually by the year 2000, estimates Richard G. Bennett, director of research and development at Du Pont's polymer products division.

Plastic membranes that are semipermeable (allowing only molecules of certain sizes to pass through) are another promising outlet for engineering resins. They are already used to separate hydrogen from refinery gases, to separate pollutants from waste water, and to desalinate seawater. They're also being adapted for citrus juice concentration, generating nitrogen for industrial purposes, and as containment media for genetically engineered microbes in commercial biotechnology. As a result, the 10 to 15 million pounds of engineering resins now used worldwide in separation membranes should grow by 15 percent annually over the next ten years, predicts Kline's Eckert.

Engineering resin makers are also keenly aware of the opportunities for plastics to replace plastics. Among the chief trends here are the growth of thermoplastics, which can be melted and remolded repeatedly without physical or chemical damage, at the expense of thermosets, which harden irreversibly when heated and thus cannot be recycled. "If you look at the possibilities for thermoplastics to replace thermosets such as epoxies and phenolics, that's a 400- to 500-million-pound-a-year opportunity in the electronics field alone," says John R. Dole, Celanese's market supervisor for electronics.—Gordon Graff.

neto-optic discs will be feasible, perhaps with some protective "passivation" coating on the surface. But according to IBM's Economy, the jury is still out.

Locking Out Germs

In the field of food packaging, plastics may prove a lightweight alternative to cans and bottles. Such firms as Du Pont, Dow, Monsanto, and Allied are vying to introduce new barrier resins that are impervious to air, moisture, and bacteria; some traditional food packagers—Continental Can and Britain's Metal Box, for example—and food processors such as Campbell Soup have also entered the fray.

Although details vary, the new resins designed for shelf-stable packaging of perishable foods are generally used as laminates, with each layer designed to do a different job; Heinz's squeezable catsup bottle, for example, has six layers. A typical barrier film might consist of a support layer of polypropylene secured to a sheet of polyethylene, which is in turn bonded to a barrier plastic such as polyvinylidene chloride or ethylene vinyl alcohol, which bars air and germs. This layer is bonded to a sheet of rigid polystyrene. The layers are assembled by a process in which each polymer is liquefied and forced through a die; the streams are then combined in a single chamber which forces the hot laminate through another die that spreads and compresses the mixture into the final structure.

Producers are burdened somewhat by the fact that barrier-resin technology is basically mature; it is already familiar in the form of disposable containers for yogurt, ice cream, and TV dinners. Industry observers thus say that further growth hinges on consumer acceptance in the all-important shelf-stable applications such as single portions of soups, stews, and other prepared foods that can be microwaved right in their containers. Many of these products, which remain stable at room temperature for up to three years, are now being test-marketed.

Choosy Membranes

Plastic films are also being designed to separate liquids, gases, and solids. During the past decade, semipermeable polymer membranes (which pass only molecules or particles of certain sizes) have been pressed into service for separating hydrogen from waste gas streams, desalinating seawater, removing toxic metals from municipal waste water, separating enzymes from microbial broths, and slowly releasing drugs through the skin. Many of these membranes, composed of such materials as triacetates, aromatic polyamides, and polypropylene, can be spun into hollow fibers to maximize surface area. Among the leaders in the separation membrane industry are Monsanto, Dow, Du Pont, and Celanese. Some specialized separation products are sold by smaller firms such as Pall Corp. (Glen Cove, New York) and Alza (Palo Alto, California).

But the action in separation membranes is more in developing new markets for existing materials than in finding new ones. In one instance, both Dow and Permea (a Monsanto subsidiary) are pushing inert gas generators—membrane systems that separate nitrogen from air or combustion exhausts to create a nonreactive, oxygen-free gas mixture. Dow is touting its unit for such tasks as blanketing inflammable chemicals or providing an inert atmosphere for treating metals at high temperatures. Permea's objective is to set up systems on offshore drilling platforms and pump the generated nitrogen into undersea oil reservoirs to prevent the platforms from sinking as the oil is withdrawn. Another proposed application: generating nitrogen that could be used to blanket fuel tanks on military aircraft to prevent fires in case of a crash.

Celanese, too, is extending the range of its separation membranes, which are based on polypropylene with microscopic pores. One major outlet for the membranes (tradenamed Celgard) will be biotechnology separations, says Henry T. Taskier, development associate in Celanese's separation products division. In one system, he envisions genetically altered bacteria growing inside the hollow fibers, fed by nutrients that seep into them; the enzymes or other microbial products would slowly diffuse out of the fibers for collection. The advantage of such a system, Taskier explains, is that the bacteria wouldn't have to be separated from the product—a costly step in many such processes.

Despite the rapidly growing applications of the new plastics, producers aren't deluding themselves: many of the ideas are, and will continue to be, a tough sell. Although separation membranes and other specialized applications face few of the ingrained prejudices against plastics, the resistance is more intense in cases where plastics are trying to edge out such well-entrenched materials as metals and glass.

One reason is that even today's tough high-performance resins are saddled with the image of cheapness that still clings to the inexpensive commodity resins. Another is the all too common reluctance of designers and engineers to chuck the familiar in favor of the new. "Many of the companies that have been around for fifty or one hundred years are very conservative," says Monsanto's Eggert. "What I tell them is that there's no such thing as a bad plastic . . . only a bad application."

Optical Fibers: Where Light Outperforms Electrons

Les C. Gunderson and Donald B. Keck

11

We are only just beginning to tap the prodigious capacity of the laser to generate optical signals. Many now think that the development of so-called "photonics" will inevitably lead to fully fledged optical computing, in which light pulses replace electrons as the basic method of transmitting information. The photonics revolution has been made possible by the development of optical fibers, tiny strands of pure glass through which light can be transmitted efficiently over long distances. This new material is already transforming telecommunications and data networks, vastly increasing our capacity to move digitized information around. The authors are both employees of Corning Glass.

A revolution has begun in the technology of transmitting information. The rebel force is optical fiber—a thread of purest glass, five-thousandths of an inch in diameter, about the size of a human hair—through which laser light of high purity and intensity can be transmitted. This radical departure from traditional electrical and electronic communications is destined to increase by a factor of hundreds our capacity to move words, pictures, and data from place to place. Today's crowded communications cableways can be relieved, complex databanks can become more easily accessible to anyone with a personal computer, and vast amounts of information can be exchanged at very modest costs.

Already over one hundred communication links using optical fibers have been installed in North America. The first segment of the largest commercial optical system in the United States went into service early this year between New York and Washington. Soon this will be part of a 1,077-kilometer optical-fiber network connecting nineteen local telephone switching systems from Boston to Richmond. The cable used in this system contains one hundred glass fibers, each of which can carry up to 240,000 telephone calls (or lesser numbers of video or data transmissions) simultaneously. The half-inch-diameter cable has as much capacity as a two-and-a-quarter-inch cable of copper wires.

Its high capacity and efficiency make optical fiber one of the most cost-effective conduits available. The small size of typical optical fibers is also a significant advantage. Another is that "signal tapping"—that is, unauthorized eavesdropping—is extremely difficult. And optical-fiber communications are free of electromagnetic interference.

The Seeds of Revolution

The first suggestion for using light to transmit sound was made about one hundred years ago by Alexander Graham Bell, inventor of the telephone. Bell was intrigued with the light sensitivity of the metal selenium—its electrical resistance varies directly with the amount of illumination—and he patented a "photophone" that exploited this phenomenon in 1880. But Bell's vision was premature; seventy years later scientists were still seeking more intense light sources and better transmission media to achieve their goal of practical optical communications.

When the laser was first conceived in the early 1960s, its potential as an effective source of the intense light required for optical communications was obvious. The scientific literature was quickly flooded with reports of methods for modulating and detecting laser light. These first lasers generated light in the visible part of the spectrum, which is affected by the same factors that impair long-distance vision—atmospheric particulates such as fog, smoke, rain, and snow. Therefore engineers proposed to transmit the light in pipes with controlled atmospheres. But the pipes, lenses, and mirrors had

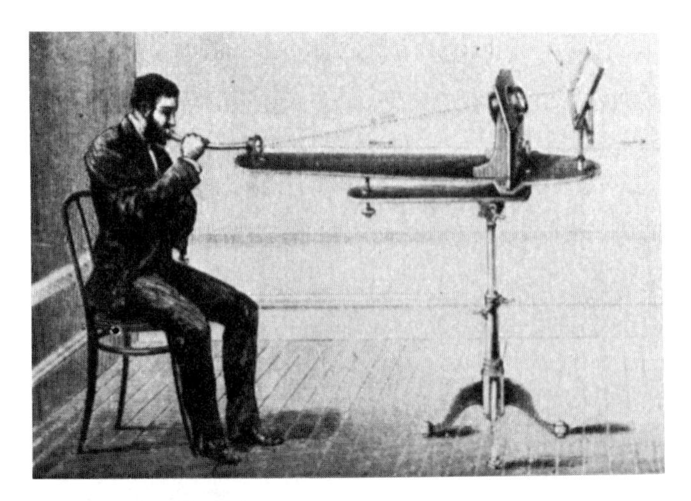

Figure 11.1 Alexander Graham Bell's photophone, patented in 1880, was the first attempt to use light to transmit sound. Sunlight, bounced off a reflector and focused by a lens (*above*), was intermittently interrupted by a mechanism that vibrated in response to sound waves. At the receiving end (*below*), the intensity of the light determined the resistance of a selenium cell and thus controlled an electrical current. A telephone receiver then recreated the original sound waves from the vibrations in this current.

to be precisely aligned, and experimenters soon discovered that instead of building transmission conduits, they had constructed some of the world's most sensitive seismometers.

In 1966 scientists at Standard Telecommunications Laboratories postulated that glass of extreme purity could be used to transmit the laser light. About four years later, scientists at Corning Glass Works demonstrated that a fiber of high-silica glass could transmit 1 percent of the light energy introduced at one end over a distance of 1 kilometer. This efficiency, although not high, was 98 orders of magnitude better than that of other fibers at that time, and was in fact considered the threshold for practical systems. If the American Revolution began with a shot heard around the world, this was the light seen round the world.

Through a Glass Lightly

Two distinct causes of loss of light potentially limit transmission through glass: absorption and scattering.

The principal cause of absorption is impurities—the ions of iron, copper, cobalt, vanadium, and chromium that have long been troublesome contaminants in high-quality optical glasses. For optical fibers, even one to two parts per billion of such metallic impurities are unacceptable; this is 0.1 to 1 percent of the amounts acceptable for most optical applications. Picture a hallway stretching five and a half times around the earth at the equator, floored with standard nine-inch-square white tile. A single row of red tile stretching across the hallway somewhere along its length would represent an impurity sufficient to absorb an intolerable amount of laser light in an optical fiber. Thus a process that makes possible extreme purity—called "vapor deposition"—has been the key to the advancement of optical-fiber communications.

Scattering, the other source of light loss, is a change in the light wave caused by a change in the density of the transmitting medium. Small irregularities that can cause scattering are very common in glass—the result of temperature variations in the mix as the glass solidifies. These irregularities can be largely, although not completely, controlled in manufacturing, so present fibers are able to conduct light for remarkably long distances. Indeed, if the oceans were as transparent as today's low-loss optical fibers, one could see the bottom of the Mariana Trench from the surface of the Pacific Ocean—32,177 feet down.

Scattering losses are inversely related to the fourth power of the wavelength of the light. This means that long-wave red and near-infrared light is more easily transmitted through fibers than blue and near-ultraviolet light. Thus a "window" of wavelengths in the near-infrared, between 600 and 1,600 nanometers, is now being used for communicating through optical fibers. (Most people cannot see light of wavelengths longer than about 700 nanometers; a nanometer is one-billionth of a meter.)

Optical Fiber: Reflecting upon Itself

The simplest form of optical fiber, known as "step-index" fiber, is based on the principle of "total internal reflection": when a beam of light strikes the boundary between two materials of different densities at a suitably low angle, the light is fully reflected back into the denser material. This is why, for example, a person swimming underwater cannot observe objects above the surface except within a 44-degree cone of the vertical. The surface of the water outside this cone has a silvery appearance. Optical step-index fiber is constructed with a core region of dense glass surrounded by slightly less dense material. Light striking the cladding (the outside layer of glass) as it moves through the fiber is reflected back into the core. Thus light zigzags through the fiber by reflecting first off one side and then another of the core-cladding interface.

The simplest optical fiber suffers from one major drawback: the transit time of a beam following the longest possible zigzag path is far greater than that of a beam traveling straight through a fiber. This means that over a 1-kilometer distance, the sharpness of the signal, which consists of a series of on–off pulses, is significantly reduced, severely limiting the system's information-carrying capacity. Indeed, step-index fiber is unable to carry more than one television channel for 1 kilometer—hardly the type of performance from which revolutions are made.

This problem has been solved by developing two other forms of optical fibers. In "graded-index" fiber, the glass in the core varies

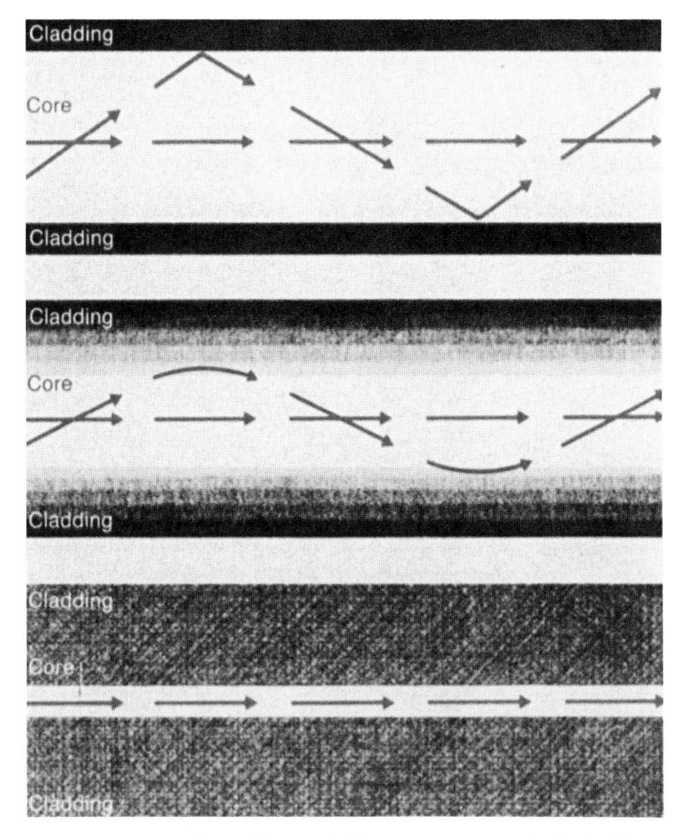

Figure 11.2 The three fibers of fiber optics. "Step-index" fiber (*top*) consists of glasses of two different densities, called core and cladding. Light zigzags down the core, bouncing off one side and then the other of the core-cladding interface. In "graded-index" fiber (*center*), the glass in the core varies in density, so the light travels in a smooth, curving path. Thus its information content is less distorted. In "monomode" fibers (*bottom*), the core is very small in relation to the wavelength of the transmitted light. The light therefore moves down the fiber in a straight line with very low distortion.

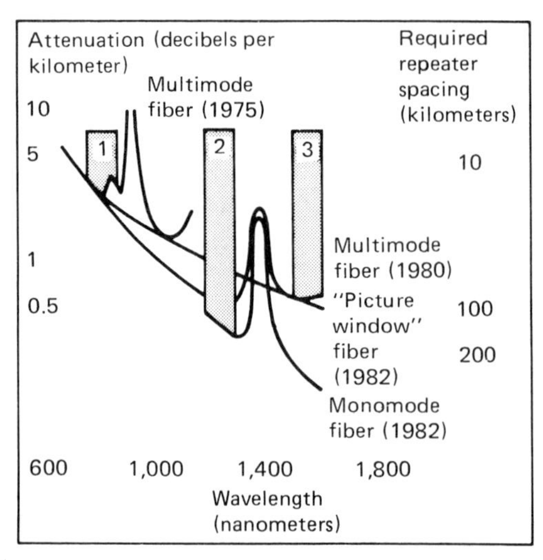

Figure 11.3 How fibers become transparent. For over a decade glass chemists have been seeking to decrease losses of light in optical fibers by increasing their purity. The goal is to find combinations of light and glass for which 80 percent of the light entering a fiber is transmitted over 1 kilometer. Three "windows" have been chosen where fiber losses are low and light sources of the right size are available. Under these conditions repeater stations are required at 50-kilometer intervals.

continuously in density—high at the core to low at the core-clad interface. As a result, instead of zigzagging abruptly, the light rays travel in smooth, curving paths, being turned gradually back toward the center as they encounter the decreasingly dense glass near the cladding. Though these curving beams travel nearly the same distance as in step-index fiber, a larger portion of their travel is in the low-density region of the fiber, where they travel faster. Their transit time is therefore more nearly that of rays that propagate straight along the axis of the fiber.

The other solution is to reduce the diameter of the fiber core to approximately the wavelength of the transmitted light—2 to 8 micrometers. In such "monomode" fiber, all light waves take a straight path, reaching their destination essentially simultaneously. Thus monomode fiber has much increased information-carrying capacity—sufficient to carry 40,000 conventional television channels over a 1-kilometer distance. Although this far exceeds the capacity of any present electrical cable, it is still about 1,000 times less than the theoretical carrying capacity of a light wave.

This theoretical capacity is determined in the following way.

Fiber-Optic Communications: Where Will They Lead?

Communication systems that send light waves through fiber-optic cables are in use in more than twenty countries, ranging from Indonesia, which just laid its first 8.7 kilometer cable, to Great Britain, which has over 4,000 kilometers of fiber in operation. These optical-fiber systems were chosen because they were more economical than the alternatives—copper wire, radio relay, and satellite. Since fiber optics is still an immature technology, scientific advances coupled with increased experience in manufacturing and operation suggest an even more promising future.

The economics of optical fibers are so favorable that Anthony Rutkowski, staff advisor to the chief scientist at the Federal Communications Commission, predicts that all telephone exchanges in the United States will be interconnected by optical fibers by the mid-1990s.

Consider the optical-fiber cable that AT&T is currently installing between Boston and Richmond, Virginia, to accommodate the steadily growing demand for telephone service. (Although demand for computer-data transmission over telephone lines is also growing rapidly, it still represents only a small percentage of total demand, according to AT&T.) Since optical fibers carry much more information than copper wires of the same diameter, AT&T will provide the additional capacity by drawing optical cables through existing conduits. By using optical fibers instead of copper wire, AT&T expects to save over $49 million in construction costs alone.

Optical fibers have another advantage. After a certain distance, signals sent on both copper and optical-fiber cable lose clarity and must be regenerated. But repeaters to regenerate the signal on copper cables are generally spaced every 2 kilometers, while the distance between repeaters on fiber-optic links is four times as long. This means significant savings in both installation and maintenance.

Indeed, it sometimes costs more to maintain copper cables than to rip them up and replace them with optical-fiber cables, according to AT&T. As better fibers and optical transmitters are developed, the distance between repeaters—and the relative economy of fiber-optic systems—will increase enormously. Already scientists at the British Telecom Research Laboratories have reported sending 140 megabits per second (equivalent to 1,920 telephone channels) without regeneration over a 102-kilometer length of cable.

The majority of communication links over 30 kilometers in length are now handled by microwaves relayed from one transmitter to the next, and fiber optics are challenging this technology too. Microwave relay service is difficult to expand in densely populated areas, where few, if any, frequency bands remain unoccupied, according to AT&T. There are no such limits on optical-fiber transmission, which will soon be as cheap as radio relay, says Malcolm Ross, a consultant at Arthur D. Little International.

The longest communication links are handled by satellites. The cost of satellite transmission, unlike that of terrestrial systems, is relatively independent of the length of the link. Such communication is especially effective between sparsely populated areas that cannot easily be connected by a telephone cable. But even here optical fibers have a role since they can carry large volumes of information more efficiently. An optical-fiber cable would be more economical for links thousands of kilometers long if it carried more than a thousand phone curcuits, according to Sir George Jefferson, chairman of British Telecom. In some cases this crossover point is already in sight. AT&T proposes to use optical fibers for the next trans-Atlantic cable, as well as for a link between the continental United States and Hawaii. Both cables are expected to be in operation by the end of the decade.

Indeed, the only place where fiber optics have few advantages is on local exchanges. The lowest-capacity optical fibers ever installed for commercial telephone use by AT&T carried 672 phone circuits per fiber pair, and many local routes require but a small fraction of this capacity. Nor is repeater spacing a factor on very short routes. So Rutkowski predicts that local service will continue on copper wire until the amount of information to be carried into homes increases dramatically.—Janet Yanowitz.

Information is typically transmitted through an optical fiber in digital form—that is, as pulses of light, much like those that might be used to transmit Morse code. The maximum frequency of such pulses through an optical fiber is set by the ratio of the speed of light to the wavelength of the light being transmitted, which works out to be about 300,000 trillion pulses per second. But so far it has not been possible to switch lasers on and off this fast. Indeed, the best lasers pulse only about 10,000 million times per second under laboratory conditions.

It's Not Like Glassblowing

Making fibers generally involves at least two distinct steps: preparing a high-purity glassy "preform" and forming fiber from it. Making glass in the conventional way—melting its ingredients together—is unsatisfactory; the heating process permits the entry of too many contaminants. Instead, glass for optical fibers is made by the "vapor-deposition" process, invented by Corning.

High-purity vapors of silicon and germanium are reacted by

applying heat to produce layers of glass. The composition of this glass can be altered by changing the proportions of these ingredients to produce the different densities required for graded-index fibers. One approach is to react the silicon and germanium within a flame, depositing the reaction products onto the outside of a rotating mandrel. Another approach is to react the materials inside a glass tube. The preform produced by either the "outside" or "inside" vapor-phase deposition technique must later be heated and pulled into fiber. The Japanese are varying the outside process by depositing the reaction products on the end of a rotating mandrel.

People usually associate glass with breakage, but the strength of pristine glass is actually greater than that of a similar piece of steel. However, if the glass surface is flawed even microscopically, the glass fractures easily. Thus, to retain their high intrinsic strength, optical fibers are immediately coated after manufacture with various polymers to protect against abrasion.

When still greater strength is required—as, for example, when fibers are to be pulled through ducts in underground communications systems or subjected to crushing forces—the coated fibers can be gathered into cables. Indeed, fifty formed into a bundle two millimeters in diameter could easily support a 200-pound person. Plastic-coated cables containing several tens of fibers are routinely fabricated just as in the conventional electrical cable industry.

New Wavelengths, New Switches

In 1970, just as Corning demonstrated the first graded-index fibers, the first gallium-arsenide lasers were emerging to provide a source of light far more satisfactory than the gas lasers first considered for optical communications. However, these new lasers still created light beams that were large compared with the optical fiber cores. But by the mid-1970s the technologies of fibers and beams converged, and optical-fiber communication reached its first plateau—lasers emitting at wavelengths of 800 to 900 nanometers projecting into graded-index fibers. These "first-window" systems were economically competitive with electrical coaxial systems for many uses requiring intermediate resolution (or bandwidth). These systems could transmit light through 10 kilometers of fiber, and repeaters (or amplifiers) placed at 10-kilometer intervals made long-distance transmission possible.

At this point the parallel evolution of fibers and lasers accelerated. Fibers were found to be more efficient—with lower losses and greater bandwidth—transmitting frequencies of 1,300 nanometers. And adding indium and phosphorus to the basic gallium-arsenide lasers made them operate in the 1,100-to-1,600-nanometer region. This led to high-performance "second-window" systems operating at 1,300 nanometers, with repeaters spaced at intervals of 15 to 30 kilometers.

At this point engineers installing first-window systems were in a quandary. The success of second-window systems assured that first-window systems would eventually be obsolete. Yet extending existing systems operating in the first window was obviously less expensive than standardizing the new second window. The dilemma was resolved when fiber manufacturers, particularly Corning, were able to design a new fiber material that operated at a wavelength between the first and second windows. These "double-window" fibers gave satisfactory performance at both wavelengths, so that first-window systems could be implemented immediately, with transmitters and receivers in the second window installed as they became available.

But events did not stop there. The latest fibers are most efficient transmitters of light at still higher wavelengths—in the range of 1,500 to 1,600 nanometers (see figure 11.2). Highly reliable lasers operating in this "third window" are now being developed that will have greater information-carrying capacity and allow for longer distances between repeaters. These systems will routinely transmit information at the rate of up to 1 billion bits per second over 50 to 100 kilometers, sufficient to carry simultaneously several video channels, high-fidelity audio, data services, and thousands of telephone calls.

Another interesting technological issue in fiber optics is the choice of the switching methods by which information is channeled to individual users. Most of today's fiber communications are point-to-point links, used to transfer information between only two terminals. But optical fibers will soon be widely used in networks, passing information among multiple users. This will require fast, dependable switching and branching functions.

For the predictable future, switching will be accomplished by conventional electronic equipment connected to the ends of optical-fiber links. Optical switching techniques are being studied but are far from ready for use. These range from mechanical devices that simply move a fiber between two or more other fibers—like a conventional relay switch—to sophisticated devices that make use of certain materials that change their density—and hence deflect the light—when subjected to an electric field.

Another technical question for the future concerns "wavelength multiplexing"—the idea of simultaneously transmitting through a single fiber signals from several lasers operating at different wavelengths. Filters are used to separate these signals—and their messages—at the receiving end. This seems to be a logical way to increase the already prodigious communications capacity of an optical fiber.

But there are complex trade-offs between multiplexing and the alternative of adding fiber links to increase capacity. Transmitting several different wavelengths simultaneously in a single fiber requires additional components that decrease the amount of light transmitted and increase complexity and therefore cost. Nevertheless, components to split and combine optical signals of different wavelengths are beginning to emerge, and a "picture-window" fiber that could accommodate light of many wavelengths is being sought.

Another important question concerns the choice between digital and analog transmission. In digital transmission, messages are encoded so they can be transmitted in terms of on–off signals. In analog transmission the laser light is continuous, with its intensity made to vary according to the information being sent—like conventional radio and television. Digital transmission has higher fidelity but requires more capacity in the transmission channel. Analog transmission requires less bandwidth by a factor of 10 to transmit a similar amount of data. Digital transmission is directly compatible with computer data, and most new telephone systems are being designed to use a digital format. However, the tremendous capital investment of the TV industry and analog's lower bandwidth requirements will spur continued work on analog optical systems. The principle technical problem is finding a light source that can handle analog information with high fidelity.

There is also competition between monomode and multimode

graded-index fiber systems, but the odds are beginning to favor the former.

We have focused on optical fibers used for long-distance telephone, video, and data transmission. However, optical communication is now competing with conventional electrical systems in many short-distance applications as well, including interconnecting computers and bringing data to computers from sensors. Requirements here are not as stringent. Transmitters can utilize inexpensive light sources, and larger fibers with somewhat less bandwidth and efficiency are adequate. The benefits of fiber in such uses include increased bandwidth, smaller size, freedom from electrical interference, and greater security.

Because of the many uses of and rapid changes in this technology, no single fiber or light source can satisfy all present or future requirements. Indeed, the systems designer already has a broad array of fibers and light sources to choose from, with more coming. However, this proliferation threatens to become too great. An important challenge to the industry is to standardize to avoid increasing the numbers of fibers, light sources, and system designs beyond those absolutely required. Without standards, the technology will be needlessly expensive and progress delayed.

Gazing into the Glass Ball

What is the place of this new technology in a future that will see an ever-increasing demand for information?

No one doubts that the growth of information services, now approximately 15 percent each year, must continue. But it may soon be prohibitively expensive—perhaps impossible—to add new transmission capacity to our existing channels. Local telephone, cable, and data-transmission lines are crowding our urban cableways, and we are running out of wavelengths on which to carry broadcast, microwave, and satellite links.

Optical-fiber communication will provide an essential and cost-effective solution to these problems. The impact will be at once dramatic and subtle—dramatic because of the very large increases in capacity that fiber-optic links will provide, but subtle because most users will be unaware that their expanded services are made possible by a technological revolution. The transition to optical communication has already occurred in some telephone trunk lines, where optical fiber now provides more cost-effective capacity than conventional alternatives. In some cases, using optical-fiber links in conjunction with satellite transmission may be the most economical way to add telephone trunk capacity to accommodate data and video as well as voice communication.

When will optical communications become a factor in local services, with each home having its own fiber? That time is moving closer because of the increasing demand for information services and the decreasing cost of optical-fiber systems. Personal computers linked to central processors and videotext systems, including newspapers, magazines, shopping services, and central video libraries; monitoring of homes by police and fire officials; direct readout of electricity, gas, oil, and water consumption; and computerized personal financial transactions are now available—or soon will be. Connecting these services to large numbers of individual homes through conventional wire and broadcast channels is almost unthinkable. But a single optical fiber linking a home to a central switching office would have ample capacity.

Such optical links into homes and offices will not be inexpensive. Indeed, this wonderful world of the future may be a hard sell. But as the total amount of information to be transmitted increases, the optical fiber network will become more attractive. Those willing to use most or all of the many services available will be able to justify the cost.

The Higashi Ikoma Optical Video Information System (HI-OVIS), an optical-fiber communication system near Tokyo, has already transmitted analog video and digital data signals to domestic subscribers. More than 90 percent of subscribers used the system to participate in interactive educational programs such as piano classes, and subscribers also reported stronger community involvement and a widened circle of friends. Indeed, the response was so positive that a second-phase system is being constructed to handle 3,000 to 10,000 subscribers by 1985.

Optical-fiber systems will soon be tested in France and West Germany. In Biarritz, France, two fibers going into each of 3,000 to 4,000 homes will be capable of transmitting two video channels, one picture phone, one stereo channel, one data channel, and one normal telephone channel. By 1984, 350 subscribers in seven West German cities will be connected by a network of 2,000 kilometers of optical-fiber cable. Two fibers into each home will provide one picture-phone channel, three switchable television channels, twenty-four channels of stereo, and thirty telephone channels. These initial forays into the use of optical fibers in a local network make it clear that this new technology will dramatically change the complexion of our society within a few decades.

Meanwhile, there are already lively markets for fiber optics in high-density communications systems. The small size, light weight, and freedom from interference of optics will increasingly dictate their use in military and other special applications.

The Irresistible Force and The Immovable Regulation?

This recital of actual and potential uses illustrates why the market for optical communications is predicted to be enormous. Yet because the future cost of fiber cables, lasers, and other devices is uncertain, the extent of this "optical revolution" cannot yet be predicted. One forecast claims that the market for North American fiber-optic hardware will be $2.8 billion per year by 1990, including fibers, cables, and components. If planning and installations costs are included, the total may be two to three times this amount.

Whether this market materializes will depend somewhat on U.S. regulatory policies during the rest of this decade. Though a single broadband optical line would provide capacity for television, telephone, and many other services, television and telephone signals cannot be transmitted on the same network in the United States except in special cases, and broadcasters are generally banned from owning cable systems. Yet using an optical channel for a combination of services may be the only way for an individual homeowner to justify the channel's cost, at least in the near term. Telephone companies, cable-system operators, and perhaps broadcasters may well vie for control of the information flow to homes in a less regulated future.

Domestic applications of optical communication are likely to come sooner in countries where the communications network is government-controlled and multiple services on one network are allowed. The current government-sponsored "wired-city" experiments in France, West Germany, and Japan will undoubtedly prove that bringing multiservice broadband networks into homes is feasible. Thus, it only can be a matter of time until the first economical systems are installed overseas.

The optical-communications revolution has begun its irreversible penetration of our lives.

Skylines of Fabric

Doug Stewart

12

While optical fibers are transforming the telecommunications industry, new materials such as high-tech fabrics are helping transform one of the world's oldest industries, the construction industry. For example, translucent Teflon-coated weaves of fiberglass strands are enabling the tent to stage a comeback. Strong enough to withstand wind, rain, and snow, the new generation of tents already cover sports stadiums, exhibition areas, airport terminals, and shopping malls. It is even suggested that many of us may be living in tents one day. The author, a former senior editor at Technology Illustrated, *is a writer based in Newbury, Massachusetts.*

The oldest human-made shelter in the world—the tent—is making a comeback. Durable new materials and clever architectural schemes are freeing fabric structures from their traditional association with circuses and Boy Scout outings. A handful of designers even believe that permanent fabric roofs may one day be a common sight on American skylines.

Tents have long been fixtures of the Arabian landscape, so it's not surprising that the Saudis have commissioned the most spectacular fabric structure to date. Completed in 1981 and shading a full 105 acres of desert, the covering of the Haj Terminal at the Jeddah airport is the largest roof in the world. Its 4.6 million square feet of gleaming white fabric shelter the hundreds of thousands of Mecca-bound pilgrims who converge on the airport each year.

The Saudi government had plenty of money—back in 1981 at least—to spend on the detailed engineering studies that such experimental, one-of-a-kind structures require, and fabric turned out to be the perfect roofing material. First, Jeddah is warm and dry, so the roof's chief function is as a sunshade. The white fabric reflects many of the desert sun's rays, and warm air can escape through the openings at the top of the roof's 210 tentlike cones. Second, an airport roof need not provide privacy for occupants, accommodate electrical wiring, or support people on higher stories. Finally, Arab culture has accepted the tent as architecture for millennia. The Haj roof fits in perfectly in Jeddah.

To most Westerners, however, fabric seems too flimsy for permanent use. And by itself, it is. A loose expanse of cloth is as limp and worthless as a flag drooping on a pole. What makes a fabric roof strong enough to resist wind, rain, and snow is tension.

The Physics of Fabric

During installation, a tent roof is stretched and twisted between an assortment of fixed high and low points—masts, arches, cables, wall tops, and ground anchorages. This carefully calculated, continuous tightening, called "pre-stressing," forces the fabric into its designed shape. The taut surface must then endure the varying, day-to-day stresses of winds or snow. Opening an umbrella is a familiar example of pre-stressing. If the spokes do not pull the fabric snugly into place, it droops, flutters, and eventually tears.

As with an umbrella, tension alone is not enough to make a tent roof hold its shape reliably: curvature is also needed. Not just any curve will do, though. Every section of a tent roof is shaped like a horse's saddle, curving up in one direction and down in the other. You can make your own saddle surface by tacking three corners of a tightly stretched handkerchief to a board and pulling the fourth corner out and slightly up.

The opposing tensions bring each point of the fabric's surface into a fixed position, just as the opposing tensions of rigged stays make a sailboat's mast stable. The fabric doesn't sway in the wind or bow under a load of snow. The more pronounced the curvature, the stiffer the surface. Without curvature, even a tightly stretched surface will yield. That's why a drum vibrates when hit.

For the architect who's tired of the same old box shapes, fabric's intriguing curves offer an escape. "The beauty of a fabric structure is that you can see where the stresses are, the way you can with a suspension bridge," says Horst Berger, a structural engineer who has pioneered the design of fabric structures. A fabric roof reveals the interplay of opposing tensions to anyone who looks upward. "With fabric," says Berger approvingly, "form *must* follow function."

Nevertheless, the ubiquitous curves in a fabric structure complicate the designer's task. Architects are trained to draw lines and angles using T squares and triangles. The hyperbolic paraboloids of stretched fabric defy such instruments. "Architects always come to me with straight lines," Berger says. "I always tell them everything must be curved."

Most of the tents used throughout history were neither highly tensioned nor highly curved. As a result few could stand up to the forces of nature for long. Primitive nomadic tents, like the North American teepee, were often simply wooden frames draped with animal skins. On a grander scale, the Coliseum in Rome is thought to have sported a huge, removable cloth awning that was hung across a series of masts protruding from the tops of stadium walls. In later centuries tents were sometimes spacious and elaborately furnished, yet they were still impermanent, sheltering military field commanders and traveling royalty.

Originally, tents were shaped like little houses, according to fabric architect Bill Moss of Camden, Maine. For instance, the traditional slab-sided tent uses fabric as a cheap substitute for a wooden roof and walls. "But when you have a flat surface on a tent, the wind will blow it down," says Moss, who now designs hardier tents for backpackers. Moss's own tents curve and twist like Chinese processional dragons.

Not all of today's fabric structures are highly curved. The balloon roofs of modern stadiums in the United States and Canada are relatively flat expanses of fabric kept aloft by air from electric blowers. These roofs are cheaper to design and build, but they are not as aesthetically pleasing as fabric tents with graceful curves. Nor are they as durable.

Bringing the Outdoors In

Efforts to exploit the physics of curvature and tension have certainly advanced the cause of permanent fabric structures. But new kinds of architectural fabrics have been even more important. In the old days there was only canvas. Heavy and dark, it sagged, caught fire, and often leaked. Then came the first wave of large, so-called permanent fabric structures—a set of fanciful, vinyl-coated polyester tents at Japan's Expo 67 pavilion in Montreal. Vinyl, however, had its shortcomings: it smoldered when burned, and it stretched over time, creating wrinkles and sags. Prolonged sunlight made its outer surface sticky and hard to clean.

The fabric of today's large roofs is usually a Teflon-coated weave of fiberglass strands. Though pliable, it barely stretches at all. It won't burn. Rain washes it clean. And the translucent Teflon coating is chemically inert, so it does not corrode and discolor.

In fact, translucency itself is a key selling point. A stadium, mall, or pavilion with a fabric roof needs no artificial lighting during the day. The octagonal stretched-fabric roof of the F.I.S. department store (formerly Bullock's) in San Mateo, California, seems to put merchandise in its best light. The sun's rays are softened to create a bright but shadowless ambiance.

Translucent roofs can also give work spaces an open feeling. FTL Associates, a New York design firm, recently stretched translucent fabric canopies under the roof that shelters 25,000 feet of office space at Chicago's Bradford Exchange, which markets decorative ceramic plates. These canopies diffuse both artificial lighting and the sun from skylights. The result is a relaxed, outdoorsy setting—"except there's no wind and there's air-conditioning," says Tom Gradel of the Exchange.

Increased computer power has also been a boon to fabric structures. A curved expanse of fabric under tension is a complex system of forces. Tightening a cable here or lifting a mast there causes mathematical pandemonium. In the early 1970s the design of the Munich Olympic stadium's tentlike roof involved as many as 200 workers at one time. Many of them spent weeks using hand calculators to determine how tightly to pull the cables and how sturdy to make the anchorages. (The roof was actually a covered mesh of crisscrossing cables, not a fabric surface, but the structural principles were the same.)

Today fabric designers rely on computer-graphics programs to calculate tautness and shape. In many cases crude tabletop models built with pins, glue, and swatches of panty-hose fabric are used only to please clients who still distrust images on a computer screen.

Computers were essential in designing the complex fabric roof of the 67,000-seat Riyadh International Stadium in Saudi Arabia. Built in 1985, the structure is architecture at its most dramatic. The twenty-four white fiberglass peaks that top it make an undulating ring 945 feet across.

Soaring above the Crowds

In the United States modern stadiums are much more likely to be covered by air-supported balloon roofs, which are less expensive. This form of fabric roof was pioneered by David Geiger, formerly Horst Berger's partner. Geiger, a structural engineer, designed the balloon roofs for the Metrodome in Minneapolis; the Silverdome in Pontiac, Michigan; the Carrier Dome in Syracuse, New York; and B.C. Place, a sports stadium in Vancouver.

Geiger's balloons resemble huge, flattish pancakes. Saddle curves and internal supports are not needed because electric blowers keep the air pressure inside just a fraction of a percent above that outside. This props up the fabric and gives it enough stiffness to withstand wind, rain, and snow.

Unfortunately, those elements sometimes win out. Geiger's domes have deflated after quick, heavy snowstorms. However, the result in each case has been far short of calamity: collapsing fabric should never be confused with collapsing steel or concrete. In fact, had a football game been underway when the Silverdome deflated a few years ago, it probably could have continued, since the roof's lowest point was one hundred feet above midfield. But part of the roof had ripped on a metal light-support tower, so a new roof was required, along with lower towers. The roof cost an extra $8 million and engendered lawsuits that are still winding their way through the courts.

Geiger says his latest air-supported roofs are designed to hold up under major snowstorms. If so, they should be ideal for stadiums. Though colossal, they are extremely light—as little as one-thirtieth the weight of a conventional stadium roof. According to Geiger and other fabric enthusiasts, this makes them less expensive overall. Money saved in construction more than makes up for whatever money may be lost on heating. The steel-topped Superdome in New Orleans, Louisiana, cost $140 million to build in 1975. That same year Geiger's Silverdome, with nearly 5,000 more seats, cost $55 million. His Carrier dome cost less than half as much per spectator seat as the steel-roofed Astrodome in Houston, Texas.

The Price of Being Different

Some fabric designers feel uneasy about any roof that needs electric blowers to stay in place. Large, unpressurized tent roofs do not need blowers, but designing their intricate curves is pricey even if computers are used. Mainly because of design expenses, the dramatic, translucent fabric roof of the F.I.S. department store cost more than a conventional flat roof would have. Even so, it is simpler and faster to install a large fabric roof than one of steel or wood, largely because fabric is flexible and lightweight. Erecting even the largest balloon roofs takes only about six weeks instead of six months.

For most architects, fear of the unknown is another major obstacle. Builders probably won't consider fabric roofs for supermarkets, gymnasiums, and schools until someone comes on the market with modular, snap-on sections. Moreover, the so-called permanence of fabric roofs is an open question. Manufacturers like to say these roofs can last "twenty-plus years," but the oldest one, at La Verne College in California, is only thirteen. No one disputes the fact that Teflon-coated fiberglass is tough under tension. A strand of fiberglass can resist a tug that would snap a strand of steel. Steel, on the other hand, can resist bending, folding, and mutilating. A fiberglass roof can be cut with a knife.

Fabric is also a lousy insulator. A single layer holds in heat as effectively as a sheet of plywood. When the climate is chilly, utility bills can soar—even if you don't need to turn the lights on during the day. Owens-Corning-Birdair, the leading manufacturer of architectural fabrics, has recently developed an expensive but workable alternative: a translucent sandwich. The outer layer is Teflon-coated fiberglass, and beneath it is a layer of woolly fiberglass insulation up to 16 inches thick. Several inches of dead air lie between the insulation and the inner layer, which is a plastic vapor barrier to prevent condensation. The sandwich both insulates and lets light through. Earlier insulated fabrics could do one or the other but not both.

In 1983 David Geiger used this new material for the roof of Canada's Lindsay Park Sports Centre in Calgary, Alberta, where subzero winters make insulation a must. Brenda Clarke, the center's general manager, says its annual utility bill is below that of a traditionally covered sports center nearby, thanks to sunlight hitting the roof. "This isn't a building, it's an environment," Clarke says. "It gives you the feeling of being outside." On dreary days, of course, that can be a disadvantage, as it is for structures with skylights. In Lindsay Park, electric lights are attached to a central arch to provide nighttime and occasional daytime lighting. Inflated domes often have lights hung from cables running along the underside, with the whole assemblage held aloft by blowers.

Geiger and other fabric champions like to say that fabric roofs have double benefits. Translucency reduces the need for furnaces and lights, while the reflective surface holds down the cost of air-conditioning. No hard-topped stadium has ever been built without air-conditioning, but theoretically a fabric-topped stadium could be. However, soon after the Metrodome's owners chose not to install any air-conditioning, they regretted it. "The Twins and their fans were boiled out the first year they had the new stadium," says Byron Engen, a heating and air-conditioning engineer for Owens-Corning-Birdair. "But that's not the fault of the roof. You put 40,000 people in a stadium, just the heat from their bodies will be more than you'll get from the sun."

Tents for the Future

Geiger's latest approach to large roofs is the cable dome, a refinement of an idea dreamed up by Buckminster Fuller. A cable dome's fabric top is supported by a complex web of tightly stretched cables that radiate from a central ring. To make the roof bulge like a dome instead of droop like a sack, the cables zigzag through arrays of metal bars. The bars themselves are suspended hundreds of feet in the air like candles on an invisible cake. The cables keep them in place.

Like a tent roof, a cable dome needs no blowers and can use heavy, insulated fabric. Like a balloon roof, it can cover large open spaces without the obstruction of masts or arches. And it won't collapse in snowstorms. Geiger has several cable domes in the works, including two stadium roofs for the 1988 Olympics in Seoul.

Geiger has other ideas for the more distant future. Perhaps most imaginative is his scheme for covering a mining town in the Arctic completely with a dome—creating a huge solar greenhouse. A permanent layer of sun-heated air at the top of the dome would keep the transparent, ultra-thin plastic skin fully inflated and keep the town warm.

Tent-maker Bill Moss's dream is to see one-story fabric structures accepted as permanent housing. As the swooping designs of Moss and others have demonstrated, fabric is sturdy and cheap enough to compete with steel, wood, and concrete for a variety of projects, big and small.

In the United States a growing number of people live in Moss's mass-produced tents year-round. In most of these cases, however, the climate and culture is Californian. At last count, for instance, some twelve families live in fabric tents in the Sausalito area. But in most cases building codes frown on residential tents.

More than restrictive codes or other practical drawbacks, the very *idea* of the tent has been blocking its widespread acceptance in the Western world. Conventional designers simply don't think fabric structures look like architecture. "To most people, tent is a dirty word," says Bill Moss. "We've got to change our whole image of what a tent is."

Materials Innovation and Substitution

IV

The Real Challenge in Materials Engineering

Thomas W. Eagar

13

Having surveyed some of the exciting materials that scientists can now design and build from scratch, it may be appropriate to lower the euphoria level and bring things back to earth. MIT associate professor Thomas W. Eagar injects a dose of reality by explaining that what is possible in the laboratory is not always practical in large-scale production. Likening the excitement over new materials to the flurry about biotechnology in the early 1980s, he cautions that it will be some time before we see their widespread use in commercial products.

In the early 1960s the United States decided to put someone on the moon by the end of the decade. To the nonscientists and engineers of the world, this was a remarkable goal. In hindsight, however, we can see that the basic knowledge needed to achieve it was available in 1960. The only thing lacking were the resources and commitment to make it happen. In contrast, researchers still have not found a cure for cancer despite huge financial expenditures over the last ten years. That's because progress in the war on cancer requires fundamental new knowledge.

Like the moon landing, the development and processing of new materials today is a problem limited mainly by the available resources. Materials used to be developed through trial and error. Researchers would make a series of alloys whose chemical composition varied slightly, and then they would try to correlate the properties of those alloys with their different atomic structures.

Today materials scientists and engineers possess enough basic knowledge about the bonding of atoms to predict and design new materials. In many cases we can decide what properties we want in a material and then make a material that has those properties. We can turn lead into gold or carbon into diamonds. We can create polymers and ceramics as strong as the strongest steel. We can perform such seemingly miraculous feats largely because we have a better understanding of quantum mechanics—the physics of how electrons behave in solids—and a clearer idea of how the structure of materials relates to their properties. New technological advances in the analysis and fabrication of materials have also contributed to this new era of materials science and engineering.

However, what is possible in the laboratory is not always practical in large-scale production—at least not yet. Some observers claim that the transition from possible to practical is only a detail, but history tells us that this detail can require decades of time and vast quantities of capital. Indeed, the processing of new materials is the most significant problem materials scientists and engineers face. U.S. industry should be devoting its considerable resources and engineering talent to learning how to process these materials for commercial use on a large scale, rather than inventing still newer ones.

The Beginnings of Materials Science

The development of new materials became a recognized science near the end of the nineteenth century after British geologist Henry Clifton Sorby discovered that he could see the crystalline structure of steel by polishing the surface and etching it with acid. The acid selectively wore away certain parts of the structure so that the structure as a whole became more distinct when viewed through a microscope.

After repeating this procedure with steels of different chemistry and heat treatments, Sorby and others were able to correlate the structure of each steel with its mechanical properties. They noted that the low-strength steels had large amounts of light gray crystals—made up primarily of iron atoms. They called these crystals ferrite (from the Latin for iron). Strong, wear-resistant higher-carbon steels had darker, somewhat shiny areas, which the researchers called pearlite. Pearlite was a mixture of pure iron crystals and of carbon-rich iron crystals. Thanks to Sorby's work, scientists began to understand the relationship between properties, structure, and processing. This relationship still forms the basis for materials science and engineering.

In the 1920s scientists found that X rays could be used to discern structures even more refined than those visible through an optical microscope. By bombarding a material with X rays and observing the angles at which they reflected off different planes of atoms, a researcher could measure the distance between the atoms in a particular crystal and determine how those atoms were arranged in space. X rays determine the three-dimensional crystal structure of a solid from a series of two-dimensional measurements in much the same way that CAT scans measure cross-sections in human body tissue, revealing three-dimensional abnormalities that are indiscernible to the naked eye.

Using X rays, scientists finally figured out why an alloy of aluminum and copper known as duraluminum could be made much stronger than the original aluminum or copper alloy. They knew that duraluminum became weak when heated and cooled suddenly from a high temperature. They also knew that heating it again later to an intermediate temperature made it five times stronger than it had been originally. The X rays showed that when duraluminum was heated and then cooled suddenly, tiny particles formed inside the crystal. When the alloy was reheated at an intermediate temperature, those particles separated from the rest of the structure and distributed themselves throughout the metal as copper-rich areas surrounded by aluminum. That "precipitation" gave the alloy its extra strength.

This was the first explanation of how precipitation hardens metals, and it is still the basis for most of our high-strength aluminum alloys today. But precipitation theory did not explain the hardening of steel, and a decade later, Edgar Bain at U.S. Steel discovered why. While the hardening of aluminum involves the formation of fine particles, steel hardens because of a distortion in the crystal itself: the atoms themselves are rearranged when steel is heated and then suddenly cooled.

A Quantum Advance

The next major gain for materials science and engineering came in the 1930s when quantum mechanics began to explain how electrons behave in solids. In a gas the electrons orbiting around the nucleus of an atom possess certain fixed energies. For instance, there are twenty-three separate energies for electrons in a sodium atom.

However, when the gas condenses and the atoms are reorganized into a solid crystalline structure, some of these energies become "smeared," according to quantum theory. What this means is that they are no longer fixed. Unlike electrons in a gas, electrons in a solid can exist only over a limited range or band of energies. These energies indicate how they are clustered around the nucleus of each atom and how the atoms bond. The bonding between atoms determines all the chemical reactions in a solid. Therefore quantum theory can tell physicists why materials behave as they do.

Quantum mechanics has shown, for instance, that electrons in solids absorb light only at specific energies. Compared with electrons in many other solids, electrons in metals can absorb light at very small energies. That's why metals are considered good conductors of electricity. Quantum theory also explains that glass is transparent because its electrons cannot absorb photons of visible light. The photons pass through the glass as if it did not exist.

Insights such as these have led to major technological breakthroughs. For instance, before quantum theory was developed, scientists realized that some materials, which had bands of energy completely filled with electrons, were electrical insulators. It was widely known that since the electrons had no room to jump around or pass from band to band, they only would respond to enormous voltages. But quantum theory revealed that other materials such as silicon and germanium are versatile: they can perform as either insulators or conductors. Under normal conditions, these semiconductor (or semi-insulator) materials behaved like insulators with completely full bands of electrons.

But if a certain number of electrons are injected into a silicon crystal, some of the electrons already present are kicked out of their original energy band into another, making the material a conductor. By regulating the number of electrons injected, we can turn the flow of electricity on and off much as a faucet turns on and off water. It was this ability to transform and control the electrical resistance of silicon that produced the transistor—the forerunner of today's semiconductor chip.

The history of the transistor demonstrates how important processing is to the success of materials development. The first transistors produced had extremely high rejection rates because the silicon and germanium crystals contained impurities that disrupted their electronic structures. Large-scale manufacturing didn't become feasible until William Pfann at Bell Laboratories showed that silicon and germanium of extremely high purity could be produced by alternately melting and freezing the materials in a process called zone refining. Later researchers at Texas Instruments took advantage of both the conducting and insulating properties of silicon by writing entire circuits of conducting silicon on a single chip, thereby producing the first integrated chip.

Looking inside Solids

The invention of techniques that analyze crystalline structures at the microscopic and even atomic level gave materials science and engineering yet another major boost. The transmission electron microscope of the 1950s enabled researchers to measure distinctions in crystalline structure 1,000 times finer than those visible through the optical microscope. In the 1960s the scanning electron microscope magnified surfaces to a degree that had never been possible before, while the electron microprobe provided a microchemical analysis of those surfaces. The Auger spectrometer was developed in the 1970s, providing an even more precise instrument for microchemical analysis of surfaces.

The latest technology—the scanning tunneling microscope—places a probe within a few atom distances of a crystal and measures the rate at which electrons jump across the gap between the probe and the crystal. That reveals both the electronic structure of the atoms and their geometric structure—how far apart they are from one another. The scanning tunneling microscope is the first instrument that can measure both these structures simultaneously.

With these and other new technologies, materials scientists can correlate structure with an expanding number of properties ranging from mechanical, electrical, and optical to magnetic, surface, and interfacial (the boundaries between crystals packed together in a material). The more we know about how different atomic structures relate to particular properties, the more precisely we can design the materials we want.

But there is more to the new era of materials science and engineering than improved characterization technologies. Materials science is no longer a knowledge-limited discipline in large part because of our ability to theoretically predict properties and then make the different materials that we conceive theoretically. The development of high-speed computers has permitted materials scientists and engineers to take full advantage of quantum theory. Now we can do more than just calculate the electron densities and properties of a hydrogen atom (which contains only one proton and one electron). We can determine the electron densities and properties of hundreds or thousands of atoms clustered together in crystals.

Such calculations involve an extraordinary amount of number crunching, which would be impossible without a computer. A sodium atom, for instance, contains a nucleus and twenty-three electrons; in order to figure out the properties of even one atom, we need to solve hundreds of quantum mechanical equations. We must solve a first equation for the nucleus and electron one, a second equation for the nucleus and electron two, and so on on up to the nucleus and electron 23. Then we must solve equations for the different combinations of electrons. To figure out the properties of the most microscopic crystal requires trillions of calculations.

Such calculations yield exciting new insights into the properties of the electrons that lie on the surface of small crystals. For example, experiments have shown that an alloy in which platinum and nickel are mixed randomly is 99 percent platinum on the outer atomic layer and only 30 percent platinum on the second layer from the surface. Now that we have this information, we can perform quantum mechanical calculations to predict—with an accuracy never before possible—the properties of the electrons on the surface of the alloy.

This information is important for a number of reasons. Surface electrons control the ability of catalysts to accelerate chemical reactions. Moreover, as we design smaller and smaller electronic circuits, their surface-to-volume ratio increases, and the properties of the surface electrons begin to dictate the properties of the entire circuit.

Building Materials Atom by Atom

There is one more essential ingredient to progress in materials science: the ability to manufacture the materials that we can theoretically predict. In the last two decades, engineers have devised methods to build materials with the desired properties atom by atom. Such methods include molecular beam epitaxy, in which streams of atoms are shot at the surface of a crystal and condense on the surface in a specific pattern. Another method, ion implantation, accelerates charged atoms to such high energies that they become embedded beneath the surface.

Furthermore engineers can now produce new materials in bulk quantities through technologies such as plasma deposition and chemical vapor deposition. In plasma deposition an electrically charged gas (as opposed to a chemical) is deposited on the surface of a material in layers to build an integrated circuit. In chemical vapor deposition a mixture of gases reacts on the surface of a material to form a solid. These processes can build up materials faster than molecular beam epitaxy or ion implantation because the gases involved put many more atoms on the surface of the materials.

Yet another innovative technology is sol-gel chemistry, which enables scientists to mix organic compounds with metals in ways that could never be done before. Chemists can "hide" a metal in an organic compound and then bake the mixture at lower temperatures than would be possible for pure metal. During this process the organic compound evaporates in much the same way that water evaporates in the baking of bread. The result is a controlled mixture of atoms that high-temperature processing could never have produced. For instance, sol-gel chemistry is responsible for high-strength ceramics that have unique properties.

What can be achieved in materials science and engineering today seems almost miraculous. We can extrude polymer fibers as strong as steel for use in bullet-proof vests and helicopter blades. We can create acoustical transmitters by designing crystals that produce sound vibrations when small voltages are applied. We can build semiconductor lasers atom layer by atom layer. The mixture of gallium, aluminum, and arsenic in such a laser gives it the properties required for fiber-optic communications. By changing the energy levels of electrons in the laser, we can tune its light to a desired frequency, and we can then transmit that frequency more than 1,000 miles without amplification through an exceptionally pure glass. In a few years fiber-optic lasers will almost completely replace copper conductors or microwaves for long-range communications.

Given such extraordinary advances, there is much talk about dramatic growth in the markets for advanced ceramics, composites, and polymers over the next few years. The ceramics and composites markets are expected to grow 20 to 40 percent annually. Many industrial leaders are excited about the potential of these materials. Technical and nontechnical magazines and newspapers herald plastics that could reduce the weight and cost of cars, ceramics that could improve fuel efficiency and lengthen the life of car engines, and electronic materials that could mean faster and larger computers.

The international community is also enthusiastic. More than half of the thirteen projects that Japan's Ministry of International Trade and Industry has identified as priorities involve materials science. I have visited major industrial labs in both Japan and the United States that are devoting 20 to 50 percent of their R&D efforts to new materials.

In some ways the excitement is reminiscent of the flurry over biotechnology in the early 1980s. And indeed, the developments in materials science do bear some resemblance to those in biotechnology. Like scanning electron microscopy, the new recombinant DNA techniques allow scientists to measure genetic structure, correlate it with genetic properties, and fabricate new structures.

But even a few years ago any biologist would have freely admitted that it would be a long time before we could predict the relationship between the structure and properties of engineered DNA. Biotechnology still lacks the theory that explains that correlation. As a result progress is slow and the glamour of biotechnology has faded. Investment in that industry peaked at $850 million in 1983 and dropped to $200 million in 1986. Materials science, on the other hand, has both the theory and the technological means to make good on its promises. Thus it may be more commercially successful.

A Dose of Reality

Nonetheless, there are limits to what we can achieve in materials science. Some of the limits are dictated by physical laws. For example, the elasticity, stiffness, and melting temperature of a material depend on the strength of its atomic bonds. Carbon–carbon bonds are the strongest, so diamond, which consists solely of carbon–carbon bonds, is the strongest material known. Diamond has the highest melting temperature as well. But even though polymer molecules also have a carbon–carbon bond, they do not have as great strength or as high a melting temperature. That's because these two physical properties are controlled in polymers by the weaker intermolecular bonds that link the carbon–carbon chain. Thus there are inherent physical limits to the strength and melting temperature of polymers, and materials science cannot alter that fact.

Scientists and engineers can now predict or explain many such differences among various materials and determine their practical limits. The problem is that not all researchers agree on what those limits are. In some cases there is honest disagreement about the proper assumptions. However, all too often the confusion stems from the distinction between what is possible in the lab and what is practical in mass production. Furthermore, just because we have the capability to produce new materials on a large scale does not necessarily mean we can convert them into usable, cost-effective products.

Take, for example, high-technology ceramics and composites. They are projected to command $10 billion in annual worldwide markets by the year 2000. This is only one-fiftieth of what the world's present steel industry commands, yet government and industry are spending much more on fine-ceramics R&D each year than they are on steel R&D. The justification is that the market for high-temperature structural ceramics is forecast to constitute up to $300 billion in the automotive industry by the end of the century. The forecast is based on the prediction that once ceramics are used in car engines, car owners can expect a 30 percent increase in fuel economy.

This is a very impressive claim, but one that many insiders question. First of all, to achieve such improved fuel economy, an engine would have to operate at a higher temperature than is now possible, and as the temperature increases, other energy losses would become greater. After all, one can hardly expect to exhaust gases from the tailpipe at 1,800°F. And very few people will want to drive a vehicle with a ten-foot-high smokestack. To cool down exhaust gases, car manufacturers would have to install a radiator, thus adding to the weight of the car and decreasing the efficiency with which it burns fuel. When the entire engine system is considered, estimates show that high-temperature ceramics will improve fuel economy by only 3 percent.

At this point ceramists cannot even produce a reliable high-temperature ceramic for less than five times the cost of a metal part. They claim that their raw materials—which include silicon, aluminum, oxygen, and nitrogen—are abundant. Yet they spend a tremendous amount of money purifying and inspecting the final product. And even so, finished ceramic parts can have rejection rates of 90 percent or more because of imperfections in their structure.

Another problem with ceramics is their brittleness. One recent technical paper on silicon-dioxide glass claimed that its resistance to fracture has doubled. However, the paper failed to note that this means the material is still more than twice as brittle as the poorest grade of cast iron. Brittle car engines, of course, are not very economical. If your engine cracks, you have no choice but to buy a new one.

Furthermore, what if researchers make the internal-combustion engine obsolete by developing an efficient, low-cost way to produce electricity directly from fuels? What will be the advantage of high-temperature ceramics then? Ceramics have much to offer in automotive applications. But to tie the future of this industry to the growth of a single material is both premature and naive.

The Complexity of Composites

Composites have problems similar to those of ceramics. Composites are materials such as fiberglass, in which a glass and a polymer are mixed to exploit the benefits of both the strength of glass and the flexibility of plastic. However, it is unlikely that the cost of composites will ever compete with the cost of metals or plastics in high-volume uses. Producing composites simply involves more steps. In addition parts made of a composite are much more difficult to join in a complex assembly.

Composites often consist of materials with vastly different types of chemical bonding. In any joining process other than putting in a simple screw, the chemical bonds of the materials being joined must match. This greatly restricts the number of joining methods available. The joint where the composite meets the larger structure must also have properties equal to those of that structure. Yet it is unlikely that the material that glues the composite to the rest of the assembly (being a noncomposite in nature) can match the properties of the composite. As a result it is very difficult to join complex composite materials together.

In a sense composites are to aircraft as aluminum is to automobile bodies. Car manufacturers could reduce vehicle weight and increase fuel efficiency by bolting aluminum hoods and fenders onto auto bodies. Airplane builders could do the same by attaching composite rudders and wing flaps to airplanes. But no one is mass-producing autobodies with aluminum or aircraft frames with composites because of the difficulties of joining these materials. Despite the claims of materials enthusiasts, most composites will remain high-performance, high-cost materials for limited markets such as the military.

The Secret Is Processing

If we look closely at the information revolution, we find that it has depended primarily on the ability to make silicon chips faster, cheaper, and smaller. Although other developments in electronic materials are very exciting scientifically, advances in silicon processing have provided the economic and technological basis for the information revolution.

Materials processing has also been crucial to the growth of the steel industry. Today Japan is readily acknowledged as the world's leading steel producer, but its secret does not lie in any special knowledge of how to make the material. Rather, Japan leads the world in steel production because it has invested more heavily in materials-processing equipment. As a result it produces steel of equivalent or higher quality at significantly lower costs than any other country.

For example, Japanese blast furnaces are more than twice the size of older U.S. furnaces, on average. With a lower surface-to-volume ratio, large furnaces lose less heat per ton of iron, which allows them to burn fuel more efficiently. In addition the amount of labor required to operate a large furnace is not very different from that required to operate a small furnace. However, as the Japanese are now learning, nations such as Korea, Brazil, and Romania can also invest in this resource-limited industry and, with even lower labor costs, compete very effectively.

In industries such as aerospace and defense, where competition is limited, companies that focus on new materials and cutting-edge technology are most successful. But in industries such as semiconductors, steel, and automobiles, where the competition is severe, companies that put more resources into materials processing have the lowest costs and are thus the most profitable.

The story of Lincoln Electric Co. shows how important processing is in maintaining a competitive edge. More than fifteen years ago, Lincoln Electric, the largest U.S. welding-electrode company, shocked the industry by introducing an electrode that did not require a shielding gas to protect the steel from the nitrogen in air. Once the shielding gas was eliminated, welding equipment became smaller and easier to use. Also since the gas had been a major expense, material costs dropped.

Within two or three years Lincoln Electric's competitors had successfully copied the new electrode in the laboratory. Yet even today they still have not made a dent in the market. Lincoln Electric now controls the entire world's use of this type of welding electrode—not because it has a technological advantage, but because it can make and sell its product for less.

How did the company accomplish this? First, it developed methods to reduce the cost of producing the alloy powder that gives the electrode its special characteristics. Second, it refined its electrode-forming equipment and quality-control procedures to produce welding wire that was of consistent quality. As Lincoln

Electric's success shows, the true competitive advantage for many companies today comes from better-quality and lower-cost processing—and not from a technologically superior material.

For this reason the hunt for new materials should remain primarily the province of academic and government laboratories. And those labs should continue to receive federal funding. However, once new materials are conceived, U.S. companies should step in and decide which are the most practical for commercial exploitation. The companies should also supply the resources necessary to develop low-cost, high-quality methods of processing these materials. After all, the successful industrial development of new materials is often driven more by market pull than by technological push.

Designing materials with curious properties is fun for the materials scientist and engineer, but it does not often yield results of major commercial or social benefit. American companies must spend their resources learning how to manufacture existing materials economically, not searching for exciting new materials. Otherwise, we will only be creating gold that is more expensive than the gold we dig out of the ground. The promise of new materials will fade, and both the materials scientists and the general public will suffer. But if we spend our resources on processing selected new products of high reliability and low cost, we will all be winners.

Innovation in Materials

George F. Ray

14

History shows that predictions of coming shortages of materials are seldom realized. In response to each crisis, new materials are discovered and new methods are introduced to process existing materials more efficiently. But, like Eagar, Ray seeks to demonstrate that the diffusion of materials innovations often takes an unusually long time, although he is confident of continued progress. This raises the fascinating question of whether it is "science push" or "demand pull" that forces the pace of change. George Ray is with the National Institute of Economic & Social Research in London.

Introduction

This paper starts with the proposition that without scientific and technological advance—in other words, without innovation—mankind would long ago have had to face severe material shortages. In the past much public anxiety has been expressed about the consequences of such shortages,[1] but new materials have been discovered and new methods introduced to produce and process them. Plastics and other synthetic materials, discussed by Freeman,[2] are cases in point.

At the time of writing, in 1985, almost all materials, both conventional and relatively "new," are available at historically rather low real prices in virtually any quantity. Although the world's nonre-

From Roy M. MacLeod (ed.), *Technology and the Human Prospect* (Frances Pinter, London, 1986). Reprinted by permission of the publisher.

newable resources are obviously finite, the lifetime of the known and proven reserves is, in general, reassuringly long.[3] Hence, one might ask, why bother? While the question seems legitimate, the problem is more complex than appears at first sight; the whole truth needs to be approached from several angles.

First, while overall physical scarcity of industrial materials is not a likely eventuality—as an OECD study put it[4]—the real problem is undisturbed access to them. World reserves data conceal the fact that the reserves and production of more than a dozen important minerals/metals are highly concentrated in a small number of countries (such as South Africa, Brazil, Zaïre, Chile, Morocco, China, and the U.S.S.R.) and any disturbance in production or transportation in any of these countries can lead to temporary shortages.

Second, there may be a sudden demand for a particular material stemming from some novel end-use. For example, the price of rhodium, one of the platinum-group metals, trebled in the twelve months to March 1985, despite the weak metal market: this was due to a demand from the automobile industry which required rhodium for its own technological development, aimed at reducing the damaging effects of exhaust fumes.

Third, in the summer of 1983, the U.K. Department for Industry set up a Materials Advisory Group (now known as the Collyear Committee) whose work resulted in the publication early in 1985 of "A Programme for the Wider Application of New and Improved Materials and Processes."[5] This report stresses the outstanding importance of such work, strongly advocating the need for government involvement in the research, development and application of new materials. Thus there is plenty of evidence indicating the topicality of the subject.

Another question that may be raised is this: What has been the contribution of basically economic studies, of analyses by economists, to progress in the materials area? Naturally, the lion's share in this sphere, as elsewhere in technical progress, goes to scientists and technologists. Nevertheless, the work of economists and historians should not be belittled: they greatly contributed to the understanding of the process of innovation in materials and the forces at work,[6] to the organization of research and development,[7] to the search for the direction of desirable research,[8] to various aspects of the economics of novel materials,[9] and to being generally better informed, to foresee and anticipate troubles. As a historian wrote: "it is surely arguable that foresight and anticipation may assist in preparing alternatives and make more likely a successful response to critical shortages."[10]

The Scope of this Paper

This study is in two main parts. The first surveys the history of some thirty materials which are relatively new or were "new" at the time of their introduction into general use. The dissemination of any innovation takes time; that of new materials often takes an unusually long time; hence this first part concerns materials that already have a history. The second, shorter, part deals with the present, that is, the progress of the area of materials. For both the past and the present the choice is very wide and therefore it was necessary to be selective.

Many of the "new" materials were discovered or developed as the outcome of a specific need under wartime or market pressure. Others were the result of spontaneous and random scientific/technological advance. The various materials are thus crudely categorized, although these definitions are frequently blurred and inevitably overlap. Because of limitations on the length of this contribution, detailed case histories of the various materials cannot be included; they may be found elsewhere.[11]

The role of science and technology, as applied to industrial materials and their use, is complex, extending well beyond the relatively simple concept of "new" materials. We therefore start by discussing briefly some of the complexities involved.

The Impact of Technology

Although important, the discovery or development of a new material is still only one of the ways in which supplies may be affected through scientific or technological progress. The supply system is complicated by interaction between miners and farmers through traders, exchanges and middle men, designers, processors and consumers to recycling. In principle, it is subject to market forces at every single stage and in practice it is further colored by state intervention. Any national system has its roots in a world market, and this gives it a truly international flavor. The technological side is equally complex: it affects not only all aspects of the material itself, from exploration to processing, but also its application in end-use.

Originally, there was only neutral matter in nature. A material became a natural resource when man began to use it for a specific purpose. Coal was a useless black rock before man started to burn it; bauxite did not even have a name before it was discovered that it could be processed into aluminium. The promotion from neutral matter to resource was itself an advance. At all later stages technology and science play significant parts and their spectrum is incredibly wide: "materials technology" (narrowly defined) makes use—particularly nowadays, in view of very rapid communications—of interventions and innovations in almost any area or discipline. Aerial photography is commonly used to assist in the exploration for mineral deposits, analysis of local flora for the existence of metals in the subsoil and techniques developed in medicine may be used in materials production (such as that used in the prevention of blood clotting applied to rubber trees to increase yield), to mention just a few of the many examples of technological cross-fertilization.[12] The vast development of mining technology is well known, including methods of strip mining, a technique introduced after, and with the experience gained from, the digging of the Panama Canal.[13]

Technology has helped to reduce production costs, and this has been of fundamental importance, for example in copper mining. In the early nineteenth century, Cornish copper ore, with a metal content of 13 percent, was the price leader; by the middle of the century, after the exhaustion of Cornish mines, Anglesey copper took over, with a metal content of about 3 percent. New mines nowadays contain ores with around 0.5 percent metal, and yet the real price of copper has decreased very considerably over time.[14] It is technological advance—and by this we mean a long chain of major and minor innovations—that has made cost-efficient production possible, using leaner ores and reducing the cost and price of the final primary metal still further.[15]

Another advance, the Thomas process, brought high-phosphoric ores into the ambit of valuable steelmaking resources. They had previously been useless. Thus, for example, the French were able to use highly phosphorous Lorraine ores for steelmaking from about 1880.[16] Similarly, around the middle of the present century, many of the major American steelworks were literally saved from extinction by the introduction of pelletized taconite as their base material, replacing the exhausted high-grade conventional iron ore from the Mesabi Range of Minnesota.[17] Both these cases represent materials that were "new" at the time. There are many other examples of technological "intervention" in conventional methods of materials production and use.

Lead Times

It is commonly held that modern communication speeds up everything and the spread of innovations is no exception. There is plenty of scope for speeding up in the field of new materials, as can be seen from the following few historical examples.

Coal was used by the Romans more than 1,600 years ago; monastic records reveal coal-extracting operations in the twelfth century in Britain and Germany. Yet the widespread use of coal had to wait for two major innovations in the eighteenth century, the steam engine and the replacement of charcoal with coal in ironmaking, after which coal became acceptable to industry.[18]

The *oil* produced by Drake's 1859 well or that from the Baku oilfield in Romania was first used as a medicine or for lighting; the real take-off for oil began much later, in the present century.

Indian *cotton* became familiar to the Greeks through the advance of Alexander the Great (323 B.C.); the Romans produced inferior quality cotton in Malta, but cotton fabrics remained a luxury until the end of the eighteenth century, when Britain's North American colonies started to cultivate it. The trade then took off rapidly, aided by a series of major innovations at all stages.[19]

After first having been isolated in 1825, *aluminium* was for a long time considered a precious metal: "silver from clay," as exhibited at the 1855 Paris Exposition, used for cutlery at court banquets, a rattle for the Prince Imperial at the time of the Second Empire in France, or a watch charm for the king of Siam. Its electrolytic reduction was developed in 1886, but processors did not know what to do with it, and wider applications had to wait until the present century.[20]

Tungsten has been known to science since the end of the eighteenth century but its use in industry did not begin until around 1900.[21]

Thus the time period from invention or discovery of a new material and its innovation to its wider diffusion, that is, commercial production and adoption by users, can be very long indeed. There are several reasons for this: the most important is probably that a "new" material is often—as Rosenberg formulated—"an example of an invention which could be produced under experimental conditions for many years before means of producing it commercially were developed."[22]

The Effects of War

The textbook example of the pressure on materials during wartime is that of the introduction on the continent of *sugar beet* during the Napoleonic wars, when France was cut off from her traditional cane sugar supplies by the British fleet.[16] Similarly, at the same time—and perhaps more appropriate to the discussion of industrial materials—the production of *alkalis* (mainly soda) was revolutionized. Widely used, from the making of textiles to gunpowder, *soda* (sodium carbonate) was originally obtained from the ashes of certain plants, grown mainly in Spain and the Canaries. Leblanc, however, worked out a method of making it from common salt, and in 1808, when the French were cut off from their Spanish sources during the Peninsular War, the manufacture of soda by this process began, remaining the main method in use throughout the nineteenth century (Britain took it over in 1823).[16]

Germany's isolation during the 1914–18 war necessitated the urgent large-scale introduction of the *nitrogen* fixation process, developed by Haber a few years earlier. Nitrogen was widely used in various forms, among other things as an explosive, and its traditional source was Chilean nitrate. By the end of the war one-half of Germany's huge production of nitrogen compounds was based on the new process. (One cannot avoid some philosophical thoughts in this context, since without this contribution the war would undoubtedly have ended earlier.)[23] Another indispensable base for the making of explosives was *glycerine;* its traditional manufacture required large quantities of fat—very scarce during World War I; a substitute was developed—again, in Germany—obtained by the fermentation of sugar residues.[13]

The shortage of *fats*—and hence of soap and candles—began to affect Europe around the middle of the nineteenth century when the supply of solid animal fats was unable to keep pace with growing requirements. Liquid vegetable oils were used as a supplement in households but this was not possible in industry. The process of hardening liquid fats by hydrogenation solved the problem. But during World War I even vegetable oils (and hence soap) became scarce in Germany and this shortage led to the development of the first synthetic *detergent.*[24] It was an inferior product, a true *ersatz,* yet it deserves attention because its later development was characteristic of all synthetics. Once the basic problem of synthetic production was solved, progress continued. In the following decades, the simple *ersatz* was developed into a sophisticated family of products, a wide range of specialized varieties that could serve particular purposes better than the original "all-purpose" ancestor. This was the case with synthetic detergents, fibers, dyes, and rubber alike.

Rubber was perhaps the most important instance of technological development during the 1939–45 war. Hitler's Germany and the United States were equally cut off from Malayan natural rubber. The Germans developed *buna,* a fairly successful substitute that covered almost all their requirements. Before the war Du Pont research in the United States had already produced synthetic elastomers with specific qualities, as had Esso for inner tubes of tires; these innovations proved decisive when the Japanese invasion of Southeast Asia compelled the United States to look for another source. The output of all the initial "new" rubber was still experimental in 1943; only four years later, however, production had reached 100,000 tons and developed into a major industry. This illustrates the rapidity of technological response when the pressure is really great and the backing equally powerful.[13,25]

Another noteworthy example of progress is that of *acetone,* used as an agent to incorporate gun cotton into nitroglycerine in explosives. Temperate hardwood provided the original source, but this became scarce during the Great War; the Germans used various chemicals to replace it, while Weizmann in the United Kingdom discovered the process whereby ordinary grain ferments bacteriologically into acetone.[13,26]

Peacetime Scarcities

Wars are obvious causes of scarcity, either because normal trade flows become disrupted by consuming countries being boycotted by their enemies, or by disturbances to the production or transportation of materials. Shortages may, however, also be generated in peacetime, through requirements widely outstripping supplies or through the manipulation of the market by agents with vested interests in achieving monopolistic gains.

Paper was originally made from fibrous textile residues such as rags, but in the nineteenth century the supply of rags was insufficient to satisfy the rising demand from paper mills. In 1841 the Halifax mill was the first to produce "groundwood paper," using a German patent. Within ten years, the chemical treatment of *wood-pulp* was introduced, making it the single base material without which the enormous quantities of paper could never have been produced.[13,16] Even so, it had to be supplemented by recycling, a process that has also gone through a chain of minor innovations, such as, for example, de-inking.[13]

Material shortages can be artificially engineered, but attempts at cornering the market may seriously misfire. This happened around 1830, when a French syndicate tried to exploit what was then a monopoly of Sicilian *sulphur,* the base material of sulphuric acid. Their speculation was overdone: it led to the invention of the process that uses sulphide ores (mainly pyrites).[13,16]

A somewhat similar case is that of *camphor,* which was mainly produced in Formosa. Having occupied that island in 1895, the Japanese attempted to benefit from the steeply rising price of camphor, the production of which had been reduced by wartime conditions; in so doing, they stimulated research activity and soon a synthetic substitute was introduced which quickly replaced the natural product in industrial use.[13,28]

The camphor incident, however, had much wider consequences, which are not only important in themselves but also present an interesting case of serendipity (a phenomenon not unusual to scientific researchers). Among others, the Belgian scientist Baekeland (working in America) was unsuccessful in his search for a camphor substitute. Although he did not entirely succeed, he did make a systematic study of certain chemical reactions and rather unexpectedly developed a hard, chemically resistant plastic material which has come to be known as *bakelite*.[29] This was important in its own right but also opened the door to significant further developments, leading eventually to today's enormous family of plastic materials.

Another outstanding factor contributing to the foundations of the later great polymer and pharmaceutical industries was the invention of synthetic *dyes*. There was no shortage of natural dyes, but by the middle of the nineteenth century it must have been fairly clear that requirements might soon outstrip the supplies then available. Research began in both the United Kingdom and Germany, and in 1856 the German patent application (by BASF) beat the British one (Perkin) by one day! The product was the first synthetic alizarine dye, and it paved the way for many other synthetic dyestuffs.[30]

Cryolite is essential in aluminium production as a flux in electrolysis. The largest deposit of it, in Greenland, was totally depleted by 1963, but somewhat earlier a perfect synthetic substitute had been developed.[31] Thus the advance of technology and science sometimes solves scarcity problems even before they become acute. Synthetic polycrystalline *diamonds* perfectly replace the rare natural stone in industrial applications;[32] man-made radio-isotopes are successfully substituted for *radium*, the great discovery of the Curies; and, to quote a more recent example, synthetic *quartz* crystals are the accepted substitute for the natural quartz crystal required in a particular quality and in steeply rising quantities by the electronics industry.[33]

There are often spin-off discoveries from research too: it was in the search for a diamond substitute that *tungsten carbide* was developed; this extremely tough material is in many respects superior to the mainly natural materials previously used in the making of dies, cutting tools, and so on.[24]

The Science "Push"

While the brief case studies in the previous sections deal with new materials that were invented or discovered under some kind of pressure, either from war or from market needs (that is, various forms of "demand pull"), there are also examples of new departures arising outside such situations, purely as a result of research. One of the best examples is that of *man-made fibers*. At the time of their initial development there was no particular shortage of the different varieties of natural fibers. Although the basic invention goes further back, commercial development started early in this century, with "artificial silk"—later to be called rayon—followed by nylon in the 1930s and by today's large group of synthetic fibers, including even "elastomeric" varieties from the late 1950s.[13] Rayon was based on cellulose, the further development of which yielded *cellophane,* a thin, nonfibrous film widely used for wrapping—originally a French product (1912) which was improved by waterproofing in the United States (1920).[30]

Numerous man-made materials that have come to be called by the blanket name *plastics* have, of course, become part of modern life. Specific properties make their wide range—from PVC through polystyrene to transparent polymers—popular in industry and elsewhere, for an endless variety of purposes.[30] In many cases the role of science has been not so much to discover new materials as to find uses for them. *Titanium*—like tungsten, already mentioned—has long been known, but its application remained a problem until a process for refining it was developed in 1936, and its high strength/weight ratio was recognized by the aircraft industry (it accounts for 9 percent of the airframe structure weight of the Boeing 747).[24] This first metal of the "rare earths" group was discovered in 1794, and in recent decades many members of this category have been used increasingly in industry.[13,27,33]

Glass fiber is a new material that has found rapid acceptance on a wide front, from insulation to boatbuilding, and its highly refined version, optical glass fiber, has started to take over the role of copper and other wires in communications systems.[13,27]

The spectacular advances in the chemical industry outside the field of industrial materials proper are also worth noting. Various types of *fertilizers* have greatly helped to raise food production.[16]

Research is well advanced with the objective of producing synthetic *proteins* for feeding animals. But it is in pharmaceuticals that scientific progress has been the most dramatic.

Napoleon's troops used to chew the bark of the willow tree to relieve pain: this bark contained a chemical, related to *aspirin,* which Bayer in Germany started manufacturing at the end of the last century.[13,29] Many people recall how *penicillin* was also discovered almost by chance.[24] More important, however, penicillin was the first antibiotic, and its later semisynthetic version was a vast improvement on the first "natural" variety, and from those early beginnings many types of antibiotics have been developed, many of them serving very specific purposes.

A historical study of the cases listed in this section leaves one with the impression that they were the result of a genuine "science push," but this elusive concept is difficult to define and delineate or to demonstrate in isolation. Let us take man-made fibers. As already stated, at the time of their introduction there was no particular shortage of natural fibers so that the new material could be treated as the result of science push, a considerable intellectual achievement by those scientists working in basic research. But one can argue just as well otherwise: the key R&D work was done within the research establishments of the large leading enterprises, such as ICI, Du Pont, and the German chemical giants; one can assume that a general idea, at least, of the potential of the hoped-for result of these particular projects (if successful) was "in the air" or perhaps more concretely formulated.

Advocates of "science push" would doubtless reply that, while this may have been so, in many other cases no further development could have taken place without the incentive—that most important first step—provided by science. In fact, while it may be of interest to theorize about market pull and science push, both theories seem so closely interconnected in modern life that one cannot but agree with Freeman that, "it is quite possible to subscribe, at least partially, to both theories."[34] (See appendix.)

Several serious scholars agree with Freeman, even if they formulate their views differently. Rosenberg says that "potential demand may exist for almost anything under the sun, and the mere fact that an innovation finds a market can scarcely be used as evidence of the undisputed primacy of "potential demand pull" in explaining innovation."[35]

Utterback approaches the question in another way: while writing that "any innovation is necessarily a combination of a user's need and a technological means to meet that need," he takes the view that "need input often takes precedence over the technology input," and this is because "basic research enters the innovation process through education and development of trained people, who later apply in meeting users' needs what they learned years earlier."[36] However, even for him this is only "often" the case, leaving Freeman's thesis still valid.

Some New Developments

All the preceding cases have a history, but progress continues and there are many new materials in the R&D pipeline. Some of them are well advanced and have already been used commercially, others are still the well-guarded secrets of their developers. Many of them face the same difficulty that hindered the rapid dissemination of the use of aluminium.

With man-made materials, widespread use depends on a number of factors, of which the most important are availability and price, properties and ease of processing. The latter is of supreme importance; without it the circle can be vicious. For example, carbon fibers encased in plastics could replace many metals (so far as properties are concerned), but processing is slow; therefore they are not available in quantity—hence the price is high, and subsequently they may not in general be a commercial proposition. This vicious circle may take some time to break.

There are composite materials that could replace steel or aluminium in the construction of motor vehicles, but their processing methods are not to hand—and therefore the change does not make commercial sense. All this refers to the long lead times of any new material. The introduction of a new material more often than not requires innovative design and may necessitate redesigning the whole product or the production system.[37] This is risky and expensive: vested interests such as existing capital investment and expertise present constraints in any case so that, in order to balance all these handicaps, new materials or allied processes are expected to offer substantial cost savings.

Despite these initial difficulties many new materials have already been commercially applied—if not generally, at least by pioneering companies. *Carbon fibers* are a case in point. Almost parallel development work in the United States, Japan, and the United Kingdom was based on the recognition that the carbon–carbon bond is the strongest of any in three dimensions. Interestingly, carbon fibers were first commercially applied in the making of sports goods (golf shafts, fishing rods, tennis racquets); only later—and rather slowly—did the new material start to penetrate the aerospace and other industries.[38]

An important group of new materials is *engineering ceramics*. These are formed from compounds, they are light, stiff, corrosion- and wear-resistant, with low thermal and electrical conductivity—properties that are increasingly attractive in engineering applications.[38]

As one example, we mention *sialon,* based on the alloys of the elements Si, Al, O, and N, which has captured a sector of the machine tool tip market and has potential for penetrating other areas.[38] The discovery of "ceramic alloying" is an example of a generic invention that may have wide applications in many fields. Ceramics are just one segment of *composite materials*—engineering materials built up from several components with different properties (the term as commonly used generally assumes reinforced materials). It is precisely this combination of different properties that represents a new material/technological approach to product manufacture, offering scope for design innovation.[5]

Thermoplastic materials are also relatively new. They are a considerable improvement on the first-generation plastics or polymer matrices that, once set, cannot be reshaped plastically by heating: thermoplastic matrices are amenable to intermediate forms because they can be heated and pressed into molds, into sheet, tube or any other complex form. This ability rapidly offered new opportunities to the mass production of plastic articles, to highly "oriented" polymers. Similar experiments with *metal matrix* composites are aimed at reducing the old danger of metal fatigue.[5]

The dramatic progress in the microelectronics industry owes much to developments in materials and their processing. We are probably far from the end of the development work in electronics, and particularly in the materials field. Parallel with work on the engineering application of the results, research is going on into *superlattices* (thin layers of semiconductor materials) whose properties promise to offer opportunities for novel devices and for tailoring the properties of material over a far wider range than is now possible. Other likely departures are device processing for special uses with materials substantially different from silicon, new chemicals—including organic semiconductors—and so on.[5]

This list of relatively new departures affecting materials could be made much longer. To conclude our tentative illustration of the continuing work in the area under discussion, it should suffice to mention some of the new manufacturing *processes* that will certainly have an impact on the search for materials, probably including new ones, that suit them best. Among them, as examples, it is worth bearing in mind:

- powder metallurgy, which is supposed to avoid the time-consuming and wasteful conventional shaping of parts and components,
- precision casting, which has become imperative in modern aerospace and other industries,
- superplastic forming and diffusion bonding, which reduce several operations into one single process,
- CAD/CAM techniques which will certainly spread from the most complex to simpler manufacturing processes and, particularly when combined with robotics, raise specific requirements of material properties,
- coating technology, which has already reached advanced stages in the fight against corrosion and other harmful effects but is presumed to go much further in surface treatment technology,
- the application of lasers to welding and other forms of joining materials.

Concluding Thoughts

In one of his many publications concerning the role of science, Freeman wrote: "Much scientific research is concerned with the exploration of the unknown. By definition we cannot know the outcome of such explorations and still less can we know its future impact on technology."[39] This general statement can be applied with some force to materials. This brief survey is intended to show how science and technology have contributed to the supply of industrial and other materials in the past and that progress has been a continuing one. History does not necessarily repeat itself; nor do the examples given in this paper of scientific and technological achievements solving materials problems provide any guarantee for the future. They do, however, provide a basis for the hope that progress will go on and future advance continue to secure a link between the demand for and the supply of materials that industry and other sectors of the economy will require.

Appendix: Demand Pull or Science Push?

Although we may theorize about these important stimuli to invention, it should not be forgotten that much of the equipment, and even food, that has become indispensable in modern life was invented at various times as the result of personal eccentricity or inconvenience. Inventiveness no doubt, but how should one classify it?

The ballpoint pen, nicknamed the "biro," was devised in 1938 by a Hungarian journalist Laszlo Biro, because he got bored with always blotting his work.[40] The first carpet sweeper was put together by Melville Bissell, a china shop owner in Michigan, because he suffered from headaches caused by an allergy to the dusty straw used for packing his wares.[40] The first zip fastener was demonstrated in 1893 at the Chicago World Fair, but it came apart easily; improved versions were then tried, but the zip did not catch on until the 1914–18 war when the American armed services started using zips everywhere since they were more convenient for the soldiers than buttons.[40] The world's first cannery was established in 1812 in Bermondsey, London; in those days the contents were introduced through a hole at the top of the can and sealed with a soldered disc. But the direction on the label said: "Cut around the top near the outer edge with a chisel and hammer." This was a considerable impediment, and the tin opener was soon invented.[40] In 1847 Fry and

Sons in Britain made the world's first solid eating chocolate. Daniel Peter, a Swiss, did not find it sweet enough and, just by intuition, added milk to it. This was the original milk chocolate—and the beginning of Swiss chocolate-making supremacy.[41]

But perhaps the oddest story is this: toward the end of the nineteenth century there were two undertakers in Kansas City. One of them was a Mr. Almon Strowger. His rival's wife was an operator on the local telephone exchange who consistently diverted Mr. Strowger's calls, and hence his business, to her husband. To overcome this handicap, Strowger developed the basic design of the mechanical telephone exchange (still named after him), and these clicking/clacking exchanges have been in operation for many decades and will remain in use until the last Strowger exchanges have been replaced by automatic ones.[42]

"Law and order" can also play a part in technological advance. The elimination of piracy in earlier centuries made it possible to dispense with the carrying of heavy metal deck-guns on merchant ships; this change not only vastly increased the freight-carrying capacity of ships but improved the safety of their design by lowering the centre of gravity.[43]

Notes

1. Apart from the classical views of Malthus and Ricardo, some more recent and also more specific works may be mentioned, such as—among others—that of W. S. Jevons, *The Coal Question,* London, Macmillan, 1865; *Resources for Freedom, A Report to the President by the Materials Policy (Paley) Commission,* Washington, U.S. Government Printing Office, 1952; J. W. Forrester, *World Dynamics,* Cambridge, Mass., Wright-Allen, 1971; or D. L. Meadows *et al.*, *The Limits to Growth,* London, Earth Island, 1972. For a thorough survey and critique of most of these works, see K. L. R. Pavitt, Malthus, and other economists in H. S. D. Cole, C. Freeman, M. Jahoda, and K. L. R. Pavitt (eds.), *Thinking about the Future,* London, Chatto & Windus, 1973, pp. 137–58; for a shorter survey, G. F. Ray, "Raw Materials, Shortages and Producer Power," *Long Range Planning* 8, 4 (August 1975), pp. 2–17.

2. C. Freeman, "The Plastics Industry: A Comparative Study of Research and Innovation," *National Institute Economic Review* 26, November 1963, pp. 22–62; "Chemical Process Plant: Innovation and the World Market," *National Institute Economic Review* 45, August 1968, pp. 29–57; and chap 3 on "Synthetic Materials" in *The Economics of Industrial Innovation,* Harmondsworth, Penguin Books, 1974, pp. 74–107.

3. George F. Ray, "Mineral Reserves: Projected Lifetimes and Security of Supply," *Resources Policy* 10, No. 2 (June 1984), pp. 75–80.

4. OECD, *Interfutures—Facing the Future,* Paris, OECD, 1974, p. 41.

5. 'A Programme for the Wider Application of New and Improved Materials and Processes', Report of the Collyear Committee, London, HMSO, 1985; engineering ceramics, pp. 19–20; composite materials, pp. 13–18; thermoplastics, p. 14; metal matrix p. 17.

6. Among others, N. Rosenberg, *Perspectives on Technology,* London, Cambridge University Press, 1976, particularly chapter 14, pp. 349–359; and *Inside the Black Box: Technology and Economics,* London, Cambridge University Press, 1982, pp. 55–80.

7. Freeman writes on the "professionalization" of R&D in *The Economics of Industrial Innovation* (op. cit., note 2, p. 48) and finds that one of the reasons for the success of the German chemical industry was that the leading companies "were among the first firms in the world to organize their own professional R&D laboratories." Elsewhere ("The Plastics Industry," op. cit., note 2, p. 32) he adds, however, that "gifted individuals still play an extremely important part in the inventive process."

8. For example, the OECD report, "Interfutures" (op. cit., p. 45) states "For asbestos, there are as yet no substitutes in many applications . . . technological advances before the end of the century in the production of synthetic asbestos substitutes would therefore be highly desirable, particularly since asbestos poses environmental problems." See also G. F. Ray, "Research Policy and Industrial Materials," *Research Policy* 8 (1979), pp. 80–92.

9. Just one example: Freeman ("The Plastics Industry," op. cit., note 2, pp. 22–23) showed that the locus of production and exports of new plastics in advanced countries was more a function of technical progress than of relative factor costs.

10. W. H. McNeill, "Coping with an uncertain future—A Historical Perspective," in C. J. Hitch (ed.), *Resources for an Uncertain Future,* Baltimore, Johns Hopkins, 1978, p. 59. (This was a collection of papers presented at a forum marking the 25th anniversary of Resources for the Future, Washington, established in 1952.)

11. Studies specifically discussing "new" materials or materials innovations in general are relatively rare (in contrast to those concerning particular materials). The four-volume *Science in History* by J. D. Bernal (Cambridge, Mass., MIT Press, 1971) discusses many of the materials now in use from a scientific angle, while D. S. Landes analyzes the development of many of them from the standpoint of an economic historian in *The Unbound Prometheus—Technological Change and Industrial Development in Western Europe from 1750 to the Present,* London, Cambridge University Press, 1969. The case histories of selected new materials are contained in J. Jewkes, D. Sawers and R. Stillermann, *The Sources of Invention,* London, Macmillan, 2d ed., 1969. A wide-ranging survey, shorter but more to the point, is in G. F. Ray, "The contribution of science and technology to the supply of industrial materials," *National Institute Economic Review*

92 (May 1980), pp. 33–51 and, covering a limited field, in C. Freeman, "Synthetic Materials," op. cit., note 2, pp. 74–107.

12. G. F. Ray, "The Wider Horizons of 'Industrial' Innovation," in M. J. Baker (ed.), *Industrial Innovation—Technology, Policy, Diffusion,* London, Macmillan, 1979, pp. 1–14.

13. Ray, op. cit., note 11. Stripmining, p. 38; nitrogen (Haber), 41; glycerine, p. 42; rubber, p. 42; acetone, p. 42; paper, p. 43; sulphur, p. 44; camphor, p. 44; man-made fibers, pp. 45–46; rare earth metals, p. 47; glass fiber, p. 47; aspirin, p. 46.

14. Sir Kingsley Dunham, "How Long will our Minerals Last?" *New Scientist,* 17 January 1974, pp. 1–3.

15. According to Nordhaus, the price of copper relative to the hourly wage rates in American manufacturing declined from around 800 (index, 1970 = 100) to 82 in 1960. S. D. Nordhaus, "Resources as a Constraint on Growth," *American Economic Review* 64 (1974), pp. 22–26.

16. Landes, op. cit., note 11; Thomas process, p. 352; sugar beet, p. 144; alkalis, pp. 110–111; paper, p. 269; sulphur, p. 203; bakelite, p. 276; dyes, pp. 274–276; fertilizers, p. 515.

17. B. Gold, W. S. Peiree, G. Rosegger, M. Perlman, *Technological Progress and Industrial Leadership,* Lexington, Heath, 1984, pp. 309–311.

18. J. Platt, *British Coal,* London, Lyon, Grant and Green, 1968, pp. 8–11. G. F. Ray, "Energy Economics: A Random Walk in History," *Energy Economics* 3 (1979), pp. 139–143.

19. C. R. Fay, *English Economic History,* Cambridge, Heffer, 1948, pp. 111–112; J. L. and B. Hammond, *The Rise of Modern Industry,* London, Methuen, 1925, pp. 179–89.

20. D. H. Wallace, *Market Control in the Aluminium Industry,* Harvard Economic Studies, Vol. 58, Cambridge, Mass., Harvard University Press, 1937.

21. K. C. Li and C. Y. Wang, *Tungsten,* American Chemical Society Monograph No. 94, New York, Reinhold, 1955.

22. Rosenberg, *Perspectives,* op. cit., note 6, pp. 213–279, gives many illustrations for materials and also for "new" processes, such as the oxygen method of steelmaking.

23. B. G. Reuben and M. L. Burstall, *The Chemical Economy,* London, Longman, 1975, pp. 17–18.

24. Jewkes *et al.,* op cit., note 11; fats, pp. 256–257; detergents, pp. 304–307; tungsten carbide, pp. 319–321; titanium pp. 314–317; penicillin, pp. 278–279.

25. F. Höllscher, *Kautschuke, Kunststoffe, Fasern,* Ludwigshafen, BASF, 1972, pp. 23–24.

26. There were important political consequences of Weizmann's development work that contributed significantly to the wartime security of the supply of acetone. Weizmann turned down offers of honor or financial reward and was

eventually recognized (as one of the leaders of the Zionist movement) by the Balfour Declaration in 1917, leading to the establishment of the state of Israel, of which Weizmann became the first president. Thus Weizmann's contribution was an important one among the many other, chiefly political, considerations that eventually resulted in the Balfour Declaration.

27. *Encyclopaedia Britannica,* 1974 ed. "Weizmann," Vol. 10, p. 605; "paper/wood pulp," pp. 13, 966; "rare earths," VIII, p. 422; "glass fiber," IV, p. 568.

28. Rosenberg gives many examples for innovative responses to materials shortages in *Perspectives,* op. cit., note 6, pp. 249–259.

29. Reuben and Burstall, op. cit., note 23; bakelite, p. 29; aspirin, pp. 17, 350.

30. Freeman, op. cit., 1974, note 2; dyes, pp. 48–51; plastics, pp. 74–83; cellophane, p. 94.

31. J. E. Tilton, *The Future of Non-Fuel Minerals,* Washington, Brookings, 1977, p. 22.

32. "The Incredible Crystal: Diamonds," *National Geographic* 195 (1979).

33. US Bureau of Mines, *Mineral Commodity Summaries,* 1983, Washington, U.S. Government Printing Office, 1983; diamonds, pp. 44–45; quartz, pp. 122–123; rare earth metals, pp. 124–125.

34. C. Freeman, "The Determinants of Innovation," *Futures* 11, no. 3 (June 1979), p. 206.

35. Rosenberg, *Perspectives,* op. cit., note 6, p. 197.

36. James M. Utterback, "The Dynamics of Product and Process Innovation in Industry," in C. T. Hill and J. M. Utterback, *Technological Innovation for a Dynamic Economy,* New York, Pergamon, 1979, pp. 46–47.

37. The importance of the second "D" in the R&D&D process is often overlooked. See C. Freeman, *Design and British Economic Performance,* London, Royal College of Art, 1983.

38. "The Fellowship of Engineering," *Modern Materials in Manufacturing Industry,* London, 1983; carbon fibers, pp. 40–46; ceramics, pp. 49–51, 62–63; Sialon, pp. 47–48.

39. C. Freeman, C. Cooper, and K. Pavitt, "Policies for Technical Change," in C. Freeman and M. Jahoda, *World Futures: The Great Debate,* London, Martin Robertson, 1978, p. 209.

40. *Reader's Digest,* April 1982, p. 75.

41. *Reader's Digest,* April 1982, p. 16.

42. *Financial Times,* 29 April 1985, p. 16.

43. H. G. Johnson, *Technology and Economic Interdependence,* London, Macmillan, 1975, p. 10.

How Critical Are Critical Materials?

Joel P. Clark and Frank R. Field III with John V. Busch, Thomas B. King, Barbara Poggiali, and Elaine P. Rothman

15

Disruptions to imports of critical materials from politically unstable countries will not shut down U.S. industry, say the authors, who are based in the Department of Materials Science and Engineering at MIT. The United States has nothing to fear from temporary shortages of the key minerals chromium, manganese, platinum, and cobalt. Like Ray, Clark and Field emphasize that shortages—or the fear *of shortages—of such critical materials merely leads to greater conservation and substitution and spurs the search for entirely new materials. Ceramics and plastics are selected as having great potential.*

How well could the United States maintain its economic vitality and military strength if cut off by international crisis from the critical metals it imports from overseas? This question is receiving a great deal of attention these days, particularly with South Africa's recent threat to stop sales of critical materials to any country that imposes economic sanctions against its apartheid regime. South Africa is a major supplier of chromium, manganese, and platinum, and some observers fear that a significant disruption in supplies of these materials would be disastrous.

The congressional Office of Technology Assessment (OTA) voiced just such a fear in a recent report on critical materials. After a detailed study, the OTA concluded that the "United States is likely

Reprinted with permission from *Technology Review,* August–September 1985.

to become more dependent upon imports of critical materials from potentially insecure sources of supply." And that dependence, according to the OTA, could put the nation in a vulnerable economic and military position—if there were significant disruptions in supply owing to a political crisis.

Each year the United States imports well over $1 billion worth of materials considered *strategic* because of their unique importance in the manufacture of defense munitions and industrial products, and *critical* because they are imported from countries with potentially unstable political regimes. The four commonly cited strategic and critical materials are chromium, manganese, and platinum, primarily supplied by South Africa; and cobalt, of which Zaire is the dominant producer. All four metals are thought to be essential to our national defense and economy since they are used in the production of steel and stainless steel, as well in the manufacture of high-temperature alloys for jet engines and gas turbines.

The U.S. government stockpiles these materials and others in varying amounts to ensure at least minimal supplies to meet defense needs in times of crisis. The Reagan administration has begun to build up that stockpile and, in fact, has made the first stockpile purchases since 1962. But it continues to rely "upon the workings of the market" to ensure the supply of such materials for civilian purposes.

Many observers advocate further action to reduce the nation's vulnerability to supply disruptions. Among proposals often made have been that:

- The federal government increase its stockpiling of critical materials to meet the needs of private industry as well as those of the defense establishment.
- The United States wield its political and economic power to bolster friendly governments in countries that supply significant amounts of critical materials.
- Government and industry support the development of alternative resources now untapped because their exploitation would be uneconomical or would cause significant environmental damage.
- The government fund research to develop new materials that can substitute for critical minerals.

In our view, the United States is needlessly concerned about the future availability of critical materials. We believe the existing stockpile, managed by the Federal Emergency Management Agency (FEMA), is adequate to deal with the threat of disruptions to national security. In the case of civilian industries, disruptions in supplies will automatically lead to conservation and recycling (a drop in demand), and to a much-expanded effort to develop alternative materials.

Thus it is neither necessary nor economically sound for the U.S. government to increase its existing stockpiles or to invest heavily in untapped resources. However, the government should fund R&D on alternative materials that not only will reduce our reliance on imported minerals but also will lead to improved performance and lower costs in manufactured products.

Reducing Our Fix on Chromium

The United States now imports about 99 percent of the manganese, 95 percent of the cobalt, 82 percent of the chromium, and 90 percent of the platinum-group metals (PGMS) that it requires. We prefer to purchase these materials at international market prices rather than exploiting higher-priced alternatives from domestic or "reliable" foreign sources.

The most important supplier of chromium is South Africa, which accounted for 35 to 40 percent of world production in the early 1980s. Next comes the Soviet Union, which is responsible for 25 to 30 percent of world production. Other important suppliers are Zimbabwe (formerly Rhodesia), the Philippines, and Turkey.

Chromium is used in metal alloys (stainless steel), chemicals (such as paint additives), and high-temperature refractory applications, such as in bricks that line industrial furnaces. About two-thirds of the chromium imported by the United States is in the form of high-grade ore used for metallurgical applications. Of this, about 70 percent is used to make stainless steel, a corrosion-resistant alloy containing a minimum of 12 percent chromium. Another 15 to 20 percent is used in other steel alloys in which chromium increases strength, heat resistance, and flexibility in shaping the metal. Less than 5 percent of U.S. chromium is used for superalloys—the nonferrous materials that are resistant to extremely high temperatures and used chiefly in gas turbines.

Metallurgical-grade chromite ores are found primarily in the Soviet Union, Turkey, and Zimbabwe. South Africa supplies the lower-grade ores used in pigments and paint additives. However, recent technical advances have also made it possible to use these ores in metallurgical applications. As a result South Africa has assumed an increasingly important position in the world market.

There is essentially no substitute for chromium in stainless steels and superalloys, since it provides resistance to corrosion and oxidation at high temperatures. According to some estimates, only 10 percent of chromium could be eliminated by replacing cobalt-based superalloys with nickel-based superalloys, which contain less chromium, or by applying protective surface coatings to alloys that have less chromium. Furthermore the costs of substitution would be very high because intensive tests would be required to certify these new materials for use in military aircraft and gas turbines. There would also be a three- to five-year lag before these substitutions could be made.

However, there are significant opportunities for "secondary substitution"—eliminating stainless steel entirely and using other steels of lower chromium content. For example, stainless steel is used for largely aesthetic reasons in architecture, furniture, automotive trim, and kitchen utensils. Aluminum, coated carbon steel, and plastics could easily be used instead, albeit at somewhat higher prices. Stainless steel could also be replaced by titanium and composite materials in oil-drilling rigs and chemical-processing facilities, again at somewhat higher prices. Copper or carbon steel encased in stainless-steel claddings, copper and nickel alloys, titanium, and plastics could be used for stainless steel in certain industrial equipment, fasteners, and automotive parts. Chromium could also be conserved in the production process.

Indeed, we estimate that a 50 percent increase in the price of stainless steel would result in an overall decrease in chromium consumption of 25 percent after one year and 35 percent after four years. The market essentially would take care of itself, which is why the existing emergency stockpile of chromium—5 million tons—is more than adequate.

Conserving the Manganese in Steel

An essential ingredient in modern steel making, manganese is used to remove the sulphur from unrefined iron. The chemical industry also uses manganese ore, especially as an electrolyte in dry-cell batteries.

The Soviet Union is the world's largest producer of manganese, but East–West trade in this metal has historically been minimal. The most important source for the noncommunist world is again South Africa, which accounts for about 36 percent of supply. Furthermore South Africa has more than 75 percent of the noncommunist manganese reserves. Brazil, Australia, and Gabon are sources of about equal importance, each accounting for about 15 percent of noncommunist supplies. The United States produces no manganese but has considerable domestic reserves of low-quality manganese, which is not now used in steel making.

Policymakers are not overly concerned about near-term supplies of this mineral because even South Africa accounts for only about 20 percent of total world production, and each of the other major producers accounts for less than 10 percent. Under these conditions it is doubtful that a supply disruption could cut off a large portion of our needs.

However, as manganese mines in Brazil, Australia, and Gabon are depleted, there may be greater cause for concern. By the year 2000 the developed countries may depend upon only two major suppliers of manganese: South Africa and the Soviet Union. On the other hand, many options for conservation and substitution are available should manganese prices escalate.

For instance, if its price were to double, steelmakers could economically reduce the amount of manganese they add to blast furnaces by about 0.05 percent. That would yield savings of about one pound of manganese per ton of steel. More savings could be achieved by installing robots that could feed manganese to the furnace in fixed, precise quantities. Manufacturers could also reduce the amount of manganese they use to remove sulphur in making some carbon and alloy steels.

Still, substitution options are more limited for manganese than for other critical metals. Thus, if the price of manganese ore should double, consumption by the U.S. steel industry would decrease by only about 10 percent in the first year, and by up to 15 percent in five years, if steel production remained constant.

How Precarious Is the Supply of Cobalt?

Like manganese and chromium, cobalt is considered strategic because of its importance in defense munitions. It is used in superalloys for high-temperature parts of jet engines and industrial gas turbines, and to take sulphur out of oil in refining petroleum. It is also used as a binder material in cutting tools and as an alloying element in steel.

Because the United States imports virtually all its cobalt from countries with potentially unstable political regimes, this metal is considered critical in a way that manganese and chromium are not. Zaire, the dominant producer, accounts for 50 to 60 percent in most years. Zambia, which like its neighbor Zaire recovers cobalt when producing copper, is the next largest producer. Canada, the Philippines, Australia, Indonesia, New Caledonia, the Botswana all recover smaller amounts of cobalt as a by-product of nickel mining and processing.

Zaire has experienced major problems in producing and exporting cobalt during the last ten years. Combatants in Angola's civil war temporarily cut the railroad from Zaire through Angola to the Atlantic in 1975. In 1978 Katangan rebels invaded the cobalt-producing Shaba province in a revolt against the ruler of Zaire. Though the damage to mining facilities was slight, virtually all foreign workers fled and the facilities fell into disrepair. In addition low copper prices have left Zaire almost bankrupt, hindering it from maintaining its transportation network and copper and cobalt factories.

However, the cobalt situation may not be as precarious as this description suggests. A recent study by Charles River Associates, a Boston-based economics consulting firm, shows there is considerable elasticity in world demand. When the price of cobalt increased dramatically from 1978 to 1980, consumption dropped sharply.

Materials such as nickel can be substituted for cobalt in many uses, including jet engines—usually at somewhat higher prices and with slight changes in performance. Increases in cobalt prices also encourage recycling and conservation.

That is indeed what occurred when the 1978 Katangan invasion of Shaba disrupted supplies. Manufacturers used alternative materials and other sources increased supplies, forcing Zaire to lower its set price from $55 per kilogram to less than $22 per kilogram in 1982. The price decreased even further to about $11 per kilogram early in 1983.

According to our estimates, if the price of cobalt were to return to the high levels of 1978, demand would drop by 50 percent in a few years. Thus a disruption of cobalt supplies from Zaire would not affect U.S. national security, especially since stockpiles could be used to cover short-term shortages.

A similar case can be made for platinum-group metals (PGMS), which include platinum, palladium, and iridium. These metals are essential in catalytic uses, petroleum refining, chemical processing, and treating automotive exhaust. For instance, platinum is the catalyst in the converter used to control automotive pollution. PGMS are also used in telecommunications switching systems (as contacts and electrodes in ceramic capacitors), as well as in jewelry and medical and dental equipment. The United States imports 49 percent of its platinum from South Africa, 15 percent from Canada, and 13 percent from the Soviet Union.

Charles River Associates recently predicted that if the price of platinum saw large, sustained increases, the U.S. government would relax pollution controls on car exhaust. If that happened, automakers would reduce their platinum consumption by at least 75 percent. In chemical processing any major increase in price would provoke major increases in platinum recycling, and consumption in this area would drop by about 30 percent. A supply disruption would also have minimal effect on the telecommunications industry, since it already plans to recycle much of its electronics equipment over the next two decades. Furthermore Charles River Associates estimates that gold would rapidly supplant platinum in dental uses if its price were to increase, although the rate of substitution would depend on the price of gold at that time.

The Chances of Calamity

What would happen if a politically unstable producer failed to supply a critical material to the United States for five years? And worse, what if a successful boycott by other countries, such as those in the Soviet bloc, cut off U.S. access to alternative resources?

To begin with, the likelihood that a disruption would continue for five years is very low because political regimes and alliances constantly change. And the probabilility of a "cataclysmic" disruption is virtually nil. The failure of the Arab nations to choke off oil exports to developed countries in the early 1970s vividly supports this point. Other punitive trading policies that failed include the U.S. embargo of grain to the Soviet Union in the late 1970s, the Rhodesian chromium embargo to the United States in the late 1960s, and the Cuban nickel embargo to the United States in the early 1960s. Each of these embargoes increased the cost of the materials, but none prevented them from reaching any country that wanted them.

For instance, even though the United States could not buy chromium directly from Rhodesia during its embargo, we simply purchased it from South Africa, which continued to buy the ore from Rhodesia. It is virtually impossible to shut off any one country from access to the materials market.

Of course, disruptions may have serious short-term effects, particularly in the form of increased costs to manufacturers. That is why we urge the U.S. government to maintain its present stockpile to protect military needs, and to cushion the shock to the private sector while it changes its consumption and production patterns. However, we see no need to greatly increase stockpiles to cover nonmilitary needs. Current supplies of platinum, for example, total 1.437 million ounces, with a market value of $350 million. But defense needs account for only 5 percent of U.S. platinum consumption. A platinum stockpile large enough to cover nonmilitary needs for five years would cost over $7 billion—an outrageous economic burden. Similarly, the cost of providing a civilian stockpile for cobalt would be $1.2 billion.

The Danger of Political Misalliances

The OTA and other analysts have urged the United States to form stronger alliances with nations that supply critical materials. This means tying such nations closer to us with trade privileges, foreign aid, weapons sales, and treaties. However, this approach presents an unsavory prospect: stronger U.S. alliances with governments whose policies are antithetical to our own.

The most compelling example, of course, is South Africa. Late in 1980 Joseph Churba, a Reagan advisor, accused the previous administration of "criminal neglect" in failing to recognize South Africa's critical importance as a supplier of strategic materials. He urged the president to end the U.S. arms embargo against South Africa and to set up a naval presence there. Although not going to that extreme, the Reagan administration has pursued a policy of "constructive engagement" in an attempt to improve relations with that country while exerting subtle pressure for reform.

Recently, however, this policy has unraveled because of the overly repressive attempts by the South African government to put down domestic unrest. In June the United States recalled its envoy to South Africa, and Congress has renewed calls for economic sanctions against South Africa. That country, in turn, has threatened an embargo on critical metals to the United States if it imposes such sanctions.

In our view this is an empty threat since the United States can obtain these critical metals elsewhere—albeit at somewhat higher prices. And South Africa must continue selling these metals to someone if it wants to preserve the health of its economy. More important, forging closer ties with a nation that practices overt racial discrimination would be disastrous to U.S. international relations. The political damage from such a self-serving policy would far outweigh the potential economic benefits.

Exploiting New Resources

Two types of unexploited resources could help this country respond to a crisis in critical materials. The first are the low-grade, land-based ores in the United States and friendly nations that cannot profitably be extracted at today's prices. The second are the mineral resources that we know exist or may discover in the deep sea. Of these, the manganese nodules in the Pacific are the best known. But like the low-grade land-based ores, extracting them is not economical at current prices and with today's technology.

Would-be producers of these resources are caught in a Catch-22 situation. On one hand, they must wait until scarcity drives up

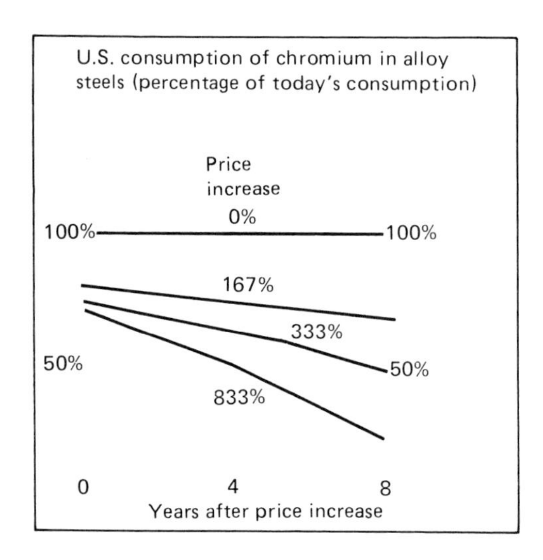

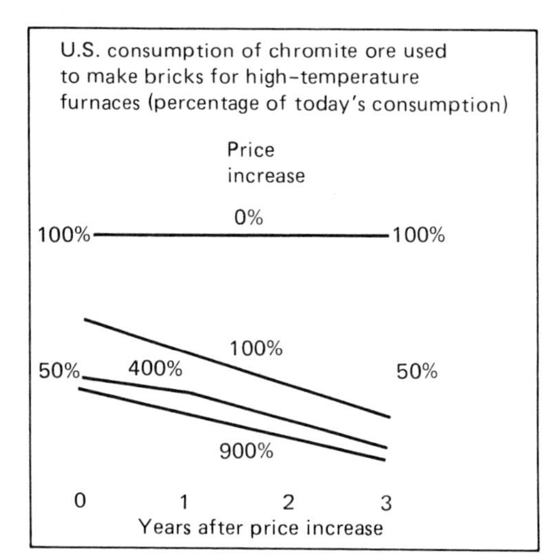

Figure 15.1 The United States imports 82 percent of its chromium. But if supplies were disrupted, industry could turn to other materials. Thus an increase in the price of chromium used in metal alloys would cut consumption (*top*). The same effect would occur for chromite ore, used in high-temperature bricks (*bottom*).

the price of strategic materials before trying to exploit them. But once they make that investment of time and money and the strategic materials reach the market, scarcity-driven prices will fall. This dilemma is particularly acute in the case of deep-sea mining projects, where large quantities—especially of cobalt—could be produced, driving the price to very low levels.

One way to overcome this dilemma, and encourage companies to invest in technology for exploiting these resources, is for the government to guarantee prices for the manganese and cobalt that would be produced. But such an approach would be very costly, since the government would be committed to purchasing materials at above-market prices.

As a less costly alternative, the government could further explore the known reserves of land-based low-grade ores in Idaho, California, and Missouri. This effort would not likely discover any "mother lode," but added domestic reserves—even if unexploited—would allow for more rational policies and a reduced sense of dependence on foreign sources. Rather than directly funding this kind of exploration, the Reagan administration has concentrated on removing the environmental constraints on companies wishing to do so.

An even better alternative would be for government to support research on new materials that could substitute for critical metals. This is not to argue that federal funds should be devoted to developing expensive technologies that would be economical only if supplies were disrupted. Such a policy would impose not one but two unnecessary drains on the economy: the direct cost of the R&D, and the loss of other investment opportunities that might have led to more profitable technologies.

It does make financial sense to support R&D programs designed to develop new materials that provide more efficient and cost-effective products. Often such developments also provide the added benefit of reducing reliance on overseas suppliers. However, replacing critical materials should not be the primary purpose of such research.

The Tough New Ceramics

Materials researchers at a number of universities are now developing ceramics, plastics (polymers), and advanced composites that meet these two criteria. None requires critical minerals either as raw materials or in the manufacturing process.

Interest in advanced ceramics is keen because they can withstand higher temperatures and more hostile environments than today's metal alloys, superalloys, and plastics. This makes them ideal for automotive and gas-turbine engines, heat exchangers, automotive turbochargers, and burner nozzles. In addition their hardness and resistance to corrosion make ceramics a promising substitute for metals in cutting tools, seals, bearings, and sandblasting nozzles. Indeed, advanced ceramics are the most likely candidates to replace our four most critical materials.

Ceramics are already cost-competitive with conventional materials in cutting tools, pump seals, and gas igniters. Ceramics are gaining commercial acceptance as cutting tools because they operate at high speeds, have excellent wear resistance, and produce smooth finishes. Since cutting tools now consume almost 40 percent of the tungsten and 20 percent of the cobalt used domestically, ceramics will be key in reducing our reliance on these materials. Overall, the use of ceramics to replace wear-resistant and heat-resistant steels and superalloys in cutting tools, seals, engine components, and magnets could reduce demand for cobalt by 10 percent and demand for tungsten by 20 percent.

Ceramic turbocharger rotors, cylinder liners, and other parts for automobile engines are close to commercialization. Such ceramic parts would replace those made of steel, specialty steels, and superalloys, eventually saving up to 10 percent of our chromium consumption. If a partially ceramic gas-turbine engine can be developed with emissions below Environmental Protection Agency standards, up to 50 percent of domestic platinum consumption could also be saved.

The Growing Market for Plastics

Direct substitutions of plastics for critical materials or metal alloys that contain them are rare. But plastics able to withstand continuous temperatures of 500°C have the stability required for use in rotor blades for gas-turbine engines.

The plastic automobile engine is even closer to commercial reality. One developer has already used a motor that is 60 percent plastic to power the winning entry in an endurance race in New York last year. The driver was able to record a lap speed of 94.8 mph with the engine attaining speeds of 14,000 rpm. Most auto engines contain only small amounts of critical materials. However, eliminating their use in the 7 million car engines produced annually in the United States would greatly reduce our vulnerability.

Plastics will play an even larger role in reducing our dependence on critical materials through secondary effects. For example, substituting them for heavier steel sheet in autobodies will make possible the use of smaller, lighter engines and components.

Similar results could be achieved by developing composite materials. Composites are composed of a matrix substance that holds together stiff fibers. These fibers—usually carbon or glass filaments—are combined with the matrix, which is often an epoxy or polyester resin, to form a solid compound. Composites made of a carbon matrix and filaments are lightweight, show great stiffness and strength at high operating temperatures—perhaps as high as 1,350° to 1,925°C for long periods—and contain no critical materials. However, the composites developed so far tend to degrade at extremes of temperature, and it is doubtful that they will play a major role in the short run. Still, industry sources anticipate that with further advances, carbon–carbon composites could replace superalloys in jet engines by the end of this decade. A similar substitution may be possible in the turbine engines that power small cruise missiles.

Scientists are developing composites made of a ceramic matrix for a variety of space and military applications. They are tougher and more resistant to rapid temperature changes than conventional ceramics, and are therefore being considered for use in gas-turbine and jet-engine components, spacecraft heat protectors, and rocket nozzles. The composites will substitute for superalloys in all these cases. If successful, ceramic composites could replace up to half the superalloys made in the United States within ten to twenty-five years.

The Reagan administration has said it wishes to support R&D on improved materials and technologies that would reduce our reliance on critical materials. Unfortunately, it has not substantially increased funds for university and industry research in this area.

Overall, however, the administration's policy of relying on the workings of the free market to ensure the supply of critical materials for civilian use is wise. In this one area, at least, the market seems to be able to take care of itself.

Material Substitution: Lessons from the Tin-Using Industries

John E. Tilton

16

The substitution of new materials for traditional materials has been cited in relation to possible shortages of critical materials and the overall transition from a low-tech to a high-tech economy. Indeed, materials substitution is a key issue in any historical or global perspective on materials. Using the example of tin, John E. Tilton looks in more detail at the substitution process and explores the links between prices, government policy, consumer preferences, and especially changing technology. The author is with the International Institute for Applied Systems Analysis, Laxenburg, Austria.

Historically, the United States and other industrialized countries have substituted relatively abundant materials for increasingly scarce materials. This has alleviated—some would even say, postponed indefinitely—the upward pressure on material costs resulting from the depletion of high-grade, readily available, and easy to process mineral deposits.

Forest products, as Rosenberg (1973) has shown, provide a dramatic illustration from the past of this benefit of material substitution. The abundance of timber in nineteenth-century America led to its widespread use, both as a fuel and as a building material. By the end of the century, however, the once seemingly endless tracts of forest land were gone, and many feared that shortages of this essen-

Reprinted by permission of Resources for the Future and Johns Hopkins University Press from John E. Tilton, editor, *Material Substitution: Lessons from Tin-Using Industries* (Resources for the Future, Inc., Washington, DC, 1983), pp. 1–11.

tial resource would soon curtail the country's economic development. The substitution of coal, petroleum, and natural gas for wood as a fuel, and iron and steel, aluminum, cement, and plastics for wood as a material, however, averted a shortage, and the country's economy continued to grow rapidly during the twentieth century. As for the future, Goeller and Weinberg (1976), Skinner (1976), and others contend that material substitution will have to play an even more critical role, if the adverse effects of dwindling mineral resources on economic growth and living standards are to be avoided.

While material substitution is an essential weapon in society's arsenal for holding the long-run problem of resource depletion at bay, it is important for other, more immediate reasons as well. Material substitution may, at least in some applications, soften the blow of abrupt and unexpected interruptions in minerals trade, whether these interruptions arise from wars, civil disturbances, strikes, natural disasters, or embargoes. During World War II Germany, the United States, and other countries managed to replace tin, tungsten, and other restricted materials with more available alternatives in many end uses (Eckes 1979, chap. 4). In recent years, interest in material substitution as a cushion against supply interruptions has grown as the political situation in south Africa and other mineral producing regions has become more turbulent.

Despite its importance we still have much to learn about the nature of material substitution. For example, just how important a factor is substitution in shaping short-run and long-run trends in material consumption? Are its effects generally evolutionary and continuous, or abrupt and discontinuous? Is the replacement of one material by another primarily motivated by shifts in relative prices, or are changes in government regulations, consumer preferences, technology, and other factors typically more influential? Where substitution does occur in response to shifts in material prices, what is the nature of the time lag between the two? What kinds of substitution can be made quickly, and what kinds require years or decades to accomplish?

Approach

To provide insights into such questions, material substitution was examined in three tin-using industries—beverage containers, solder, and tin chemical stabilizers used in the production of plastic pipe. In these case studies, which are found in part II, material substitution is defined to encompass a number of different types of events that may substantially alter material use.

The first is material-for-material substitution, where one material is used in place of another. Examples include the use of an aluminum beer can for a glass bottle, the installation of plastic rather than copper pipe, and using aluminum rather than copper-brass radiators to cool automobile engines.

The second is other-factors-for-material substitution, where consumption is reduced by increasing nonmaterial inputs, such as labor, capital, and energy. The hand soldering of radios, televisions, and other electronic products, for example, requires less solder than more automated production using printed circuit boards. As indicated in part II, however, producers have favored printed circuit boards despite their higher material intensity because they lower labor costs.

The third is quality-for-material substitution, where material is saved by reducing the quality or performance standards of the final product. For example, the lightweight, nonreturnable bottles produced during World War II had relatively high breakage rates. These bottles could have been strengthened by using more glass to make them thicker.

The fourth is interproduct substitution, whereby a change in the mix of goods used to satisfy a given need alters the demand for one or more materials. Television may reduce the need for movie houses, public transportation for private automobiles, and the telephone for letters or visits. This type of substitution, unlike others, does not affect the manufacturing process or the materials used in the production of individual goods, but instead influences material use by changing the composition of the goods and services produced. It includes what Chynoweth (1976) and others call functional substitution, which occurs when a product (along with the materials it embodies) is replaced by an entirely different means of achieving the same end. The use of satellites instead of underground cables to transmit long-distance telephone messages is an example.

The fifth is technological substitution, in which an advance in technology allows a product to be made with less material. The introduction of electrolytic tinplating during and after World War II greatly reduced the tin requirements for beer and soft drink cans, and provides a good example of this type of substitution. New technology can also create new products, as the aluminum can illustrates, and thus increase the opportunities for interproduct substitution.

In short, material substitution may result from the introduction of new technology, from shifts in the composition or quality of final goods, and from changes in the mix of factor inputs used in producing these goods. This is clearly a broader and more encompassing definition than the common perception of material substitution, which often is restricted solely to material-for-material substitution. A broad definition, however, is more appropriate and useful if ultimately one is concerned with alleviating material shortages or forecasting future mineral requirements.

There are, of course, other ways to study material substitution than the case study approach followed here. One possibility entails the estimation of production or cost functions for entire industries, economic sectors, or even the economy as a whole. Such efforts, which Slade (1981) has recently reviewed, generally treat materials as a single factor of production, and assess the extent to which they can be replaced by other inputs such as capital, energy, and labor. Their findings vary greatly, depending on how the production or cost functions are specified, how technological change is treated, and how other issues are resolved. They also tend to suffer from the use of aggregate input data and from their static nature (Kopp and Smith 1980; Slade 1981).

A second alternative involves more micro investigations that consider individual materials and often focus on specific end uses. These efforts specify and estimate formal models, the most common of which are econometric supply and demand models, such as the Fisher, Cootner, and Baily model of copper (1972) and the Woods and Burrows model of aluminum (1980).

These models typically assess the effects of material substitution by including in their demand equations price variables for close substitutes, and so implicitly assume that material prices are the primary driving force behind substitution. The specification of such models also normally presumes that the relationship between demand and material prices adheres to a particular lagged structure that remains fixed over time, and that this relationship is continuous and reversible. The latter implies that the demand lost when a material's price increases is recovered when price returns to its original level. Although such assumptions are perhaps plausible in certain instances, they are open to question. Even more important, in making them, these models are assuming away many of the interesting questions associated with material substitution.

The decision to concentrate on a limited number of in-depth case studies was taken in the belief that this was the most promising approach given the current state of understanding and knowledge about the material substitution process. Like other approaches, it has its limitations. In particular, since material substitution occurs constantly in literally thousands of products throughout the economy, only tentative conclusions about the general nature of material substitution can be drawn on the basis of such a limited sample of actual situations.

The focus on tin-using industries was motivated by two considerations. First, the price of tin rose substantially during the 1970s, both in real terms and in relation to the prices of most of the alternative materials with which it competes. In this regard there is some concern that the depletion of known, high-quality deposits and the failure to discover new deposits of comparable quality will continue to force prices up in the future. Second, the United States and other industrialized countries rely almost entirely on imports from developing countries for their primary tin requirements. These imports come in large part from Malaysia and other countries in Southeast Asia. The lengthy sea routes over which they travel could easily be interrupted during military or other emergencies.

Although all of the case studies examine uses of tin, they are not concerned exclusively with tin. Indeed, by its very nature one cannot study material substitution without considering more than one material. In its many applications tin competes with a variety of materials, including chromium, steel, glass, aluminum, cast iron, copper, antimony, lead, plastic, and silver. In certain applications steel, plastics, lead, and other materials are complementary products in the sense that they are used with tin in producing a product. As a result a reduction of their price tends to increase the demand for tin.

The three sectors of the tin industry selected for study—beverage containers, solder, and tin chemicals used in the manufacture of plastic pipe—were chosen for different reasons. As every consumer of beer or soft drinks knows, the beverage container market has been a lively battlefield for competing materials over the last two decades. The sturdy returnable glass bottle, the lighter one-way bottle, the tinplate can, the aluminum can, the tin-free steel can, and recently the plastic bottle have all fought for this market. The variety of competing materials and the dramatic speed with which their fortunes may rise and fall make the container market of particular interest.

In contrast, it is widely assumed that little material substitution has occurred in the solder market, and that little is possible. Indeed, solder by definition is an alloy of tin and lead, with other materials added in minor amounts in certain applications. Moreover, for technical reasons the tin content of solder cannot be varied without seriously affecting its performance in certain uses, particularly electronic equipment. So solder provides an opportunity to examine the extent and nature of material substitution in applications where substitution is generally presumed difficult. As solder is consumed in a multitude of products, this case study concentrates primarily on its uses in can seams, new motor vehicle radiators, and fillers for automobile bodies. Other important applications of solder in the electronics and plumbing industries are considered in less detail.

Tin chemicals, and particularly organotin chemicals used as stabilizers in the production of plastic pipe, have grown substantially over the last two decades. In contrast, as table 16.1 shows, most sectors of the tin industry have been stagnant or decreasing. Consequently an investigation of tin chemicals provides an opportunity to examine why material substitution may not always reduce the demand for a commodity whose price is rising rapidly. Due to the many uses of tin chemicals, the focus is restricted to organotin chemicals and their use in plastic pipe production.

The scope of the case studies is further limited in three respects. First, although many of the trends in U.S. tin consumption are also found in other countries, in a number of instances important differences do exist between the United States and other countries. Although some of these differences are noted in passing, the emphasis is on material substitution in the United States.

Second, the case studies do not cover all the important tin consuming industries. Table 16.1, which shows the major end-use sectors for tin in the United States for the years 1955 and 1978, indicates

Table 16.1
Tin consumption in the United States by end-use sector, 1955 and 1978

End-use sector	1955 Thousands of tons[a]	1955 Percent	1978 Thousands of tons[a]	1978 Percent
Tinplate	34.1	37	17.3	27
Solder	22.6	25	18.3	29
Bronze and brass	20.0	22	10.4	17
Chemicals[b]	1.9	2	7.6	12
Other[c]	13.3	14	9.5	15
Total	91.9	100	63.1	100

Source: American Bureau of Metal Statistics (various years).
a. Metric tons are used throughout this study.
b. Includes tin oxides and miscellaneous.
c. Includes terne plate, babbit, collapsible tubes, tinning, pipe and tubing, type metal, bar tin, miscellaneous alloys, and white metal.

that over two-thirds of the country's supplies are used to make tinplate, solder, and bronze and brass. Beverage containers are part of the tinplate sector, but so are fruit and vegetable, meat, soup, and other cans. Beer and soft drink cans have accounted for only between 10 and 20 percent of total U.S. tinplate consumption. The case studies also examined important uses of solder and chemicals, but as noted already, coverage of these sectors is far from complete.

Third, the studies concentrate on explaining the past, primarily the period since 1950, rather than predicting the future. The past can be documented with data, and the factors causing material substitution identified more or less. The future is far more uncertain, and detailed information on material consumption, prices, and other relevant variables is not available. Thus the analyses carried out here are designed to enhance knowledge of the nature and determinants of material substitution by improving our understanding of events that have taken place. Of course, it is hoped that this information will ultimately prove useful in assessing the role that material substitution may play in the future.

The case studies are similar in that they all follow three analytical steps. In the first, the pounds or tons of tin consumed in specific applications, such as soft drink containers, are quantified by the years covered in the analysis.[1]

In the second step, what are called the apparent determinants of tin consumption are identified and measured. These determinants are directly related to tin consumption by an identity, and basically indicate how tin usage has changed over time. For example, the amount of tin in the solder used to produce new motor vehicle radiators has varied over time as a result of changes in (1) the number of new motor vehicles and hence radiators produced; (2) the proportion of radiators made from copper and brass, rather than aluminum, and so requiring soldering; (3) the pounds of solder used per copper-brass radiator; and (4) the tin content of the solder used for this purpose. Any change in tin consumption from one year to another in radiator solder must result from a change in one or more of these apparent determinants.

Although the apparent determinants vary from one end use to another, they contain in all instances one variable that reflects the change over time in the output of the end use. In the preceding example, it is the number of new motor vehicle radiators. This apparent determinant may change over time for reasons other than material substitution. The number of new motor vehicles and in turn radiators produced in the United States, for instance, may increase simply because population and per capita income are growing. Changes in all the other apparent determinants, however, are the result of one or more of the five types of material substitution identified earlier. Moreover, while the apparent determinant reflecting changes in the output of the end product can change for other reasons, it too may be affected by material substitution, particularly interproduct substitution. The growth of mass transit systems in metropolitan areas, for example, may slow the growth of new automobile radiators.

By identifying the apparent determinants and empirically assessing their effects, one can dissect the change in tin consumption into its component parts. The reasons why these parts change can then be assessed.

This is done in the third step, which identifies and evaluates the major underlying factors responsible for the changes in the apparent determinants, and thus ultimately in tin consumption. These factors include the price of tin, the price of alternative materials, technological developments, government regulations, and a host of other considerations. At this step the analysis cannot be as empirically rigorous. Assessing the relative importance of the major underlying factors involves some judgment, even after weighing the available information from industry and other sources. Surprisingly, however, in many instances, due largely to the level of disaggregation on which the analyses are conducted, the important underlying factors are readily apparent.

Findings

In all of the end uses studied—beer and soft drink containers, can seams, motor vehicle radiators, automobile body solder, and chemical stabilizers used in the production of the PVC plastic pipe—material substitution greatly affected tin consumption over the longer run, a period of ten years or more. Moreover, in many instances, it sharply altered or reversed trends in tin usage even in the short run.

The tinplate can, for example, after years of increasing its share of the beer container market, abruptly found itself during the late 1960s being pushed out of this market by aluminum and tin-free steel cans. Even with solder, often considered immune to material substitution, the introduction of low-tin alloys, produced by substituting lead and minute amounts of other metals for tin, substantially reduced the need for the latter in can seams and automobile body solder. More recently, the trend away from large automobiles has decreased the size of the average motor vehicle radiator and in turn the need for solder and tin in this use.

Two types of material substitution are especially prevalent. The first is material-for-material substitution. The beverage container market in particular has experienced substantial changes over time in the number and quantity of materials it consumes. A wide variety of materials also compete for the pipe market. Indeed, much of the growth in tin chemicals for this industry has come about as a result of the substitution of plastic for copper, cast iron, steel and other kinds of pipe. In solder, material-for-material substitution has occurred on a modest scale with the appearance of aluminum radiators, and on a more significant scale with the widespread adoption of low-tin alloys for can seam and body solder.

Technological substitution is the second type often encountered. New technology reduced the tin content of the average size tinplate beverage container by 93 percent between 1950 and 1977. Similarly, in the plastic pipe industry, the introduction of second and third generation organotin stabilizers caused their tin content to fall by nearly 50 percent.

Other types of material substitution, though apparently less common, also occur. The growing distribution of soft drinks in bulk containers, due in large part to the rise in fast food chains, slowed the penetration of the tinplate can in this market. Another example of functional substitution is the electronics industry, where the miniaturization of components greatly reduced the number of electrical connections requiring soldering.

The case studies also strongly suggest that intermaterial competition and substitution are becoming more intense and prevalent over time, as the number of materials increases and the properties of existing materials are enhanced, allowing them to penetrate new markets. This tendency is clearly found in the beverage container market. In the early 1950s, the glass bottle basically monopolized both the packaged beer and soft drink markets. The only competition came from the tinplate can, in which a modest amount of beer was shipped. Starting in the 1960s and continuing in the 1970s, however, the bottle encountered increasing competition from the tinplate can, the aluminum can, the tinfree steel can, and recently the plastic bottle. While the glass bottle continues to be the most popular packaged container (if the returnable and one-way bottle are considered together), its share of the soft drink market dropped from 100 to 62 percent between 1950 and 1977. In beer it fared even less well, retaining only 40 percent of the market by 1977.

Growing material competition is found in other sectors as well. Polyvinyl chloride (PVC) plastic did not enter any of the pipe markets examined until the early or mid-1960s. Antimony stabilizers as an alternative for organotin first appeared in the late 1970s. Solder, once required to seal the side seams of all cans, had its monopoly of this market broken by the development of welded and cemented seams, and by the introduction of the two-piece can, which has no side seam.

The factors responsible for material substitution are numerous and their significance tends to vary over time and by end use. However, three factors—relative material prices, technological change, and government regulations—are of particular importance in all of the case studies. In examining their influence, it is useful to start with material prices, since economists and others often assume that they are the principal motivation or incentive for material substitution. On this, the evidence is mixed; or more correctly, it supports the proposition that material prices are important but that several qualifications or corollaries to this conclusion are necessary.

There is no doubt that the high and rising price of tin has discouraged its use. As table 16.1 indicates, total tin consumption in the United States, in contrast to that of nearly all other metals, has declined over the postwar period. Moreover it is not difficult to document specific examples of substitution away from tin that is motivated at least in part by its price, as is illustrated by the development of second and third generation tin stabilizers, low-tin alloy solders for can seams and automobile body fill, and the declining tin content of tinplate.

Still, material substitution has in many instances increased, and increased substantially, the use of tin. Such substitutes take place, despite the high price of tin, for two reasons. First, in some uses tin provides superior quality or performance that outweighs its higher cost. For example, electronics manufacturers, after experimenting with solders containing 50 percent tin and 50 percent lead, reverted back to 63 percent tin solders because the latter's lower melting temperature resulted in less damage to printed circuit boards during the manufacturing process.

Second, substitution takes place on many levels. In the pipe industry, for example, it occurs in the production of stabilizers, plastic compounds, and pipe.[2] Since tin constitutes between 18 and 35 percent of the final cost of producing organotin stabilizer, producers are strongly motivated to reduce or eliminate their use of tin. However, at subsequent stages of production, tin becomes an increasingly smaller fraction of total cost, finally accounting for 1 percent or less of the final price of PVC plastic pipe. At this stage the price of tin has much less of an effect on the type of pipe purchased.

So it is not surprising that the price of tin has its greatest impact on material substitution at relatively early stages of production, where tin costs constitute a significant portion of total costs. It is here that substitution has with considerable consistency reduced tin usage. In plastic pipe, for example, the tin content of stabilizers has dropped substantially as a result of the development of second and third generation tin stabilizers and the recent appearance of antimony stabilizers. This contrasts with the material substitution at later stages of production, which has significantly increased the use of tin. At the compound stage, polyvinyl chloride, which requires a stabilizer, has replaced other types of plastic; and at the pipe stage, plastic has replaced pipe made of copper, cast iron, and other materials. At both of these stages the price of tin has had a small impact on final cost, and has been easily offset by other considerations.

When tin prices do stimulate material substitution, three possible time dimensions or lagged responses are found. First, where existing technology and equipment permit the use of one material for another in the production process, substitution can respond immediately to changes in material prices. In such situations, when the price of a material rises above a certain threshold level, producers switch away from it. When the price falls below this threshold, they switch back again. Not only is the response fast, but it is reasonably predictable. The case studies of containers, solder, and tin chemical stabilizers uncovered a few instances of this type of response. The dual canmaking line, introduced about 1976, can substitute between aluminum sheet and tinplate in four hours. Also plastic pipe producers can switch from one plastic resin to another quite quickly since the same equipment is used. However, the opportunities for such a rapid response to changes in material prices or other conditions are limited.

Second, where the technology for substitution exists but equipment must be altered or completely replaced, material substitution occurs only after some delay. Moreover the lag normally exceeds the minimum period required to build new facilities or modify old ones, for producers hesitate to make such changes, in light of the costs, until they are reasonably certain that the change in material prices is not temporary. The expense of switching also means that the threshold price at which substitution occurs is higher than would otherwise be the case, and that once a change is made, the price of the replaced material must fall appreciably below the original threshold level before producers will switch back again. This type of substitution has occurred from time to time in the beverage container and pipe markets. Though not all that prevalent, it is more frequently encountered than the first type of response.

Third, where new technology must be introduced before substitution occurs, the response to a change in material prices takes even longer. In this case there is also greater uncertainty and variation regarding both the size of the response and the length of its time lag. Once substitution away from a material does occur, a decline in its price is even less likely to result in the recapturing of a lost market than in the second situation.

It is this third type of response that has consistently had the greatest impact on tin usage. In beverage containers, for instance, tin prices have encouraged the introduction of electrolytic plating, differential and lighter tin coatings, the tin-free steel can, and the tin-free steel bottom used on all tinplate cans—all major developments reducing tin usage. In solder the experiments that led to the use of low-tin alloys for can seams and automobile body finishing have been particularly important. In the pipe industry tin consumption has been cut by about 50 percent from what it would otherwise have been by the development of superior tin stabilizers and the introduction of antimony stabilizers. Thus the relatively high price of tin, when it has effected material substitution, has often done so indirectly by stimulating new tin-conserving technologies.

This conclusion is part of a broader finding. In all of the end uses examined, technological change appears as the dominant factor affecting material substitution. The use of tin in beer and soft drink containers, for example, was made possible by the development of the tinplate can. Its competitiveness over time has been enhanced by electrolytic plating, the easy-open aluminum top, the cheaper tin-free steel bottom, two-piece production techniques, and a host of other innovations. Conversely, the development of the one-way bottle, the aluminum can, the tin-free steel can, and subsequent improvements in these containers have reduced tin consumption in this market.

The use of solder in electronics was first stimulated by the introduction of printed circuit boards, and then reduced by the trend toward miniaturization. In cans, it declined substantially after welded and cemented seams were developed and two-piece canmaking technology was introduced. In motor vehicle radiators, the appearance of the sweat-soldered tube and corregated fin core for the dip-soldered cellular core used in the early postwar period substantially reduced solder and in turn tin usage.

The use of tin as a stabilizer for the production of PVC plastic was originally made possible by new technology. The rapid penetration of PVC in the pipe market that followed was stimulated by advances in extruder techniques, the development of superior tin stabilizers, and other innovations. After this rapid ascent tin consumption in PVC plastic pipe may decline as quickly as it rose due to another new development, antimony stabilizers.

These are but a few of the many developments identified in part II. Technological change has been a powerful force shaping the use of tin in all of the applications examined. Moreover its impact has often been abrupt and uneven, at times stimulating and at other times curtailing tin use. This random and discontinuous character of technological change makes it difficult to foresee its effects.

Government regulations have also influenced material substitution. Although less important than technological change, the influence of this factor appears to be growing over time. Recent container deposit legislation, for example, has favored the returnable glass bottle; and among metal containers, the aluminum can, whose homogeneous composition makes it cheaper to recycle. During military emergencies the government has restricted the use of tin containers in the domestic market, and encouraged tin-conserving new technologies, such as electrolytic plating and low-tin solder alloys. Tin consumption in solder for cans has been influenced by the regulations of the Food and Drug Administration, in automobile radiators by the fuel efficiency standards of the Environmental Protection Agency, and automobile body finishing by the lead exposure standards of the Occupational Safety and Health Administration. In the pipe industry the ubiquitous building codes imposed by thousands of local authorities have encouraged the use of tin by inhibiting lead stabilizers and slowing the growth of antimony stabilizers. They have also impeded tin consumption by delaying the use of plastic pipe.

Although government regulations, material prices, and particularly technological change are the more prevalent and influential factors responsible for material substitution, other considerations have also been important in certain applications. For example, changes in customs, causing the decline of the local tavern and the rise of fast food chains, have altered the mix of packaged and bulk containers in the beer and soft drink markets. The popularity of the vinyl-roofed automobile has reduced the consumption of body solder, while the desire for quieter plumbing systems has favored heavier materials over plastic.

Implications

The findings described in the preceding section point toward several general conclusions regarding the nature of material substitution. First, such substitution may, and often does, substantially alter material requirements. This is particularly true over the long run, but applies at times in the short run as well. Second, material-for-material and technological substitution are the easiest types of substitution to identify and appear more prevalent than other types. Third, a major cause (perhaps the major cause) of substitution is technological change. Fourth, changes in material prices typically have little effect on the mix of materials in the short run because producers are constrained by existing technology and equipment. Over the long run they have more of an impact. Indeed, the major influence of material prices apparently is exerted indirectly over the long run by altering the incentives to conduct research and development on new material-saving technologies. Finally, government activities and regulations often motivate material substitution.

These conclusions, of course, are tentative. As pointed out earlier, material substitution is taking place continuously in a multitude of products throughout the economy, and so there are dangers in drawing general conclusions on the basis of only a few cases. It is important to keep this caveat in mind in examining the implications of the findings.

The Demand Curve

The downward-sloping demand curve, a conceptual tool widely used by economists, business analysts, and others, indicates the quantities of a commodity that the market will demand at various prices over a particular period, such as a year, on the assumption that all other determinants of demand remain unchanged. Although the curve is formally derived in microeconomics from the theories of the firm and consumer behavior, the nature of the curve and in particular its downward slope seem reasonable. As the price of a good goes up, less will be demanded, first because consumers will substitute other goods whose prices have not risen (the substitution effect) and second because consumers will have less real income to spend on all goods and services (the income effect).

Materials are rarely desired for their own sake, but rather their demand is derived from the demand for final goods and services. Moreover the proportion of total costs contributed by any particular material in the production of most finished products is small. This means that changes in material pieces generally do not produce major shifts in the output of final goods and services. Nor do they cause significant changes in real consumer income. As a consequence the reduction in material demand resulting from an increase in price comes about entirely, or almost entirely, as a result of material substitution.

As commonly drawn, the demand curve presumes that the functional relationship between price and demand is reversible. Frequently, however, commodity analysts claim, if a material loses a particular market, that market will be lost forever. Such statements imply that an industry may not be able to recapture a market lost during a price rise even if its price subsequently returns to its previous level. In other words, if a commodity moves up its downward-sloping demand curve, it may not be able to reverse itself and move back down the same curve, as the conventional demand curve implies. Such concerns, it is even suggested, help explain the restraint that molybdenum, nickel, aluminum, and other material producers exercise in raising their prices during boom conditions.[3] As is well known, many material producers charge less than the market-clearing price at such times.

In assessing the reversibility assumption, it is useful to distinguish among short-run, medium-run, and long-run demand curves. The short-run demand curve indicates how demand responds to a change in price during a time period that permits neither plant and equipment nor technology to change. The medium run covers a period sufficiently long to allow plant and equipment to change, and the long-run demand curve for technology as well as plant and equipment to change.

The assumption of reversibility seems most plausible for the short-run demand curve. The material substitution that can take place in the short run, such as the use of aluminum sheet for tinplate on a dual canmaking line, involves changes that can be made quickly with little cost or disruption. When these conditions govern the switch from one material to another, they are also likely to hold for a switch back to the original material. Although the number of such substitution opportunities is limited, causing the short-run demand curve to be relatively steep, the possibilities that do exist and that can be quickly exploited tend to be easily reversible.

Reversibility appears more questionable over the medium run. Material substitution is in such instances likely to entail considerable expense. New equipment may have to be ordered, personnel re-trained, and production lost during the changeover period. Once a firm has incurred such conversion costs, it will not find it worth-while to switch back to a material unless the latter's price drops con-siderably below the level at which the original substitution became attractive.

In the long run the assumption of reversibility appears particu-larly doubtful. Within this time frame, price-induced innovations may substantially alter the underlying technical and economic con-ditions governing the use of a material in a number of its applica-tions. The introduction of better products can prevent a resurgence of consumer demand for older, traditional products even when the raw material costs of the latter fall sharply. Who, for example, would return to mechanical calculators even if the cost of their em-bodied materials were zero? Such uncertainty surrounding the gen-eration of new technology makes it most unlikely that the demand induced by a fall in price will exactly offset the demand lost by an equivalent rise in price.

The demand curve is also often assumed to be continuous. The importance and nature of material substitution, however, suggest that this too may not be very realistic, particularly for those mate-rials whose use is concentrated in a few major applications. Price may rise over a range, with little or no effect on demand. Then, once a particular threshold is crossed, making the use of an alternative material attractive, demand may fall sharply. Such discrete jumps may be found in both short- and medium-run demand curves. They are even more likely to distinguish the long-run demand curve, since innovations, such as the antimony stabilizer or the aluminum can, by their nature are discrete events that either do or do not occur. When they do occur, their impact can be substantial.

A third questionable but common assumption concerns the stability of the relationship between demand and price over time. Although the demand curve is often assumed to shift in response to changes in income, the prices of substitute materials, and other factors affecting demand, once the influence of these variables is taken into account or controlled for, the relationship between the price of a material and its demand is presumed to be reasonably stable from one period to the next. Efforts to estimate demand curves, or entire demand functions, on the basis of time series data must make such an assumption, at least implicitly.[4] The use of demand functions, however, estimated, for forecasting or assessing future market conditions also requires this assumption.

The importance of price-induced technological change, however, suggests that such stability may not exist. The occurrence of innovations inherently involves a certain random or chance element. In addition their effect on demand varies greatly. In short, there is not a stable relationship between price and the number of induced innovations, or between the number of induced innovations and their cumulative effect on demand. Yet both of these conditions are needed if the response of demand to a change in price is to remain stable over time.

Since the demand curve is defined as the relationship between the demand for a commodity and its price, assuming all other factors remain unchanged, one way of trying to preserve the assumptions of reversibility, continuity, and intertemporal stability is simply to exclude the effects of price-induced technological change. By definition, one might argue, such developments involve a change in one of the other factors (technology) which the construction of a demand curve can presume remains constant. This approach is certainly feasible, and it increases the plausibility of the three assumptions. Unfortunately, if the principal effect of a change in price on material demand occurs indirectly via induced technological change, as the studies here suggest, this approach reduces the demand curve to a sterile academic concept with little practical use.

Indeed, by ignoring the major impact of price, it may even be misleading.

Material Shortages

In assessing the role that material substitution can play in mitigating or eliminating shortages, two very different types of shortages must be distinguished. The first is due to the depletion of high quality mineral deposits, and imposes upon society the necessity of procuring its mineral needs from increasingly costly sources. This type of shortage comes about slowly with considerable warning, as the real price of a material climbs persistently over time and in the process gradually constricts demand. Although examples are hard to find because the real costs of most materials have actually fallen over the last century, there is concern that this type of shortage may become a serious problem some time in the future.

The second type of shortage, in contrast, is quite common. It tends to be temporary, rarely lasting for more than three to five years, and often arises quickly with little warning. It can be caused by a variety of factors, including war, embargoes, cyclical surges in demand, strikes, accidents, natural disasters, and inadequate investment in mining and processing. It may manifest itself in the form of sharply higher real prices, or where a few major firms maintain a relatively stable producer price, in the form of actual physical shortages.

The case studies suggest that material substitution can make a major contribution toward alleviating the first type of shortage, but much less of a contribution toward the second type. Substitution has substantially reduced tin consumption in a number of end uses, and the high and rising price of tin appears to have been a significant factor, at least indirectly, in motivating this shift. However, the response to changing material prices is typically small in the short run, for rarely is one material substituted for another while the same equipment, the same technology, and the same production processes continue in use. At a minimum, material substitution normally involves at least the replacement or modification of existing equipment, and more commonly, the development of new technologies. Since such changes take time, the major impact of a price rise on the consumption pattern of a material is not realized for a number of years. By then, the second type of shortage is usually over.

Market Power and Producer Cartels

In assessing the market power of the major material producers, most research has emphasized market concentration, which reflects the number and size distribution of the firms in an industry. Since many metal and other material industries are fairly concentrated, this approach often leads to the conclusion that firms do possess market power and the ability to earn excess profits. Moreover in these oligopolistic industries the major producers frequently quote or set a producer price, rather than simply accept a price determined on a competitive exchange. This behavior is often cited as collaborating evidence for the conclusion that these firms possess market power.

During the 1970s, concern over the possible exercise of monopoly power expanded to encompass the governments of mineral producing countries, as well as the multinational mining corporations and other major producing firms. The success of OPEC in raising the price of oil in 1973 led many to conclude that other mineral-producing countries would also attempt to form producer cartels and artificially raise material prices.

Material substitution, however, can severely limit market power even in highly concentrated industries. Collusive efforts by established producers to raise prices substantially are likely to stimulate new technological activity, and eventually end in failure, with markets irretrievably lost. Although higher prices may have little effect on material substitution in the short run, permitting a cartel to succeed for a while, the adverse consequences over the longer term are likely to far outweigh the short-term benefits. This discourages collusive activities except where firm managers or government officials have unusually short time horizons or are ignorant of the long-term consequences.

This conclusion, coupled with the findings that intermaterial competition is growing more intense over time as new materials are developed and traditional ones expand into new markets, suggests that the conventional view of market power in the material industries, based primarily on considerations of market concentration and pricing behavior, may be twenty years out of date. It also suggests that Schumpeter (1950, pp. 84–85), writing in the early 1940s, correctly described the nature of competition in the material industries today, if not then:[5]

. . . It is still competition within a rigid pattern of invariant conditions, methods of production and forms of industrial organization in particular, that practically monopolizes attention. But in capitalist reality as distinguished from its textbook picture, it is not that kind of competition which counts but the competition from the new commodity, the new technology, the new source of supply, the new type of organization (the largest-scale unit of control for instance)—competition which commands a decisive cost or quality advantage and which strikes not at the margins of the profits and the outputs of the existing firms but at their foundations and their very lives. This kind of competition is as much more effective than the other as a bombardment is in comparison with forcing a door, and so much more important that it becomes a matter of comparative indifference whether competition in the ordinary sense functions more or less promptly; the powerful lever that in the long run expands output and brings down prices is in any case made of other stuff.

It is hardly necessary to point out that competition of the kind we now have in mind acts not only when in being but also when it is merely an ever-present threat. It disciplines before it attacks. The businessman feels himself to be in a competitive situation even if he is alone in his field or if, though not alone, he holds a position such that investigating government experts fail to see any effective competition between him and any other firms in the same or a neighboring field and in consequence conclude that his talk, under examination, about his competitive sorrows is all make-believe. In many cases, though not in all, this will in the long run enforce behavior very similar to the perfectly competitive pattern.

Forecasting Material Requirements

Government agencies, private firms, international organizations, independent consulting firms, and others forecast future material requirements. These forecasts are needed to formulate sound public policies in the resource field, as well as for private investment decisions.

Although many different types of forecasting procedures are used, these techniques can be separated into three generic groups. The first encompasses a variety of statistical procedures that differ greatly in their complexity but basically involve analyzing past trends and projecting them into the future. One such technique is simply extending into the future the linear trend of past consumption. Far more complicated is the Box-Jenkins procedure, which employs a mathematical model to identify various functional trends present in past consumption and to estimate the parameters of these trends. It then uses these estimated trend functions to project consumption into the future.

The second group of forecasting methods is comprised of models that specify causal or behavioral relationships. Traditional supply and demand analyses fall into this category. In these models demand is typically assumed to be a function of the price of the commodity, an income or activity variable, and perhaps the prices of complementary and substitute products as well. Since the full effect of a change in price on demand may not occur immediately, various types of lag responses may be built into such functions. Once the causal equations are identified, their parameters must be estimated. This is usually done on the basis of past data and behavior using econometric techniques, though parameters may be determined on the basis of known technical relationships or other a priori information.

The third group includes forecasting techniques that are qualitative or judgmental in nature, and less quantitative. The Delphi method in which the views of a number of experts are solicited and then integrated to produce a forecast falls into this category. So too do forecasts based on the informed judgment of analysts who have adjusted current trends in consumption or the predictions of causal models for the effects of anticipated changes in public policies, consumer preferences and habits, technology, and other factors whose future influence on consumption cannot be measured with precision.

Material substitution greatly complicates the task of forecasting mineral requirements, and is likely to render any technique regardless of the category to which it belongs vulnerable to wide margins of error when forecasting over the longer term, ten to twenty years into the future. In the short run the effects of material substitution may be small or relatively continuous, and so more predictable. Yet even this is not certain. Within a year the strong upward market penetration of a product can be accelerated or completely reversed by material substitution. Moreover the growing intensity of material competition over time makes this type of volatility increasingly likely.

The first group of forecasting techniques implicitly assumes that material substitution is either a relatively unimportant factor shaping consumption, or that it is continuous and evolutionary so that its future influence can be projected from its past effects. Neither of these assumptions is reasonable over an extended period of time, and as just noted, may not even be valid in the short run.

The second group of forecasting techniques, when it considers material substitution explicitly, generally assumes that it is motivated or caused by a shift in relative material prices. This is the case, for example, when the demand functions within a supply and demand model include price variables for major substitute commodities. Moreover, although such models do not necessarily assume that a change in material prices immediately affects demand, as they can specify a lagged response, they do normally assume that the structure of this response follows a stable pattern over time and that it reflects a continuous and reversible functional relationship between prices and demand.

These assumptions can be questioned for reasons that have already been discussed. In particular, since the major impact of changes in material prices occurs indirectly through the development of new technology, neither the magnitude of the response to a change in price nor its lagged structure remains constant over time. On some occasions the response is negligible, while on others price changes provoke innovative activity that drastically alters consumption patterns. In addition complete reliance on the use of material prices to capture the effects of substitution is certain to miss the impact of those substitutions that are motivated by considerations of performance and quality or that occur at later stages of production where material costs are trivial. In tin-using industries such substitutions have often been important, and have at times substantially increased the use of tin despite its rising price.

Thus only the last of the three groups of forecasting techniques appears to offer any hope of adequately accounting for the effects of substitution in predicting future mineral requirements, at least over the longer term. This is the only approach that can take full account of the abrupt and inconsistent, yet major, effects of material substitution that are caused not only by shifts in material prices but also by changes in technology, government regulations, and other factors as well. Even here, in light of the inherent uncertainties involved in predicting the future course of new technology, government regulations, and the other factors affecting substitution, wide margins of error should be expected.

To some extent this pessimistic assessment can be modified when forecasting the requirements for a material in all of its end uses together, rather than requirements in specific end uses. Aggregation in this situation may lead to better results because the large distortions in consumption caused by material substitution in individual end uses may to some extent cancel out. This is particularly likely to be the case for materials used in many applications. Still, given the pervasive influence of material substitution, the reliability of forecasting techniques that do not explicitly consider material substitution and take account of its discontinuous and abrupt nature must be questioned.

In summary, an examination of various uses of tin over the past several decades in the United States suggests that substitution, when defined broadly, is a major force shaping and altering material consumption patterns. Substitution, in turn, is driven primarily by technological change, and to a somewhat lesser extent by relative material prices and government actions. Material prices influence substitution largely by encouraging the development of new technologies that conserve or replace materials whose price is high and expand the uses of materials whose price is low.

These findings, to the extent that they reflect the nature of material substitution in general, have a number of implications. They call into question the common assumptions regarding the reversibility, continuity, and intertemporal stability of the demand curve, and suggest that the relationship between the demand for a material and its price is much more complex than is often assumed. In this regard they raise the possibility that some of the methods now being used to estimate material demand functions and to analyze commodity markets may not be appropriate.

In addition the findings imply that substitution greatly complicates the forecasting of future material requirements. If reliable long-term predictions are possible at all, they require that the future effects of material substitution be explicitly assessed. This cannot be done on the basis of past trends of historical relationships between material prices and demand. The findings also indicate that material substitution is a major constraint on the exercise of market power.

Substitution can force firms in even highly concentrated industries to behave in a competitive manner. It also undermines the viability of cartels that the governments of mineral exporting countries might wish to create, and thus reduces the prospects for such collusive efforts. Finally, while cautioning that substitution may respond too slowly to changes in material prices to alleviate temporary shortages due to war, embargoes, cyclical demand surges, and strikes, the findings suggest that material substitution has a major contribution to make in the long-run struggle to prevent persistent shortages caused by the depletion of mineral resources.

Notes

1. Metric tons are used throughout this study.

2. The production of plastic pipe involves the melting and extruding of a plastic compound, which is composed of a plastic resin, such as polyvinyl chloride (PVC) or acrylonitrile-butadiene-styrene (ABS), plus several chemical additives. The latter include lubricants to prevent the compound from sticking to the extruder, impact modifiers to increase resilience, and flame retardants. In addition compounds made from PVC resins must contain a stabilizer to prevent decomposition and other undesirable effects caused by heat during production. Organic tin chemicals or organotins are the most commonly used stabilizers, though lead and antimony stabilizers are also available. The stages of plastic pipe production are described in more detail in part II.

3. The argument that producers keep prices below what the market will bear during economic booms to prevent substitution is not by itself very convincing. Since demand exceeds supply when price is held below the market-clearing level, producers must allocate or ration their customers. This means that once a firm has received its quota, the price of additional supplies is infinitely high. This presumably should encourage customers as much or more to search for possible alternative materials. The argument becomes more plausible if producers while restraining prices also discriminate in their rationing in favor of those customers with the greatest substitution possibilities. The extent to which producers actually engage in such discrimination, however, is unknown. Normally, it appears they allocate supplies on the basis of their customers' past purchases.

4. The demand function indicates the relationship between demand and all factors influencing demand. It is thus more encompassing than the demand curve which portrays the relationship between a commodity's price and its demand, assuming that income and other factors affecting demand remain constant at some prescribed level.

5. Reproduced with permission from J. E. Schumpeter, *Capitalism, Socialism, and Democracy*. Harper and Row, New York, 1950.

References

American Bureau of Metal Statistics. Annual. *Non-Ferrous Metal Data*. New York.

Chynoweth, A. G. 1976. "Electronic Materials: Functional Substitutions." *Science* 191, no. 4228, 725–732.

Eckes, A. E., Jr. 1979. *The United States and the Global Struggle for Minerals*. Austin: University of Texas Press.

Fisher, F. M., P. H. Cootner, and M. N. Baily. 1972. "An Econometric Model of the World Copper Industry," *Bell Journal of Economics* 3, 568–609.

Goeller, H. E., and A. M. Weinberg. 1976. "The Age of Substitutability." *Science* 191, no. 4228, 683–689.

Kopp, R. J., and V. K. Smith, 1980. "Measuring Factor Substitution with Neoclassical Models: An Experimental Evaluation." *Bell Journal of Economics* 11, 631–655.

Rosenberg, N. 1973. "Innovative Responses to Materials Shortages," *American Economic Review* 63, no. 2, 111–118.

Schumpeter, J. A. 1950. *Capitalism, Socialism and Democracy*. 3d ed. New York: Harper & Row.

Skinner, B. J. 1976. "A Second Iron Age Ahead?" *American Scientist* 64, 258–269.

Slade, M. E. 1981. "Recent Advances in Econometric Estimation of Materials Substitution." *Resources Policy* 7, 103–109.

Woods, D. W., and J. C. Burrows. 1980. *The World Aluminum-Bauxite Market*. New York: Praeger, pp. 184–230.

New Frontiers in Materials

V

Looking to the Future in Ceramics

W. David Kingery

17

In the first of four forward-looking articles, David Kingery, professor of ceramics at the University of Arizona, describes the diversity of high-tech ceramics and shows how their very existence is dependent on the prior creation of synthetic raw materials. Looking to the future, he predicts a tenfold growth in the market for high-tech ceramics over the next decade and argues that user needs will continue to drive the development of new, improved devices like sensors, insulators, cutting tools, and optical systems, which are being utilized in a wide variety of industries. Emphasizing the leverage effect that high-tech ceramics has on the profitability of many products, the author concludes that nations that lag behind in this area will lag behind in everything else. This article first appeared in Ceramics and Civilization: High-Technology Ceramics, Past, Present, and Future *(vol. 3), published by the American Ceramic Society.*

A new era for ceramics began to gain momentum about forty years ago with the burgeoning use of chemically prepared powders. Among the first was Bayer alumina, used for improved spark plug insulators required for high-compression engines demanded by automobile and aircraft developments that revolutionized our transport system. Soon came the use of barium titanate for capacitors and synthetic ferrites for magnetic materials; this was the dawn of what has come to be known as modern high-technology ceramics, or fine ceramics. In each case synthetic materials were developed to exploit the special properties of a particular composition and crystal structure to give improved performance in an existing application by replacing a less pure, less controlled mineral-based ceramic or a metal. In most cases new materials led to or required the development of new processing methods, such as the use of isostatic pressing for spark plug insulators. Once available, device engineers began using these materials for new purposes. Beginning in the 1950s, the relationship of the properties of these fine ceramics to their crystal structure and microstructure, and the development of new processes to obtain these structures, became the focus of a few university ceramic science and engineering programs; more recently this approach has become widespread.

The traditional ceramic industry had been and remains based in large part on the fact that clay and silicate mineral raw materials are widely distributed and inexpensive. Improvements in their manufacture and properties have led to worldwide high-volume production of that branch of the silicate industry producing clay, concrete, and glass products. These materials have transformed twentieth-century architecture and automobile design, two of the material aspects of our culture with which all of us have intimate daily contact and which we see everywhere around us.

In contrast, the newer fine ceramics are less visible—often invisible—high-value-added componenets of even higher-value devices and assemblies. A better functioning, more efficient multilayer ceramic packaging system for computer chips, which allows higher-speed computation, can double the efficiency of a device having a value a thousand or more times greater than the component itself. A low thermal expansion ceramic with sufficient toughness and strength for use in machine tools can enhance the precision of the entire machining operation, increasing the value of the machine tool by much more than the cost of the ceramic component. It is this leverage in the value of ceramics through their impact on device performance that explains the imperative for continued ceramic research and development.

The electronic and magnetic functions of ceramics shown in table 17.1 illustrate the exciting state of product development typical of high-technology ceramics. A principal application of ceramics is as insulators for integrated circuit substrates, for integrated circuit packaging, substrate wiring, and substrates for resistors and for interconnection electronics, along with spark plug insulators. Ceram-

Table 17.1
Electronic and magnetic applications of high-technology ceramics

Electric functions	Insulation materials (Al_2O_3, BeO, MgO)	IC circuit substrate, package, wiring substrate, resistor substrate, electronics interconnection substrate
	Ferroelectric materials ($BaTiO_3$, $SrTiO_3$)	Ceramic capacitor
	Piezoelectric materials (PZT)	Vibrator, oscillator, filter, etc.
		Transducer, ultrasonic humidifier, piezoelectric spark generator, etc.
	Semiconductor materials ($BaTiO_3$, SiC, ZnO-Bi_2O_3, V_2O_5), and other transition metal oxides	NTC thermistor (temperature sensor, temperature compensation, etc.)
		PTC thermistor (heater element, switch, temperature compensation, etc.)
		CTR thermistor (heat sensor element)
		Thick-film thermistor (infrared sensor)
		Varistor (noise elimination, surge current absorber, lighting arrester, etc.)
		Sintered CDS material (solar cell)
		SiC heater (electric furnace heater, miniature heater, etc.)
	Ion-conducting materials (β-Al_2O_3, ZrO_2)	Solid electrolyte for sodium battery
		ZrO_2 ceramics (oxygen sensor, PH meter fuel cells)
Magnetic functions	Soft ferrite	Magnetic recording head, temperature sensor, etc.
	Hard ferrite	Ferrite magnet, fractional horsepower motors, etc.

Source: Adapted from Yano Research Institute Report by G. B. Kenney and H. K. Bowen, *Am. Ceram. Soc. Bull.* 62(5):590–596 (1983).

ics compete primarily with plastics and porcelainized metals and are dominant where good heat dissipation, good resistivity, and high reliability are required. About 20 percent of the tens of billions of integrated circuits—those of highest quality—manufactured each year utilize ceramic substrates. The principal material is aluminum oxide, though beryllium oxide has been used for special high thermal conductivity application, and recently an insulating silicon carbide with additions of beryllia to give a good combination of high thermal conductivity and high electrical resistivity had been developed; new innovations such as silicon nitride, mullite, aluminum nitride, and crystallized glass are being evaluated. Another major application of ceramics has been for capacitors, about one-third of which are made with ceramic dielectrics. In response to an increasing frequency range and desires for miniaturization, ferroelectric materials are the dominant dielectrics; multilayer capacitors and grain-boundary capacitors have been developed to meet these requirements.

Piezoelectric materials are those in which the application of a mechanical force generates an electric charge, or the application of an electric field creates a change in dimension. They have traditionally been used as transducers for sonar devices, but a variety of new materials, mostly related to lead zirconium titanate (PZT), have been used for applications such as the piezoelectric spark generator for lighting stoves as well as a whole variety of new products such as vibrators, oscillators, filters, and ultrasonic humidifiers. As these materials replace more bulky components, it is expected that their production will expand significantly.

Ceramic oxide materials are mostly insulators, but include a variety of semiconductors such as barium and strontium titanate with appropriate additions—silicon carbide, vanadium oxide, iron oxide, zinc oxide, and others—which have many different applications in products such as thermistors with a controlled change in resistivity with temperature and varistors which have a controlled change of resistance with the applied electric field. Thermistor materials, mostly based on barium titanate, are used for temperature sensors, temperature-controlling heater elements, and switches. A positive-temperature coefficient thermistor can be used as a heating element which shuts off when it reaches a controlled temperature. The resistance suddenly increases so that it automatically maintains a constant temperature without complicated control circuitry. Varistor materials, such as those based on zinc oxide, show a sudden increase in electrical conductivity as a critical applied voltage is passed, making them valuable and widely used as surge current absorbants and lightning arresters. For high-temperature heating elements necessary to withstand furnace environments, resistive heaters such as silicon carbide or tungsten embedded in silicon nitride are becoming widely used.

Special materials being developed for energy storage and battery applications are solid electrolytes in which the ionic electrical conductivity is very high. Beta aluminum oxide is employed in the sodium-sulfur battery. Another ion-conducting material, zirconium oxide, has electrical transport entirely by oxygen ions, and is useful as an oxygen sensor as well as a fuel cell constituent. Oxygen sensors for control of the combustion process are used with many internal combustion engines. Finally, in many different devices, magnetic functions of both soft ferrites and permanent hard ferrite magnets are used in applications ranging from recording heads and temperature sensors to fractional horsepower motors and other permanent magnet applications.

The wide variety of functions provided to the rapidly growing electronics industry, by new high-tech ceramics having many different crystal structures and chemical compositions, illustrates a trend toward product specialization and the focusing of properties for particular applications which seems certain to continue.

Table 17.2 shows some of the nuclear, optical, and mechanical functions for which high-technology ceramics are being developed. Those in the nuclear industry are mature applications. The optical functions include nearly transparent aluminum oxide and mullite which are used for the high-pressure sodium-vapor lamps that have revolutionized outdoor lighting, and for special-purpose infrared transmission windows. Ceramics are used for laser hosts. New developments related to light memory elements, video display systems, and light shutters are at the very beginning of a growth pattern which seems certain to become a major industry.

One of the most exciting new possibilities is the application of ceramic materials for mechanical and structural products. Cutting tools of aluminum oxide, titanium carbide, and silicon nitride are now in use and constitute a significant fraction of the throwaway

Table 17.2
Nuclear, optical, and mechanical applications of high-technology ceramics

Nuclear functions	Nuclear fuels (UO_2, UO_2-PuO_2)	
	Cladding material (C, SiC, B_4C)	
	Shielding material (SiC, Al_2O_3, C, B_4C)	
Optical functions	Translucent alumina	High-pressure sodium vapor lamp
	Translucent magnesium, mullite, etc.	For a lighting tube, special-purpose lamp, infrared transmission window materials
	Translucent Y_2O_3-ThO_2 ceramics	Laser material
	PLZT ceramics	Light memory element, video display and storage system, light modulation element, light shutter, light valve
Mechanical functions	Cutting tools (Al_2O_3, TiC, TiN composite SiC whiskers in Al_2O_3, others)	Ceramic tool, sintered SBN
		Cermet tool, artificial diamond
		Nitride tool
	Wear-resistant materials (Al_2O_3, ZrO_2)	Mechanical seal, ceramic liner, bearings, thread guide, pressure sensors
	Heat-resistant materials (SiC, Al_2O_3, Si_3N_4, composite SiC fibers in glass and recrystallized glass, composite Al_2O_3 fibers in Al, composite SiC whiskers in Si_3N_4, others)	Ceramic engine, turbine blade, heat exchangers, welding burner nozzle, high-frequency combustion crucibles

Source: Adapted from Yano Research Institute Report by G. B. Kenney and H. K. Bowen, *Am. Ceram. Soc. Bull.* 62(5):590–596 (1983).

cutting tools appropriate for numerically controlled machine tools and other automatic production systems which make tool reuse less practical. Demands for higher cutting speeds and the emergence of ceramic materials, cermets, and silicon nitride-based materials are increasing at the expense of tungsten carbide. Similarly, wear-resistant materials such as aluminum oxide, zirconium oxide, and silicon carbide are becoming used as mechanical seals, liners, bearings, thread guides, pressure sensors, and so forth, as the toughness is increased and products less prone to mechanical failure are developed.

In terms of the potential markets, one of the most significant new applications is the use of ceramic materials, principally silicon nitride and silicon carbide, as heat-resistant components for ceramic engines, turbine blades, heat exchangers, and so forth. The capability for operating in air at higher temperatures than are possible for metals using materials based on inexpensive and widely available constituents is very attractive. As the operating temperature of an engine is increased, its efficiency goes up rapidly so that, in an age of uncertain, highly variable fuel costs, the combination of high-temperature capability and low weight is particularly effective. Some products, such as swirler compartments for diesel engines, are already in use. It seems likely that commercialization of silicon nitride trubine blades for automotive superchargers will occur in the coming year, an application in which the lower density of the ceramic is particularly important.

Applications of ceramics in chemical and biological functions are just beginning, but almost certainly will show high growth rates (table 17.3). Alumina ceramics and hydroxyapatite glasses are being used as artificial tooth, bone, and joint material, having an appreciable influence on medical technology. Of greater commercial importance are chemical functions such as gas sensors and humidity sensors, which are becoming the eyes and ears of computer-controlled devices of the next generation. For applications such as gas leakage alarms, automatic ventilation processes, and a host of other gas detectors, ceramic oxides such as zinc oxide, iron oxide, and tin oxide are being developed. We can anticipate the use of sensors for other applications and other materials. In chemical operations, ceramic materials such as electrodes and catalyst carriers are being widely used in processes ranging from enzyme carriers in or-

Table 17.3
Biological and chemical applications for high-technology ceramics

Biological function	Alumina ceramics implantation	Artificial tooth root, bone, and joint
	Hydroxyapatite bioglass	Gas leakage alarm, automatic ventiliation fan, hydrocarbon, fluorocarbon detectors, etc.
Chemical functions	Gas sensor (ZnO, Fe_2O_3, SnO_2)	
	Humidity sensor ($MgCr_2O_4$-TiO_2)	Cooking control element in microwave oven, etc.
	Catalyst carrier (cordierite)	Catalyst carrier for emission control
	Organic catalyst	Enzyme carrier, zeolites
	Electrodes (titanates, sulfides, borides)	Electrowinning aluminum, photochemical processes, chlorine production

Source: Adapted from Yano Research Institute Report by G. B. Kenney and H. K. Bowen, *Am. Ceram. Soc. Bull.* 62(5):590–596 (1983).

ganic catalysts to catalyst carriers in automobiles for emission control. Titanates, sulfides, and borides are being proposed for various electrochemical and photochemical processing procedures.

Thus we see that a characteristic of high-technology ceramics is their great diversity. The field comprises the application of dozens of different functions based on diverse properties employed for hundreds of separate applications. This diversity makes ceramics an interesting profession and guarantees that no individual, no laboratory, no company, can be at the forefront of all ceramic materials and processes. New opportunities are always arising.

As we have seen for aluminum oxide insulators for high-compression engines and UO_2 fuel for nuclear reactors, most of the developers of new contrived compositions have been outside the traditional industry. This has always been true; the material called Egyptian faience is important in the history of ceramics as the first of these contrived nonclay bodies. It was first developed about 6,000 years ago and by 4,000 years ago was being widely produced in the Near East and eastern Mediterranean region as a white ground for blue and green colors in imitation of lapus lazuli and turquoise which were highly valued, had strong symbolic importance, and could not be made with clay ceramics. Its development began with attempts to glaze quartz and steatite and occurred in the stonework-ing industry independent of traditional pottery making. The desire to be able to form inexpensive substitutes for expensive semiprecious stones and to be able to produce larger and more complicated objects led to the development of a process in which pure quartz pebbles were crushed and mixed with alkali and copper ore to produce a clay-free bonded ceramic having the special optical effects desired, a result that could not possibley have been achieved with the traditional clay-based ceramic technology of the time. Thus in this very first example of nonclay ceramics we see the importance of (1) a market demand for particular properties and (2) the development of new processing methods outside the clay-working industry.

More recently, during the seventeenth century, Chinese and Japanese porcelains were being imported into Europe in large amounts. At this time, along with the centralization of authority and the establishment of autocratic rule after the model of Louis XIV in France, chemistry and physics experiments were exploring new phenomena. There was new exploitation of natural resources, and new manufacturing methods were being developed with the support and encouragement of the central government. This was true in Saxony, as with other monarchies of the time. But Saxony was a special case; Augustus the Strong, ruler from 1694 to 1735, had a special interest in porcelain: his personal collection was the largest in Europe. His active patronage of a Saxon porcelain research effort—and the commitment of government to support manufacture—was a new element and essential to its success. An important participant was Count Ehrenfried Walter von Tschirnhaus, scion of an ancient Bohemian family, who had studied mathematics and physics at Leiden. After an extended scientific tour of Europe he carried out numerous experiments on materials behavior using high temperatures achieved by focusing sunlight in a solar furnace. This research was sufficiently appreciated that he was made a member of the French Royal Academy in 1683. He studied Saxon mineral resources and set up a laboratory where the principal investigator, Johann Friederich Böttger, undertook a program to develop porcelain based on von Tschirnhaus' research findings that a mixture of silica and lime melted at a lower temperature than either component alone. This was among the first experimental observations of eutectic melting applied to ceramics. Using this principle, various mixtures were tested for application to a variety of potential products such as artificial gems, stoneware, and white porcelain.

An essential constituent to achieving success was the development of furnaces reaching the high temperatures required (above 1,350°C). This level was much higher than had ever been achieved in commerce. With the support of royal patronage, the concept of partially fluxing a white clay with lime, the experimental testing of suitable compositions with von Tschirnhaus' burning lenses, and achieving the high kiln temperature necessary, a new European porcelain was successfully developed with the white body, translucency, and sonorous ring of the Chinese wares. Over the next several years its composition was modified such that Meissen porcelain was fully competitive with the Chinese and Japanese products. This first European success in ceramic science and engineering not only created a new product but also transformed our views of how to go about ceramic research and development. It still stands as a model of successful ceramic research.

The development of European porcelain owed much to experiments with von Tschirnhaus' state-of-the-art solar furnace. Similarly, in our time new science and innovative ceramic developments required the invention and use of new methods for probing into structure, microstructure, and microchemistry. New precision and new limits of resolution will increasingly lead to innovations derived from molecular and structural engineering rather than empirical methods.

If we turn to the future of ceramics, there will certainly be new inventions and discoveries which we cannot anticipate but which, on the basis of past experience, we may be sure will occur. There will also be markets, such as those for synthetic lapus lazuli and turquoise in ancient Egypt and white sonorous porcelain at the beginning of the eighteenth century, for which innovations are actively sought. To anticipate the general future of ceramics, it is probably best to consider the overall direction of changes in device development and the opportunities for existing and new ceramics associated with trends in our material culture that are just beginning, or are even well underway, but have not fully affected ceramic production.

Low-cost, widely available, relatively low-energy-expenditure materials such as glass, clay products, and concrete will continue to be employed in increasing amounts in the future. We may anticipate that improved strength and toughness obtained by residual stress control, the addition of additives to use these materials as a composite base, and particularly more sophisticated use of chemically bonded ceramics such as the macro defect-free cement developed by ICI and the use of new manufacturing methods such as sol-gel techniques to form particulates, fibers, and surface coatings will broaden and extend their application.

Another area for ceramic applications will be associated with *automatic feedback-controlled processing,* manufacturing, and robotics. Developments that seem certain will require sensors of all kinds: sensors for motion (the sensitive gyros for missile and spacecraft control require fine ceramic components), displacement, sound, touch, gas, liquid, pressure, radiation, magnetism, electricity, temperature, and so forth. Many of these sensing systems will depend on ceramics for critical parts, as is already true for automobile combustion control devices. These systems will also be used for all sorts of ceramic processing and will permit reliable commercial use of metastable materials such as epitaxially grown diamond surfaces, the use of ultrahigh-temperature plasma processing, rapid solidification, reproducible thin-film ceramics, and other mehods of shaping and controlling the structure of ceramics that are now possible in the laboratory but difficult to control.

Composite materials with controlled structures on a nanostructure level achieved by vapor deposition, rapid solidification, and plasma processes will be used in various devices. There will be composite constructions using fibers for strengthening, but also integration of passive electronic components into substrate packages, more extensive application of effective metal-ceramic bonding, better control of grain-boundary and surface properties, and even the lamination of electronic circuitry and display systems into an integrated whole. These developments should give new performance capabilities that will permit innovative designs and more effective devices.

The nature and extent of optical information handling, manipulation, and transfer are just starting to be developed. *Optical systems* utilizing lasers, optical information storage and sensing, laser-read disks, light pipes, control of photon energy levels, and transmission are all feasible and are being developed and used to an increasing extent. As photonic innovations combine with, and perhaps even overtake, electronic innovations, new and increasing markets of polycrystalline ceramics, glasses, and crystals will develop as device components.

The development of carbides, nitrides, and sulfides has reached commercial proportions, but the range of compositions, processing alloys, and applications of these untraditional, *nonoxide ceramics* is still growing rapidly. The first commercial applications have been for abrasives and cutting tools, but we can be confident that a whole family of materials will develop for structural properties, refractories, metals handling, insulators, and electronic applications. Alloying of silicon nitride with alumina to form the Sialons, which are a whole series of solid solutions with Al partially replacing Si and O partially replacing N, is but one example of further development sure to come.

Along with these new applications and new ceramic materials we shall surely see an expansion of *novel methods of powder preparation* involving monodisperse powders and sol-gel methods; an increased control over organic-inorganic interactions; greater application of sophisticated *feedback-controlled automatic processing;* an increased range of temperature and pressure for ceramic fabrication; more extensive use of vapor-phase reactions and thin-film constructions; controlled crystallization from glass, vapor, and liquid; and formation of metastable compounds in a controlled way such that the range and extent of ceramic processing will hardly be recognizable to a ceramic engineer of traditional training.

Summary

Based on the examples given, and many more that could be cited, it is estimated that the sum of old and new requirements for precision high-technology fine ceramics will lead to a ten-fold increase in the total world market for these materials in the next decade. This will still be an insignificant part of world commerce; its importance lies in the fact that it will mostly consist of critical components of devices valued at a hundred or thousand times that figure, or more. In any evaluation of the future of high-tech ceramics, this leverage in value must be taken into account. Companies or nations that do not remain at the forefront of ceramic developments will find it impossible to remain at the forefront of the device developments that will play a dominant role in the future of our materials culture.

In summary, the driving force for the development of new ceramic innovations is the leverage derived from their criticality to the effective functioning of a whole range of devices to which our world's culture is dedicated. We can foresee powerful new requirements related to chemically bonded ceramics, automatic feedback control systems, composite materials, optical systems, nonoxide ceramics, and chemically based automatic-feedback-controlled processing and powder preparation methods for the formation of sophisticated new materials, including many new metastable materials. The variety of these device developments and new ceramic requirements is such that an adequate return for investment in ceramic innovation seems assured.

The exciting variety of innovative high-tech ceramic processes and products now being developed is a new element that is transforming the very nature and structure of ceramic science, technology, education, and manufacturing. As a result, though individual companies may be well advised to focus their efforts on a narrow range of processes and products, successful national programs will require a broad spectrum of excellence and innovation, including core programs to develop the underlying science necessary for taking advantage of new opportunities as they arise.

Note

I am indebted to the following for comments and suggestions: R. E. Charles, A. R. Cooper, T. L. Francis, M. Goto, K. H. Jack, P. E. D. Morgan, R. E. Newnham, R. Roy, B. Schwartz, and R. R. Tummula. Research was supported by the U.S. Department of Energy under Contract No. DE-AC02-76ER02390.

Materials Processing in Space

Peter Marsh

18

The zero gravity of space makes possible the creation of special alloys, crystals, and perfect spheres that are literally out of this world. Chemical reactions and biological processes take on a different form and cheap solar energy is more abundant above the earth's atmosphere. All this suggests that "space factories" will soon get into orbit, says journalist Peter Marsh, and in the next century we will be buying goods stamped "Made in Space." Since the editor of this volume wrote a similar piece on space processing back in 1978 (see bibliography), I feel it incumbent upon me to warn that too much hype has accompanied the minor moves toward factories in the firmament—but it's nevertheless an exciting prospect.

Introduction

Products made above the atmosphere will go on sale in the early years of the next century. They will be manufactured in large orbiting complexes which capitalize on the special characteristics of space—low gravity, vacuum, relative freedom from impurities, and abundant energy easily obtainable by turning the sun's rays into electricity with solar cells. After the initial razzmatazz, people will think it no more unusual to buy goods stamped "Made in Space" than they do now to buy those produced in the Philippines or Taiwan.

From *The Space Business* (Penguin Books, London, 1985), pp. 130–137.

The trends that will bring about industrial parks in space are already well established. The Soviet Union is ahead in this area. It has put into orbit seven small space stations, the last two of some sophistication. Such a station can dock with other spacecraft to produce an orbiting complex up to 50 meters long, an outpost in the heavens the length of Nelson's Column. The U.S.S.R.'s Salyut space bases can hold, at a squeeze, five cosmonauts at a time. People have lived and worked in the stations for up to eight months in one stretch. The Soviet Union is keen to develop new, much bigger orbital complexes and site in them small factories that produce goods difficult to make on earth.

Meanwhile the United States plans to build in the 1990s a space station that would emulate the Soviet efforts. The $8,000 million base, in which people would live permanently, would be built in modules taken into orbit by the space shuttle. Western Europe, under the auspices of the European Space Agency, may decide to help. The work would follow from the agency's development of Spacelab, the world's first reusable orbiting laboratory, which made its maiden mission, on board a U.S. space shuttle, at the end of 1983. Spacelab promises to provide a "shirt-sleeves" environment in which scientists can do experiments that may lead to new industries above the atmosphere in the next century. The laboratory, however, stays in space for no more than ten days or so at a time inside a shuttle's cargo bay, whereas a space station, with its own propulsion system and life-support equipment, may have a life of several years.

Space stations are useful to industrialists and scientists primarily because of the exotic conditions that space has to offer. In a spacecraft in a state of free fall in orbit around the earth, strange things happen—or, to be more precise, don't happen. The gravitational force experienced by people or objects inside the spacecraft is between one-thousandth and one-millionth of the figure on earth. As a result many of the physical effects that we have grown used to on our planet are simply nonexistent.

For example, substances will mix far more evenly because their weight is no longer relevant in determining how they diffuse when shaken up with other liquids or solids. Thus in space you would have relatively little trouble ensuring that, for instance, sugar is stirred into your tea. A bartender mixing drinks in the cocktail lounge of a space station could do his job with a lot less effort than on earth.

Other familiar events would no longer occur, for example the sedimentation of particles in the oil in the gear box of an engine. Instead, the particles would be distributed uniformly throughout the liquid. In a similar way it would be supremely simple to make fruit jellies in which the lumps of pineapple, pear, and so on, are evenly dispersed. They would not follow the annoying trend in earth-bound kitchens and end up on the bottom of the container.

As a result of the absence of what is called hydrostatic pressure, viscous materials would not tend to collapse under their own weight. Hence slowly solidifying castings would be no wider at the base than at the top. And anyone who lights a candle in space would find that the steady trickling-down the stem of oil, a source of mild concern to any earth being who has suffered a power cut, would be a legacy left behind on the launch pad.

New Products and Processes

In manufacturing terms, the near absence of gravity holds out remarkable prospects. It may be possible in space stations to make a bewildering array of materials difficult, or downright impossible, to produce on earth. Researchers in the United States, U.S.S.R., Japan, and Western Europe have drawn up a list of the materials in this category. They include alloys, superconductors, new types of glass, semiconductors, detectors for infrared light and nuclear radiation, catalysts, enzymes, polymers, superplastics, magnetic materials, corrosion-proof coatings, solar cells, and high-strength composites or special steels.

The specific processes that are important, according to the researchers, include crystal growth, casting of metals, containerless processing, chemical reactions (such as polymerization), and biological separation through mechanisms such as electrophoresis, in which materials are isolated from mixtures according to the different electric charges on their constituent particles.

In addition to these areas of technology, any process that requires a lot of energy may be suitable for space factories on account of the abundance of energy available from the sun. By the time the radiation from this source reaches ground level, much of its power is absorbed by the molecules in the atmosphere. Above the mantle of gas that separates us from the rest of the universe, it would be an easy matter, using solar cells, to turn the full power of the sun into electricity to drive industrial processes. Among the energy-intensive operations that scientists think may take place in space factories are metals smelting, welding, and cutting with electron beams, joining up pieces of metal by brazing, and the sintering of powders to produce solids.

Crystals and Solid Objects

Of all the disciplines that appear promising for space industrialization, crystal growth holds out perhaps the most exciting possibilities. On earth scientists generally make crystals either by separating them from a solution or by techniques of vapor deposition. In the latter a solid lattice is "grown" by depositing gas onto a cool surface. Gravity interferes with both methods. Because the different parts of the crystal have a specific weight, the crystal can bend or sag, in the same way that a building suffers from stresses arising from different loads. Even the minute deformations that result from these forces may be enough to cause cracks or other structural defects that can change the performance of the crystal.

In the absence of gravity, however, it should be possible to produce crystals that are near perfect. Thus space factories may turn out semiconductors, such as gallium arsenide, that have much better electronic properties than anything obtainable on earth. Alternatively, semiconductor crystals may be grown with ease to an enormous size. In conventional factories electronic circuits made from silicon are normally produced from circular "wafers" no more than 10 cm in diameter because of the difficulty of producing pure silicon in larger dimensions.

Other crystal products could have particularly good magnetic or optical properties; materials such as compounds of mercury, tellurium, and cadmium that are useful as infrared detectors could be made. Scientists also believe that defect-free crystalline substances made from gallium and bismuth, for example, could turn out to be good superconductors.

The company that has shown, in public at least, the strongest enthusiasm for growing crystals in space is Microgravity Research Associates, based in Florida. The company has signed an agreement with NASA to cooperate in experiments on shuttle flights. Bob Pace, vice-president of Microgravity Research Associates, says he hopes, by 1990, to produce 40 kg a year of gallium arsenide made in space. The material would be sold to electronics companies requiring particularly pure specimens of the substance for ultraefficient microchips. Companies in this position would have to pay a premium for material manufactured in this way. Pace plans to sell his gallium arsenide at $1 million a kilogram, roughly one hundred times the 1984 price of gold. Gallium arsenide made by conventional terrestrial means costs $20,000 a kilogram. By 1997, says Pace, annual production would increase to 1,200 kg. By the end of the 1990s the figure would reach 3,000 kg, at which point the price would have dropped to $500,000 a kilogram.

Solid objects cast from molten metals or plastics in earth-bound forges often suffer from imperfections. These are due to particles migrating through the fluid as a result of convection currents induced by gravity. Such currents are a particular problem in composite materials where substances in a liquid form are solidified around fibers, of carbon or plastic, for example. Alternatively, liquid metals can be made to harden around particles of, for instance, tiny spheres of aluminum oxide (alumina). In both cases the extra materials lend strength to the finished product, but the effect can be ruined if the structure contains imperfections.

The presence of convection currents due to gravity may mean that some substances do not mix at all, or do so only under special conditions. This may preclude scientists from making castings or alloys from materials that, in theory, would produce unusual properties, for example, in terms of strength, when combined. One example is the difficulty of making compounds in earth-bound laboratories from gold and germanium; lead, tin, and antimony; or aluminum and antimony. In space, researchers would have virtually carte blanche to try their hand at mixing substances irrespective of what can or cannot be done in terrestrial workshops. Hence a new range of alloys and compound materials could become possible, with applications in many industries from consumer goods to vehicles to construction.

John Deere, the U.S. manufacturer of farm equipment, is another company that intends space experiments. It is interested in testing how carbon and iron mix in zero gravity. The interreaction is a crucial one in ironmaking, and, with gravity "turned off," the company may be able to gain a new understanding of the mechanism which would help in its operation of terrestrial iron foundries.

Liquids and Electrophoresis

In the third broad area of low-gravity techniques in which researchers are interested, factory workers should be able to handle liquids without containers. In space, liquids have no weight and so are held together as spheres by surface tension. One result is that processes to make highly sensitive chemicals that can be harmed by impurities will make ideal candidates for manufacturing plants in the heavens. Chemicals in this category include certain kinds of catalysts. Such reactions may be difficult in ordinary chemical plants because the containers in which they take place introduce contaminants. In space the liquids in the reaction would float around surrounded by nothingness. They would be held in place long enough for chemical or physical reactions to occur, by electrical or magnetic fields.

A specific application for containerless processing is the manufacture of glass. The material present in many millions of window frames around the world is, in reality, rather special. Glass is best described as a supercooled liquid. The atoms in it are frozen in position very quickly, before they have time to arrange themselves in the regular lattice-type structure of a solid. This is the basic explanation for the specific optical and structural properties of glass. However, although the manufacture of some types of glass, the sort made from silicon, for example, is fairly easy, glass made of other substances may be very hard to produce. With these materials the sudden cooling from the liquid phase is difficult to achieve. One way around the problem could be to cool the liquid in space without a container. On earth the receptacles in which the cooling takes place inhibit the ability to remove heat quickly, as they act as a point of nucleation around which the liquid can start to solidify, nullifying the whole idea of producing glass. Containerless processing could, then, lead to new kinds of glass with unusual properties, for example, for use as lasers or to form a conduit for electrical signals in strands of optical fiber.

Polymer chemists are also interested in experiments in the heavens. With no gravitation to worry about, researchers may be able to build long chains or molecules of a kind that, on earth, would be too unwieldy or which would snap under their own weight. In particular, scientists think they could produce new kinds of polymers known as latex spheres, which are tiny balls of polyvinyl latex. Latex spheres formed in the same reaction are of near-identical dimensions—of the order of a few micrometers—and so can be used as a calibration standard in microscopy. In space researchers think they will be able to vary the dimensions of these balls rather more easily than on earth, for example, to build up bigger ones than are possible in conventional laboratories.

The final area that is, according to scientists, ripe for exploitation in orbiting complexes concerns a way of separating chemicals from mixtures in which they are naturally present. Electrophoresis is a technique tried and tested on earth. An electric field channels substances in the mixture into different compartments depending on their electric charge. With this technique workers can produce high concentrations of specific chemicals from, for example, human or animal secretions in which the substances are naturally present. For many important substances, though, the process is painfully slow, and as a result it is often uneconomic for ordinary factories. In space, however, the lack of gravitational force speeds up the technique possibly by several hundred times, depending on the chemicals involved. Substances that could be produced in spaceborne electrophoresis laboratories include many materials useful in medicine. Examples are beta cells, useful in treating diabetes; interferon, a drug that may be able to treat cancer among other diseases; growth hormones, for stimulating bone development in children; urokinase, which can prevent blood clots; erythropoietin, for kidney failure; and epidermal growth factor, used to heal burns.

Space electrophoresis is probably closer to commercialization than any of the other ideas for orbiting factories. McDonnell Douglas and Ortho Pharmaceuticals (a division of Johnson and Johnson) say they want to put into orbit, by 1988, a prototype electrophoresis facility. This would probably be on an unmanned platform and would be lifted into space, perhaps 300 km above the earth, by the space shuttle. The two companies have already tried several experiments in space with small-scale electrophoresis hardware, the first of which was on the space shuttle's fourth flight in June 1982.

Other North American firms that have expressed interest in capitalizing on specific aspects of manufacturing above the atmosphere are TRW (production of turbine blades); GTI (castings and metallurgical furnaces); Du Pont (catalysts); Battelle (biotechnology, for example the growth of collagens); Westech Systems (large silicon crystals); Boeing (pharmaceutical products); General Electric (latex spheres); 3M (new materials); Kodak (fluid physics and production of new chemicals for film processing); and Inco, a Canadian nickel producer (electroplating techniques). These firms have either signed agreements with NASA to operate experiments in space over the next few years or are doing research in terrestrial laboratories to investigate the prospects for space manufacturing.

The Soviets

The industrial interest in making things in space is probably as great in the Soviet Union, which is behind the West in many conventional manufacturing activities and so has still more reason than the United States to start processing operations above the atmosphere. With the know-how and the products so obtained, the U.S.S.R. could compensate for, at least partially, its deficiencies in other spheres of industry. The Soviets have undoubtedly done more practical work on this subject than the United States in their succession of space stations. Concrete details of what the U.S.S.R. is planning, however, seldom leak out.

Most of the important Soviet activity in space experiments has been on Salyut 6 and Salyut 7, launched in 1977 and 1982. The United States has also made strides in materials processing in studies in their first (and so far only) space station, Skylab, to which astronauts were sent in 1973. Spacefarers from both countries have also carried out isolated experiments in short trips in space capsules. Such studies in weightless conditions have supplemented the experiments that, for twenty years, have taken place with sounding rockets and aeroplanes. When such a vehicle accelerates to the top of a ballistic trajectory and then suddenly dives, the physical forces acting on the craft are such that a brief period of weightlessness is obtained. But such spells last only a few minutes at the most and are not much use to researchers who are serious about convincing people of the merits of spaceborne processing. It is safe to assume that we will see many more advances in this area of technology in the new, more sophisticated, orbiting platforms that should be launched over the next few years. For a couple of decades the work will stay firmly in the research phase, leading to fully fledged space factories some time after the year 2000.

A permanent platform in space has attractions not just to the hard-nosed industrialist. Scientists engaged in the purest, most non-commercial forms of research are interested in space laboratories. They think that with gravity "switched off" they may obtain fresh discoveries about a great range of phenomena.

Ever since men and women started systematically to study their surroundings, so the argument runs, they have had to contend with gravity. It is an ever-present force that continually "modulates" scientific results. If, over the past couple of hundred years, scientists had been able to alter gravity as easily as they can modify conditions of temperature and pressure, then our understanding of the physical and biological world would probably be enormously improved.

In the near absence of gravitational attraction in space laboratories, scientists think they will be able to shed new light on many of the basic properties that affect materials, how substances conduct electricity or how their molecules link together, for example. Biologists are also interested in low-gravity work. In workshops above the atmosphere they may be able to investigate how human cells grow and evolve, or the mechanisms that affect bodily stress and heart disease. Experiments in space could produce an understanding of the nature of the gravity-sensing organs of the body, for example, the vestibular apparatus of the inner ear. Experiments may also be devised to examine the basic questions of how people and animals have evolved to cope with gravity, such as the way bone structures grow.

Seabed Materials

James M. Broadus

19

From space to "aquaspace." Vast material resources lie on the seabed. Some, such as hydrocarbons and tin, are already exploited. But others, like polymetallic sulphides, lie untouched. Scientific advance has increased our knowledge of seabed materials and is creating the technology for recovering them. Although commercial realities dictate that most potential reserves of seabed materials will not be exploited in the near future, James M. Broadus is optimistic about the prospects for more R&D. The author is the director of the Marine Policy Center at the Woods Hole Oceanographic Institution.

The optimistic response to recurring concerns about economically debilitating secular exhaustion of material supplies emphasizes the ability to substitute among materials and to extend available resources through technological innovation and exploitation of previously uneconomic sources [1, 2]. Although mining of materials directly from the seabed occurred as early as the sixteenth century and materials of seabed origin have been exploited for millennia, only in recent decades has the inventory of promising unconventional sources been substantially augmented by seabed materials [3]. Some of these seabed sources, such as hydrocarbons and tin, are already making large contributions to the world's supply of materials. Others, such as the massive polymetallic sulfides in deep-sea hydrothermal vents, probably will not be exploited for many decades or even centuries.

From *Science,* vol. 235, 20 February 1987, pp. 853–860.

This article summarizes the current position of seabed materials on their long and uncertain way to market. The perspective adopted here is based on the modern economic interpretation of the social process by which natural resources are identified and called into use [4]. In this view, naturally occurring materials become resources only when they have been brought within reach of practical exploitation, typically through the costly application of human effort and ingenuity [5]. Materials are valued for the attributes and services they can provide, and the demand for these factors is derived from the demand for the things they are used to make. The "best" (easiest to find, develop, extract, and use) resources tend to be used first, with the use of lower quality, more costly resources postponed until the better ones have been depleted. Increasing reliance on more costly resources tends to raise the price of the extracted materials, prompting substitution, conservation, recycling, and exploration. New resources are thus created through a combination of (1) increased attractiveness of previously uneconomic sources, (2) discoveries of previously unknown sources, and (3) increased accessibility through technological advances.

Because of the unconventional, difficult, and clearly distinct nature of their physical setting, seabed materials provide an attractive case study of this resource-creation process in action. Questions for the seabed materials case are how abundant are better sources and how rapidly are these better sources being depleted. A related question is how quickly are potential seabed sources closing the quality gap through technological gains (including understanding and know-how) and through identification of other advantages. Consideration of the latter question involves some attention to broader issues of public policy, international organization, and long-term industrial strategy.

Seabed Deposits and Material Commodities

Most material prospects from the seabed are minerals, in the broad sense of that term, although living precious coral is harvested from seabed habitats for its $50-million-per-year market. Of the sixty-five material resources they examined, Goeller and Zucker listed fifteen that can be vastly extended with oceanic supplies, but eleven of these are seawater extracts [1]. Of seabed materials, only four contained in seafloor nodules are listed by Goeller and Zucker as extending resources (reducing projected depletion by the year 2100 from 120 to 18 percent for manganese, 152 to 35 percent for nickel, 150 to 36 percent for cobalt, and a moderate reduction in copper depletion) (table 2, in [1]). Seabed deposits eventually may provide additional resources for at least twenty-two other materials (table 19.1), although the magnitude and timing of these additions remain open to question and study.

Offshore oil and gas resources are in a class by themselves among seabed materials, making an important contribution to the world's economy. Even at recently lowered prices, [6], offshore hydrocarbons generate about $80 billion of annual revenues; this sum is comparable to the total world markets for all the other materials with potential seabed sources combined (about $90 billion). Offshore deposits within continental margins account for nearly a third of the world's estimated oil and gas resources, and evidence suggests that additional resources may someday be found on the deep-ocean floor [7, 8].

The world's nonfuel seabed mineral prospects (figure 19.1) can be classified into (1) minerals that may be obtained from deposits in relatively shallow coastal waters (less than 200 m), including aggregates such as sand and gravel, shell, calcium carbonate, phosphorites, placer deposits of heavy minerals or gems, barite, and subseabed sulfur deposits; and (2) deep-sea deposits including the abyssal (3,500 to 5,500 m) manganese nodules [3, 9, 10], the richest deposits of which have been found in the 13-million-km^2 Clarion-Clipperton zone of the Pacific, cobalt-enriched crusts on the flanks of seamounts [11], marine polymetallic sulfides (MPS) precipitated around hydrothermal vents at crustal spreading centers [12], and perhaps certain deposits of marine phosphorites [13].

None of the deep-sea prospects are anywhere near production, but some minerals have been produced from nearshore deposits for many years and now generate revenues of over $600 million per year. Their contribution to the supply of minerals is small, however, compared with the more conventional onshore sources of the same commodities. Only offshore tin, dredged for nearly a century from shallow waters in the "tin belt" of Southeast Asia, supplies more

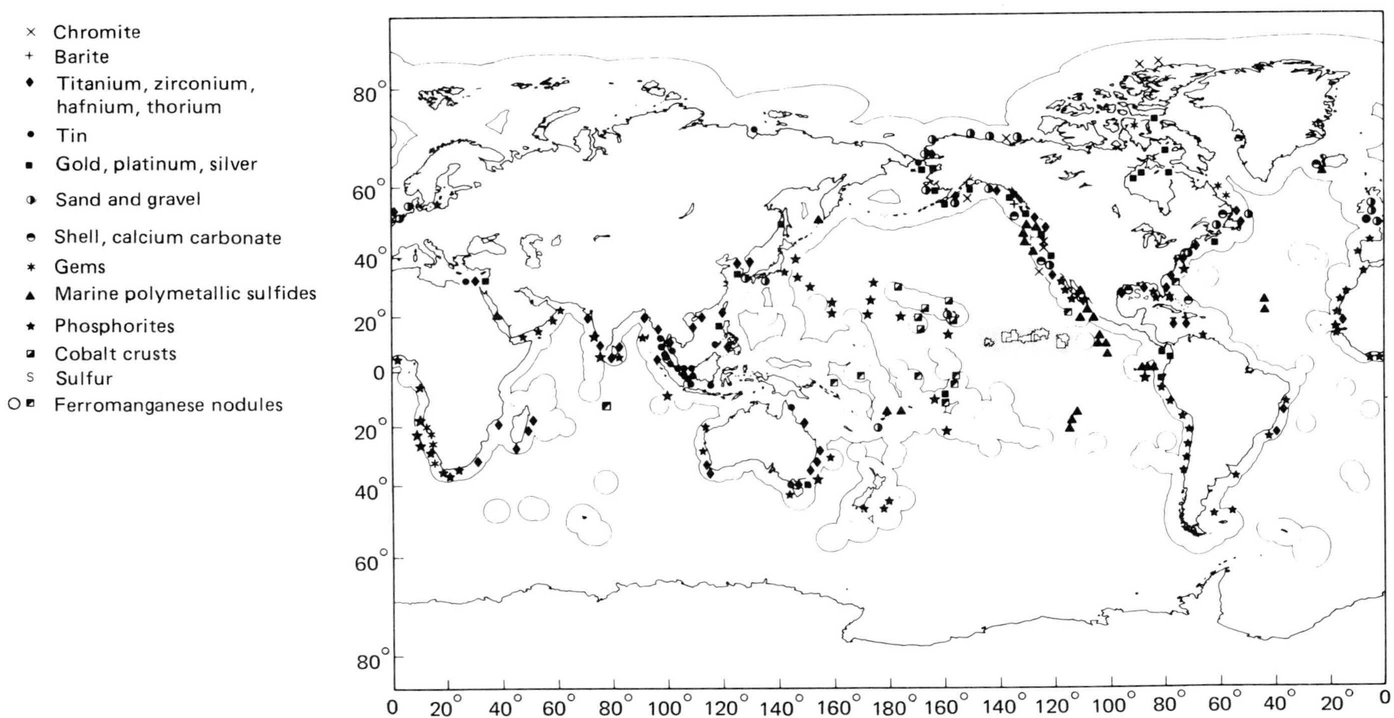

Figure 19.1 Approximate locations of major identified nonfuel seabed deposits, showing the extent of existing or potential 200-nautical-mile Exclusive Economic Zones and publicly disclosed license areas for seabed mining activities under domestic laws in the nodule-rich Clarion-Clipperton zone (stippled area).

Table 19.1
Seabed materials in world perspective (abbreviation: MT, metric tons)

Seabed deposits	Material commodity	Seabed production (10^3 MT)	World mine production (10^3 MT)	Estimated average price ($ per MT)	Seabed revenues[a] ($ in millions)	World revenues[b] ($ in millions)
Hydrocarbons[g]	Crude oil	788,834	2,788,913	70	55,218	195,224
	Natural gas	246,670	1,296,405	95	23,434	123,158
Sand and gravel	Sand and gravel	112,300	7,620,480	3	334	22,861
	Industrial sand		181,440	14		2,540
Shell	Calcium carbonate	16,667	1,666,667	6	100	10,000
Sulfur	Sulfur	381	54,000	105	40	5,670
Barite	Barite		5,652	31		175
Phosphorite	Phosphate rock		159,000	24		3,816
Mineral placers	Tin	28	201	6,614	185	1,329
	Rutile		356	364		130
	Ilmenite		4,187	49		205
	Titanium[i]		90	12,236		1,101
	Zirconium		709	182		129
	Hafnium		<<1	231,483		17
	Yttrium		<1	35,020		14
	Thorium		2	35,850		72
	Chromite		9,616	42		404
	Gold		1	10,600,000		10,600
	Silver		12	206,667		2,480
	Platinum		<<1	9,000,000		1,980
Nodules and crusts	Platinum[j]					
	Cobalt		32	25,353		811
	Nickel		745	5,026		3,744
	Manganese		23,406	141		3,300
	Copper		7,805	1,475		11,512
Massive sulfides	Copper[k]					
	Zinc		6,560	893		5,858
	Lead		3,350	419		1,404

a. Seabed production times estimated average price.
b. World mine production times estimated average price.
c. Seabed revenues times 100, divided by world revenues.
d. Seabed reported potential resources times 100, divided by world onshore resources.
e. World onshore resources divided by world mine production.
f. From [32], based on low growth case for developing economies.

Seabed share of world revenues[c] (%)	Seabed reported potential resources (10^3 MT)	World onshore resources (10^3 MT)	Seabed comparison to world resources[d] (%)	"Resource life" index[e] (years)	Projected onshore depletion by year 2030[f] (%)
28	>61,429,000	181,857,000	34	65	185
19	>60,000,000	228,214,000	26	176	45
1	665,778,000	Very large	Small	Long	
	Large	Very large	Small	Long	
1	90,000,000	Very large	Small	Long	
<1	27,125 [h]	5,000,000	<1	93	120
	2,087 [h]	453,600	<1	80	
	7,939,000	129,500,000	6	814	12
14	2,500	34,500	7	172	105
	13,060	181,440	7	510	
	230,500	907,200	25	217	
					40
	29,040	54,432	53	77	
	290	544	53	7,452	
	{3,450}[h]	172		430	
	5,168		2,584		
	30,158 [h]	32,659,200	<1	3,396	1
	<1 [h]	72	<1	72	443
	743}[h]		62	295	
	<<1 [h]	99	<<1	446	13
	2–3		2–3		
	6,000–24,000	10,886	55–220	340	
	35,000–131,000	129,730	27–101	174	77
	706,000–2,600,000	10,886,400	6–24	465	17
	29,000–108,000	1,600,000	2–7	205	86
	5,000–216,000		<1–14		
	11,000–518,000	1,800,000	<1–29	274	47
		1,400,000		418	46

f. From [32], based on low growth case for developing economies.
g. Hydrocarbons in metric tons of oil equivalent.
h. Seabed estimate for the United States only; the number in braces (3,450) is for U.S. seabed monazite deposits containing yttrium and thorium.
i. Titanium resources are included in rutile and ilmenite resources.
j. See numbers directly above in mineral placers for platinum.
k. See numbers directly above in nodules and crusts for copper.

than 1 percent of its world market. Offshore deposits now represent about half the tin resources for Indonesia and Thailand [13] and a growing proportion for Malaysia. Sand, gravel, and shell, amounting to nearly half the value of the world's offshore nonfuel mineral production, are pumped or dredged for construction uses, primarily in Japan and Western Europe. Practical recovery depths are <50 m, and high transport costs limit these construction materials to distribution in local market areas, with great variety in price among markets geographically. Calcium carbonate, whose artisanal mining from nearshore reefs has led to coastal erosion problems in several developing nations, accounts for almost a fifth of the total value of global offshore production. Offshore sulfur production is now limited to a single operation off the coast of Louisiana in the United States, and recovery of waste sulfur from pollution control equipment may replace this source entirely by the year 2000. Altogether, annual revenues produced offshore from the various nearshore sources of nonfuel minerals are <1 percent of the annual value of offshore oil and gas production.

In view of their relatively trivial economic contribution, nonfuel seabed minerals have generated a surprising amount of interest and activity. Nonfuel seabed minerals were a major stated cause in the refusals of the United States, Great Britain, and West Germany to sign the Law of the Sea (LOS) Treaty, were the central source of contention throughout those negotiations, and appear to have been a rationale in the United States for the creation of a 200-nautical mile Exclusive Economic Zone (EEZ) in 1983. Most of this interest centered on the potential of deep-sea deposits. No sooner, it seems, are the economic limitations of one seabed mineral prospect recognized than another prospect is brought forth to be touted. Partly through this pattern of shifting attention, much has been learned about seabed resource potential in the past two decades. Several overviews have appeared within the last five years [3, 13–16]; slightly older but still useful sources are also available [8, 17, 18].

As population and economic growth lead to greater demand for materials, seabed deposits will tend to move closer to expanded exploitation. Changing patterns of materials consumption, environmental restrictions, or higher value alternative uses for resource-bearing lands could work to similar effect. However, the seabed potential resources will have to push their way past both conventional and more speculative rival sources onshore or wait their turn until those sources have been consumed.

Resource Estimates, Scarcity, and Depletion

The best advice with which to identify the position of various seabed materials as emergent resources is a long-run supply function for each material. These functions would describe the amounts that could be obtained economically at different levels of incremental or unit cost, given a consistent set of assumptions about prospective changes in technology and costs over time. Ideally, these supply functions would include a description of an estimated probability distribution for each point on the curve. The relative position of each potential source of a material, both onshore and seabed, could then be compared concisely by reference to its "availability" or expected share of output at each cost level. For most materials one would expect a period of exclusive production from successively costlier onshore deposits until a cost level is reached at which the least-cost seabed deposits, as with oil, gas, and tin, join into total production. Beyond that point, the division of total output between onshore and seabed sources would depend on their respective available quantities for each increment of elevated cost. Reliable estimates of such functions have yet to be developed [19]. Combined geological, engineering, and economic estimates of potential material flows organized in a way similar to this are beginning to emerge for some materials on the seabed list, but none has yet been attempted for any seabed material as such [20].

Therefore we are limited to imperfect stock estimates, which cannot convey the flow of new resources into the line of supply over time. However, it is still useful to compare the reported quantities of seabed materials that have been proposed as potential resources to the stock estimates of existing conventional resources (table 19.1). Consistent estimates are available for onshore resources, and "Identified Resources" are reported in nearly every case here [21].

In the absence of further exploration and sampling, however, virtually all quantity estimates for the seabed materials must be taken as speculative [22]. The basis for these estimates differs across commodity and deposit type as well as by investigator and region.

A variety of geostatistical approaches to mineral resource estimation have been developed [23], and some of the most powerful have been used to estimate remaining resources of oil and gas. Even with a rich body of production experience and extensive offshore search and sampling for these materials, uncertainty about estimates remains high. This was manifest in 1984 when the U.S. Department of the Interior revised downward by nearly fifty percent a 1981 estimate of undiscovered outer continental shelf oil and gas resources [24]. This revision was equivalent to a reduction of perhaps hundreds of billions of dollars in the estimated asset value of federal oil and gas holdings [25].

Quantity estimates for nonfuel seabed materials are typically based on more limited data and less sophisticated techniques. Geologic information and scientific reasoning are used liberally to condition the estimates because of the typically wide scatter of sample data and the geologic controls on distribution. Resource estimates for sand and gravel tend to be based on a volume obtained by multiplying a defined area by the average thickness of its deposit. In most cases cutoff depths and distances from shore are not reported. Estimates of manganese nodule resources are based on apparent areal concentration within study areas defined by the average grade of samples [9]. A large body of global information on nodules is available from years of oceanographic sampling, and more densely spaced data have been obtained by commercial exploration efforts.

"Resource" estimates arrived at in a similar manner have been reported for cobalt crusts in certain areas of the central Pacific [26], but these are based largely on limited sampling or hypothetical grade-concentration combinations. For MPS deposits, geological and geochemical inference provide the only basis for estimates of potential quantities in-place because of the limited number of observations (fewer than fifty sites have been sampled to date) and the absence of data on deposit thickness. The Red Sea brines are the most thoroughly investigated, and Mustafa et al. [27] report resources of 2.4 million metric tons of zinc, 0.6 million metric tons of copper, and 5.2 thousand metric tons each of silver and cobalt. I have reported [28] speculative extrapolations of materials in vent deposits on the mid-ocean ridge, on the basis of Mottl's geochemical model of the depositional process of these materials.

With few exceptions the seabed estimates summarized in table

19.1 equal surprisingly large proportions of onshore resources. However, the seabed estimates have not been determined on a consistent basis and cannot be read as equivalent to the estimates of identified resources reported for the onshore materials [29]. All estimates are taken from published reports, are limited to occurrences with concentrations comparable to ores currently worked in onshore deposits, and are presented only as a rough catalog of potential seabed material resources. With further geological exploration (or even broader search of the literature) the reported amounts will expand, although not necessarily their resource potential.

At prevailing rates of consumption, current onshore resources would not be consumed for many years. If we assume broadly that identified onshore resources for each commodity are of higher quality and will be used before any potential seabed sources, this "resource life index" may suggest a waiting time before exploitations of the seabed materials. The life indexes for materials in the deep seabed deposits are high, from 174 years for nickel to nearly 500 years for manganese. Barsotti uses mineral availability studies of existing capacity and reserves to reach a similar conclusion that resource limitations alone will not be sufficient to draw nodules into production until many years to come [30]. Of course, these life indexes are naïve in not accounting for increasing materials consumption with population and economic growth [31]. Neither, however, do they consider the influence of additions to resources over time from previously undiscovered or uneconomic conventional deposits onshore.

Simulation results from Leontief's input–output model of the world economy are shown in table 19.1 to help account for projected growth in consumption [32]. Projected percentage depletion of world resources, based on United Nations low-case assumptions about growth in developing economies, have been reported [32]. These depletion rates, however, tell little about the rate at which new resources will be added or the cost of doing so. In addition, the input–output technique is not well suited to capture the effects of recycling, conservation, and substitution over time. With the possible exception of phosphates in fertilizer, good economic substitutes exist for all of these materials in most of their uses [21], and other substitutes emerge as economic alternatives when a material becomes scarcer and relatively more expensive.

In the absence of a more precise measure of increasing scarcity of natural resources [19], the long-term behavior of market price is probably as good a signal as any [33]. If the relative scarcity of a material is increasing, we would expect its real price to rise. In accordance with Slade [33], trends in real price indices for principal seabed deposit types are thus examined here by using both linear and quadratic fits (figure 19.2). From this limited evidence we cannot conclude that price behavior is signaling increased relative scarcity of these materials. In recent years the composite prices of all the deep seabed materials have been well below the fitted trends and have been falling throughout the decade. Further, from the evidence on resource size and projected rates of consumption, depletion effects alone would not prompt a resort to most seabed materials for at least several decades. With sufficient cost advantages, however, and with relative gains in technological progress, more seabed materials might be used before some conventional resources.

Comparative Costs and Pace of Technological Change

In the quiet shallow waters of protected bays or estuaries, dredging costs for sand and gravel, placer minerals, or phosphate may be comparable to those onshore. Indeed, this case is little more than an extension of conventional onshore production. For more exposed, high-energy (weather and waves) offshore environments much greater throughput costs can be expected. Mining costs are controlled by numerous factors and tend to be case-specific, but industry sources suggest as a rule of thumb that seabed dredging for these materials would cost at least three to five times more than inland dredging. When one mining technology is more costly than another for a given level of throughput, it can still be competitive if the ore grade is rich enough to compensate with higher metal yield or if the deposit is large enough to spread fixed costs over greater levels of output. Offshore oil and gas are good examples. Although average drilling and equipping costs tend to be three to four times greater offshore, the large size of the seabed deposits that are in production allows them to compete.

Similarly, other seabed deposits would have to offer compensating grade or size premiums to be competitive. Under some local

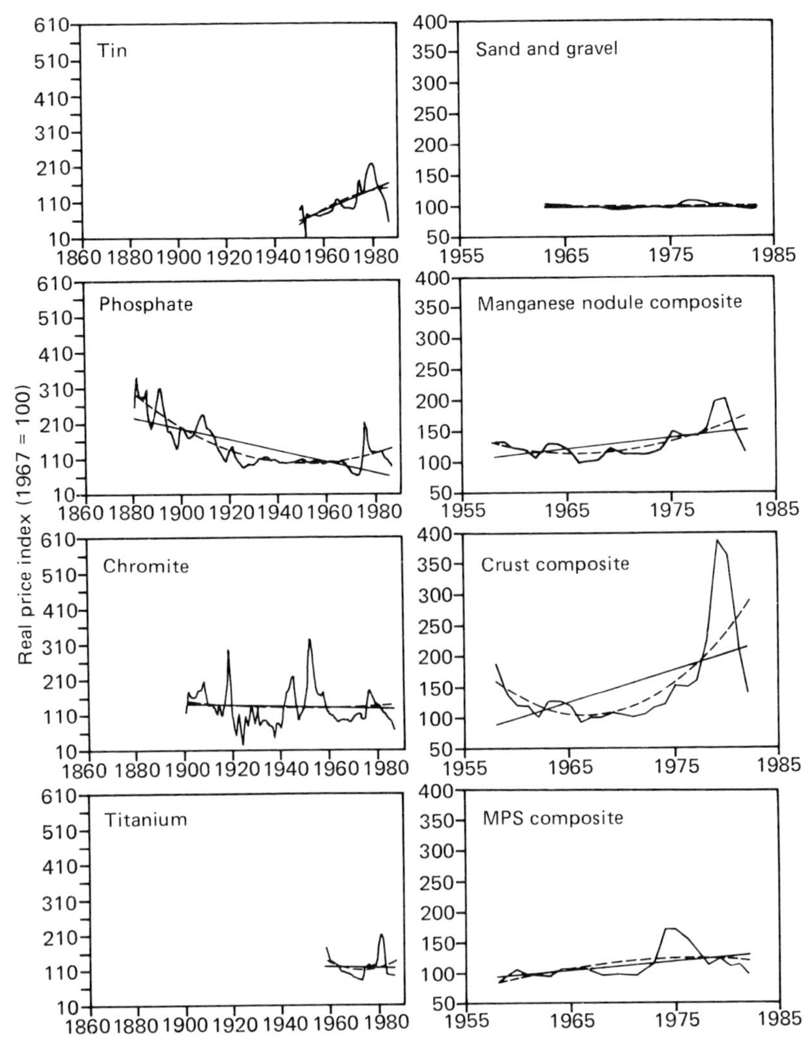

Figure 19.2 Behavior of real price indexes (1967 = 100) for selected commodities with potential seabed sources, including grade-weighted composite prices of materials contained in typical polymetallic deposits (manganese nodules are 1.28 percent nickel, 0.24 percent cobalt, 1.02 percent copper, and 25.4 percent manganese, based on [11]; crust composite is 0.47 percent nickel, 0.73 percent cobalt, and 23.06 percent manganese, from [11]; MPS is 32.3 percent zinc and 0.81 percent copper, based on [12], and showing linear and quadratic fits. Note differences in scale.

conditions generating locational or deposit-size advantages in delivered cost, offshore sand and gravel overcomes the usual cost differential [18]. Seabed placer minerals so far reported do not seem to exhibit much larger size or higher grade than their onshore rivals, and Emery and Noakes have shown that strong physical constraints will generally limit the distribution, grade, and accessibility of marine placers relative to those onshore [34]. McKelvey reported that development of new borehole mining methods may enhance access to subseabed phosphates [3], but this technology also favors expansion of deep rival resources onshore.

A number of cost estimates have been attempted for deep seabed nodule mining; these are usually based on detailed engineering scenarios [35]. The most recent estimates have shown total capital costs that range from $1.3 billion to $1.8 billion and annual operating costs from $224 million to $440 million. The conclusion of their analysis is that commercial nodule mining is unlikely for "the foreseeable future" [35]. Several earlier estimates were compared by Dick [36] who concluded that seabed nodule mining costs would be roughly comparable to the costs of production from newly developed nickel laterite deposits onshore. Dick did not, however, assign the seabed cost estimates a penalty to account for the uncertainty about operating conditions, engineering problems, and unanticipated costs that might be encountered. Analysis of cost histories in various other pioneer projects reveals that early cost estimates for commercially unproven technologies are not only typically biased low but are often so uncertain that they cannot be relied upon at all [37].

Attempts at this stage to characterize potential mining costs for MPS and cobalt crusts might be especially prone to this shortcoming. No technologies are known for breaking, sorting, and lifting these hard-rock deposits at such great depths, and only the most preliminary mining concepts have so far been presented [38]. No method is known by which the crusts can even be selectively sampled in quantity without obtaining much barren substrate material, and practically nothing is known about the thickness (size) of the MPS deposits. Development of techniques, such as a hard-substrate drill, to overcome these shortcomings is a priority in seabed minerals exploration.

The rate of technical progress in exploration and discovery may be one means by which the seabed deposits are gaining on their conventional onshore rivals. Advances in deep-sea exploration technology, such as multibeam sonar, underwater photographic and electronic imagery transmission, robotics, and deep submergence vehicles, permitted the firsthand verification and continuing refinement of geophysical theories that for several years had predicted the occurrence of the hydrothermal MPS deposits at oceanic crustal spreading centers. Not only were the theoretical results largely exogenous to the search for commercial seabed mineral deposits, but the advances in exploration hardware were too. Technology developed to support offshore oil and gas operations has made a major contribution to the study of seabed nonfuel minerals, and spillover benefits are also provided by investments for military and national security purposes. The work financed by the *Glomar Explorer* submarine recovery effort of the mid-1970s is an obvious example, as more recently is the U.S. Navy-sponsorship of the development of the *Argo/Jason* system at the Woods Hole Oceanographic Institution.

Marine scientific research will continue to provide an exogenous "input subsidy" for potential seabed material resources. Meanwhile, real discovery costs onshore have been rising, perhaps doubling in the past eighty years [19]. Reliance on scientific theory to target search for onshore deposits is increasing, and continuing study of marine deposits may also help focus the onshore search. Oceanographic knowledge has already been used successfully in locating onshore occurrences of marine phosphorites, and some scientists expect that observation of deep-sea MPS deposits will eventually help locate commercial analogs on land.

Commercial Activities, Strategic Industrial Behavior, and "Strategic" Materials

Even with substantial exogenous contributions to technological advance, the pace of seabed materials developed will be determined mainly by investments directed purposely at seabed materials. The incremental extension of conventional production of sand and gravel, placer minerals, and phosphates to offshore resources can be accomplished in the early normal course of business by incumbent producers of those commodities and by dredging contractors. Exploitation of the deep seabed deposits, however, requires the development of entirely new industrial capabilities.

Since the early 1960s more than $650 million (constant 1982) has been spent to develop technologies and explore for deep seabed manganese nodules [39]. The time profile of investments by the international industrial consortia that mounted this effort reveals a sharp decline from the 1978–79 peak spending levels of nearly $100 million (constant 1982) per year (figure 3). These spending estimates have been reconstructed from a fragmentary published record and spotty clues from industry sources. Nonetheless, they give an accurate general impression of the scale and time profile of industry efforts, as confirmed by reference in the public record to the annual number of patent grants [40].

Swayed by growth rates in metals consumption before the embargo of the Organization of Petroleum Exporting Countries and persuaded by the entrepreneurial leadership of seabed mining enthusiasts, parent firms may simply have invested mistakenly in a wasteful, losing venture. Half-billion dollar industrial mistakes are not rare, and a few companies have since withdrawn from their consortia. Certainly, earlier expectations about the scale and pace of development of the nodule resource were greatly overblown. Firms that have already invested in seabed mining R&D may have gained "first-starter" advantages, including learning and skills, patents, and increasingly secure claims to exclusive exploration and mining areas on the seabed [41]. If the investment in these uncertain assets was a mistake, for most of the firms it was a relatively small one. My estimates of spending behavior indicate that even during the five-year peak spending period from 1976 through 1980, only one of the twelve companies whose spending was examined devoted more than 7 percent of its average annual exploration and R&D budget to seabed mining (and most spent less than 5 percent).

Economists increasingly are interpreting R&D activities, including those aimed at unconventional extractive technologies, as a form of strategic behavior, in the sense that a parent course of action is chosen both in anticipation of and to influence the future behavior of rivals for future market rents [42]. Although their market structure is dynamic, both the history and concentration of the markets for nickel and cobalt suggest that monopoly rents may be earned [43]. The deep seabed deposits themselves could be extensive

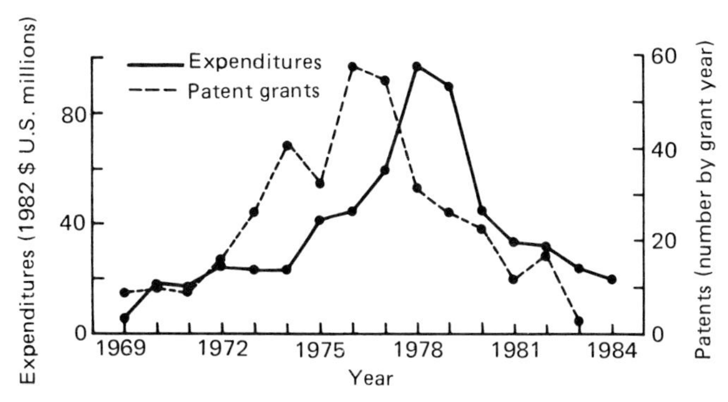

Figure 19.3 Estimated seabed mining expenditures and patent activity from 1969 to 1984

enough to provide the basis for a broad-scale, sustained market penetration. A comparable restructuring was witnessed in recent years as large, low-grade nickel laterite deposits were brought into production or, in historical experience, as porphyry ores in the southwestern United States remade the copper industry early in the century [33]. Potential entrants, or aggressive smaller sellers wishing to expand their market share, might rationally seek to establish a good technical basis for entry by means of seabed mining [44]. Major incumbent producers may try to preempt such innovative entry by demonstrating with R&D their own commitment to and capabilities in seabed mining. The combined result can be "premature" or excess capacity creation, "sleeping" patents, and idle mine sites. Where the game is played in rounds, a spiral of responses in turn can lead the players to levels of commitment that, but for the observed or expected action of rivals, they would rather avoid.

A high degree of governmental involvement adds further to the strategic dimension of seabed mining investment. Most of the consortia have at least some participation by national governments or state companies, and some are largely or entirely governmental operations. Governments account for more than a third of cumulative investment to date. Their sponsorship appears in most cases to have been motivated more by interest in exploration of alternative long-run sources of materials supply and of advanced technology development than by prospects for commercially generated profits from the production and sale of metals. For the time being at least, while some governmental "catch-up" programs go forward, the commercial consortia, after completing their first-phase planning objectives, have gone largely dormant except for protracted legal maneuvering to secure mine site claims in the Clarion-Clipperton zone.

Although overlaps still exist among the sites sought for exclusive development by various industrial and government entities, the international legal situation is becoming increasingly clear. Credible estimates of the number of suitable "first generation" nodule mine sites range from nine to forty [10]. Exploration licenses to define some of these sites further have been issued to four of the international consortia by the United States, West Germany, and the United Kingdom (figure 19.1). Licenses to other sites are being sought under the laws of other governments and through the LOS system. Two international systems have thus emerged to resolve the problem of conflicting claims and to grant security of tenure. One system, of reciprocating arrangements based on the domestic laws of certain seabed mining pioneer states, made a significant advance in 1984 with the signing by the United States, Belgium, France, West Germany, Italy, Japan, the Netherlands, and the United Kingdom of a Provisional Understanding Regarding Deep Seabed Matters (PU). The LOS process embodies another system for resolution of conflicting claims, and strong objections to the PU arrangement have been voiced by the Soviet Union and other LOS participants. Some parties to the PU (Belgium, France, Italy, Japan, and the Netherlands) also are involved in the LOS procedures through which further progress has been made.

In the United States much of the interest in seabed minerals has shifted from the international area to the EEZ. Reducing dependence on "strategic material" imports is often cited as a rationale for a greater effort to develop seabed materials [1, 15]. President Reagan referred to "recently discovered deposits" that "could be an important future source of strategic minerals," as he declared the EEZ [45]. The federal office promoting leasing and development of nonfuel minerals in the EEZ is called the Office of Strategic and International Minerals. Many people disagree about exactly what materials are or are not "strategic" [46]. A careful attempt to sort out the issue recently narrowed the list to only four "first tier" commodities: chromium, cobalt, manganese, and platinum group metals [47]. All four are seabed materials, but the Office of Technology Assessment has concluded that a number of options based on substitution, conservation, or production from alternative conventional sources are superior to seabed mining as approaches to reduced dependency on imports.

In recent months major studies of EEZ minerals exploration and policy have been undertaken by the National Academy of Sciences, the Office of Technology Assessment, and the Bureau of Mines. The issue arousing the most attention concerns the best arrangements for governmentally assured access by private parties to explore and develop minerals potential in the EEZ. Questions have been raised about the adequacy of the bonus-bid leasing provisions in the Outer Continental Shelf Lands Act administered by the Department of the Interior. Representatives from some commercial firms, environmental groups, and coastal states have expressed a preference for a licensing system modeled after the provisions of the Deep Seabed Hard Mineral Resources Act, which governs seabed mining activities by U.S. firms beyond the limits of national jurisdiction. Public discussion of this issue will raise questions about the importance of EEZ minerals development, the effects of exclusive licenses of leases on the conduct of scientific research, and fundamental public goals for seabed materials in the EEZ.

Conclusions

An optimistic outlook for long-run materials supply is reinforced by the presence of and increasing knowledge of seabed materials. Knowledge of the resource potential of these materials has grown as a by-product of basic scientific research and with the dedicated efforts of bureaucratic promoters and industrial entrepreneurs. The eventual realization of the resource potential of seabed materials will be determined by their relative economic accessibility compared to rival, onshore resources. For most seabed materials, superior sources are abundant enough to meet projected usage for at least several decades. Furthermore, although increasing consumption of current, conventional resources signals greater promise for seabed materials, it also triggers economic mechanisms that will expand onshore resources (through price effects and discoveries), while moderating consumption (through higher cost, conservation, recycling, and substitution).

More of the nearshore sources of materials will be exploited on an isolated basis in the relatively near future. This exploitation involves scanning and probing for opportunities embodying some combination of deposit grade, size, and locational advantage. Some of this type of exploitation is already taking place, and the important offshore oil and gas resources are being extended to much deeper regions. All the deep-sea marine mineral prospects, including manganese nodules, MPS, and the cobalt crusts, have long-range potential at best. They are attended by great uncertainty and will require considerably more study and investment before they can contribute to materials supplies.

Substantial progress has been made toward bringing metals from deep-sea manganese nodules into the stream of supply, and a higher level of understanding and practical know-how has been achieved for the nodules than for other deep-sea prospects. Strategic behavior in seabed minerals development implies that investment in exploration and R&D could proceed on a larger scale and at a more rapid pace than might be expected solely on the basis of the apparent commercial potential of the deposits. Even so, the commencement of production might not occur as soon as suggested by the pace of preproduction activity, since posturing may be a component of that activity.

The process of improving our understanding of marine minerals can contribute to scientific progress in general, and technological advances achieved in the process can have beneficial applications beyond marine minerals development. Close study of MPS mineralizations, for example, may also foster improved inferences about such basic mysteries as geotectonic processes and the earth's thermal and geochemical dynamics. Further, development of technologies to aid such scientific inquiries will have the spillover effect of generally advancing human capability to function in a hostile environment.

Notes

1. H. E. Goeller and A. Zucker, *Science* 223 (1984), 456.

2. Materials are the "stuff that things are made with," including minerals, forest and paper products, plastics, ceramics, and nonfood agricultural products such as fibers and oils [F. P. Huddle, ibid. 191 (1976), 654].

3. V. E. McKelvey, *Subsea Mineral Resources* (Bulletin 1689, U.S. Geological Survey, Denver, 1986).

4. R. M. Solow, *Am. Econ. Rev. Pap. Proc.* 64 (1974), 1; P. S. Dasgupta and G. M. Heal, *Economic Theory and Exhaustible Resources* (Nisbet, Cambridge, United Kingdom, 1979); D. R. Bohi and M. A. Toman, *Science* 219 (1983), 927.

5. D. B. Brooks and P. W. Andrews, *Science* 185 (1974), 13.

6. "Revenues" in table 18.1 are the product of production multiplied by an average price for the marketed commodity. Different degrees of processing and transportation costs are contained in the various commodity prices, so the revenue estimates are best used for a rough comparison of the size of markets supplied rather than for exact comparison of value of the raw materials. For a conservative estimate of offshore oil revenues, a price of $10 per barrel is used here.

7. *Oil and Gas Technologies for the Arctic and Deepwater.* (OTA-0-271, U.S. Office of Technology Assessment, Washington, DC, 1985).

8. K. O. Emery and B. J. Skinner, *Mar. Min.* 1–2 (1977), 1.

9. Ocean Economics and Technology Branch, United Nations, *Assessment of Manganese Nodules: Final Report* (Charles River Associates, Boston, 1981).

11. F. T. Manheim, *Science* 232 (1986), 600.

12. J. L. Bischoff et al., *Econ. Geol.* 78 (1983), 1711.

13. GERMINAL (Group d'Etude et de Recherche de Minéralisations au Large), *Proceedings of the 2nd International Seminar on Offshore Mineral Resources* (Brest, France, 1984).

14. K. O. Emery and E. Uchupi, *The Geology of the Atlantic Ocean* (Springer-Verlag, New York, 1984).

15. U.S. Department of the Interior, National Oceanic and Atmospheric Administration (NOAA), the Smithsonian Institution, *Symposium Proceedings of the Exclusive Economic Zone: Exploring the New Ocean Frontier* (U.S. Department of Commerce, Washington, DC, 1985); U.S. Department of the Interior, *A National Program for the Assessment and Development of the Mineral Resources of the United States Exclusive Economic Zone* (Circular 929, U.S. Geological Survey, Reston, VA, 1983).

16. D. S. Cronan, ed., *Sedimentation and Mineral Deposits in the Southwestern Pacific Ocean* (Academic Press, London, 1986); P. A. Rona, *Nat. Resour. Forum* 7 (1983), 329.

17. F. C. F. Earney, *Petroleum and Hard Minerals from the Sea* (Winston, New York, 1980); D. S. Cronan, *Underwater Minerals* (Academic Press, New York, 1980); G. P. Glasby, *Endeavour* (new series) 3 (1979), 82; U.S. Department of the Interior, *Program Feasibility Document: OCS Hard Minerals Leasing* (PB 81-192544, National Technical Information Service, Springfield, VA, 1979); D. M. Leipziger and J. L. Mudge, *Seabed Mineral Resources and the Economic Interests of Developing Countries* (Ballinger, Cambridge, MA, 1976); M. J. Cruickshank and R. W. Marsden, eds., *SME Mining Engineering Handbook 2* (American Institute of Mining, Metallurgical, and Petroleum Engineers, New York, 1973), section 20.

18. D. B. Duane, in *Marine Sediment Transport and Environmental Management,* D. J. Stanley and D. J. P. Swift, eds. (Wiley, New York, 1976), pp. 535–556.

19. D. P. Harris and B. J. Skinner, in *Explorations in Natural Resource Economics,* V. K. Smith and J. V. Krutilla, eds. (Johns Hopkins Press, Baltimore, 1982), pp. 247–326.

20. *Minerals Availability System Program Appraisals* (U.S. Bureau of Mines, Washington, DC) [lead and zinc (IC 9026, 1985), cobalt (IC 9012, 1985), nickel (IC 8995, 1984), phosphate (IC 8989, 1984), manganese (IC 8978, 1984), chromium (IC 8977, 1984), copper (IC 8930, 1983), and platinum (IC 8897, 1982)]. The Bureau of Mines has recently undertaken a study of the potential of selected EEZ minerals, which could result in Minerals Availability Appraisals expressly for some seabed materials in the United States.

21. *Mineral Commodity Summaries* (U.S. Bureau of Mines, Washington, DC, 1986); *Mineral Facts and Problems* (U.S. Bureau of Mines, Washington, DC, 1980). Exceptions from Identified Resources in table 19.1 follow: for platinum group metals and copper, the resource estimates include identified plus undiscovered resources; for manganese, the resource estimate is limited to the reserve base. Proper definitions of "resource" and "reserve" categories, including Identified Resources are given in these Bureau of Mines reports. Because of the long time perspective adopted in our discussion, the category "reserves" is not explicitly considered. See [5] and [19].

22. Except for commodities or deposit types with historically established production parameters, such as tin, sulfur, or sand and gravel, reports of seabed "reserves" or "resources" must be viewed with particular skepticism.

23. D. P. Harris, *Mineral Resources Appraisal* (Clarendon, Oxford, 1984).

24. A. R. Solow, in *Proceedings of NATO Advanced Study Institute on Statistical Treatment for Estimation of Undiscovered, Economically Recoverable Oil and Gas Resources for the Outer Continental Shelf as of July 1984* (OCS Report MMS-85-0012, U.S. Minerals Management Service, Washington, DC, 1985).

25. M. J. Boskin *et al., Am. Econ. Rev.* 75 (1985), 923.

26. A. L. Clark *et al., Resource Assessment: Cobalt-Rich Manganese Crust Potential* (OCS Report MMS-85-0006, U.S. Minerals Management Service, Long Beach, CA, 1985).

27. Z. Mustafa et al., in [13], p. 528.

28. J. M. Broadus, ibid., p. 574.

29. The basis for the estimates varies by deposit type and often by commodity. Some consist of estimates for only a single deposit (barite or sulfur) or region (heavy mineral placers or nodules and crusts), whereas others are global extrapolations from scant evidence or reasoned inference (nodules and crusts or massive sulfides). Because the economic potential of most seabed sources is at best speculative, a more balanced comparison might be to total onshore resources, including hypothetical and speculative undiscovered resources, which are typically several times as large as identified resources.

30. A. F. Barsotti, *J. Resour. Manage. Technol.* 13 (1984), 85.

31. But, see W. Malenbaum, *World Demand for Raw Materials in 1985 and 2000* (McGraw-Hill, New York, 1978).

32. W. Leontief et al., *Techniques for Consistent Forecasting of Future Demand for Major Minerals Using an Input–Output Framework* (NSF/CPE-82002, Institute for Economic Analysis, New York University, New York, 1982).

33. M. E. Slade, *J. Environ. Econ. Manage.* 12 (1985), 181.

34. K. O. Emery and L. C. Noakes, *Tech. Bull. ECAFE* (Economic Commission for Asia and the Far East), 1 (1968), 95.

35. Australian delegation to the Preparatory Commission on the Law of the Sea, *The Enterprise: Economic Viability of Deep Sea-Bed Mining of Polymetallic Nodules* (LOS/PCN/SCN.2/WP.10, United Nations, New York, 1986); C. T. Hillman and B. B. Gosling, *Mining Deep Ocean Manganese Nodules: Description and Economic Analysis of a Potential Venture* (IC 9015, U.S. Bureau of Mines, Washington, DC, 1985).

36. R. Dick, in *The Economics of Deep-Sea Mining,* J. B. Donges, ed. (Springer-Verlag, Berlin, 1985), pp. 2–60.

37. E. W. Merrow, K. E. Phillips, C. W. Myers, *Understanding Cost Growth and Performance Shortfalls in Pioneer Process Plants* (R-2569-DOE, Rand Corp., Santa Monica, CA, 1981).

38. J. E. Halkyard & Co., in *Draft Environmental Impact Statement, Proposed Polymetallic Sulfide Minerals Lease Offering* (U.S. Minerals Management Service, Reston, VA, 1983), pp. 503–545.

39. Large companies from the United States, Canada, the United Kingdom, Japan, West Germany, France, Italy, Belgium, and the Netherlands have been involved in these efforts [J. M. Broadus, *J. Mar. Resour. Econ.* 3 (1986), 63]. Since the abandonment of some earlier approaches, all the consortia members appear to have converged on the hydraulic lift recovery technology, and a nine-year Japanese R&D program with an $80-million budget seeks to advance this approach by 1989. Seabed mining exploration and R&D programs by the governments of India, China, South Korea, and the Soviet Union are also under way.

40. P. Hoagland, *J. Resour. Manage. Technol.* 14 (1985), 211. That the peak in patents precedes the peak in spending is to be expected, since inventive work leading to the patents took place before the expensive at-sea tests of the resulting systems.

41. The consortia learned through their field work that (1) areal coverage by the nodules is patchier, (2) the seafloor terrain is less uniform and holds more obstacles, and (3) weather conditions are more of a factor than initially expected.

42. J. E. Stiglitz and G. F. Mathewson, eds., *New Developments in the Analysis of Market Structure* (MIT Press, Cambridge, 1986).

43. R. Rafati, in *The Economics of Deep-Sea Mining,* J. B. Donges, ed. (Springer-Verlag, Berlin, 1985), pp. 62–112 and 253–335.

44. Advantages of entry cannot be addressed without some attention to failure. See A. Glazer [*Am. Econ. Rev.* 75 (1985), 473] and P. Dasgupta (in [42], p. 519). An executive in one of the British parent companies explained that his group's objective was not to be first entrant into seabed mining, but to watch and learn from the likely failure of first entrants. He likened the contest to a "slow-speed bicycle race" in which the winner is the last rider to cross the finish line without falling over.

45. "Statement by the President on the Exclusive Economic Zone of the United States," accompanying Proclamation No. 5030, *Wkly. Compil. Pres. Doc.* 19 (no. 10) (14 March 1983), 383.

46. The term "strategic materials" has been used in the United States since at least 1922, but its meaning has changed often and still is not precise. A comprehensive review is provided by L. H. Bullis and J. E. Mielke, *Strategic and Critical Materials* (Westview, Boulder, CO, 1985).

47. *Strategic Materials: Technologies to Reduce U.S. Import Vulnerability* (OTA-ITE-248, U.S. Office of Technology Assessment, Washington, DC, 1985). The designation was based on the criteria that quantities "required" for "essential" uses exceed "reasonably secure" supplies and that timely substitution seems unlikely.

48. I gratefully acknowledge the assistance with data compilation and interpretation by P. Hoagland and A. R. Solow and helpful comments from K. O. Emery, J. Padan, and R. M. Solow. Prepared with funds from the J. N. Pew, Jr., Charitable Trust and the Department of Commerce, NOAA Office of Sea Grant, grant NA84AA-D-00033. Woods Hole Oceanographic Institution contribution 6352.

The Coming Era of Nanotechnology

K. Eric Drexler

20

We round off our anthology with a speculative piece which some might say is more in the realms of science fiction. We include it, nevertheless, as a stimulus to thought. Drexler argues that the old style of technology is bulk technology, where we handle atoms in unruly herds. Nanotechnology, on the other hand, will allow us to handle individual atoms and molecules, so we can build up complex structures one atom at a time. Nanotechnology, says Drexler, will completely transform information technology, biotechnology, and materials science, enabling us to build self-replicating engines of abundance, engines of healing, and engines of destruction. The author is the founder of the MIT Nanotechnology Study Group. From Whole Earth Review, no. 54, Spring 1987.

We all copy one another's successes and try new things. As people seeking wealth, health, and power we have come up with better medicines, missiles, seeds, socks, and video games. To create these, we invent. Indeed, our most powerful invention was the method of invention.

Yet we cannot "uninvent" this meta-technique short of smashing our civilization. Thus in our diverse and competitive world, our technological inventions evolve toward the limits of the possible. So to outline our future, we must outline the possible.

Whatever is, is obviously possible. Life is. Therefore that dem-

Reprinted by permission of the author and the *Whole Earth Review*, 27 Gate Five Road, Sausalito, CA 94965. Subscription price $20.00 per year.

onstrates the possibility of molecular machines able to build other molecular machines—the essence of both life and a new method called nanotechnology ("nano" meaning billionth, because it uses parts measured in billionths of a meter).

Whatever obeys natural law is also possible. Science now understands the laws of ordinary matter and energy well enough for most engineering purposes. Nanotechnology will enable us to build new kinds of things. Physical laws let us calculate what some of those things will be able to do.

The basic idea of nanotechnology is straightforward. We live in bodies made of atoms on a planet made of atoms, and how those atoms are arranged makes all the difference. Atoms are objects. They have size, shape, mass, and strength. A hammer is a large collection of atoms; a molecule is a small collection. Both are held together with the same forces. Molecular machines are simply machines made of molecular-scale parts having carefully arranged atoms.

Chemical reactions happen when two reactive molecules bump together in the right orientation, making atoms rearrange to form new molecules. Nanotechnology assemblers will be molecular machines that grab reactive molecules and bring them together in a controlled way, building up a complex structure a few atoms at a time.

Today, genetic engineers reprogram the molecular machinery of living cells to make new molecules. They work at the scale of proteins—thousands of atoms. Eventually, nanotechnologists will build new, smaller machinery, and program it to make almost any pattern of atoms a designer might specify, atom by atom. Imagine an industrial robot arm, directed by a computer. It can build complex things by putting parts together, one at a time. To picture nanoassembler, imagine that the arm is made of the smallest possible parts, each containing a couple to several thousand atoms. This makes it less than a millionth the size of the industrial arm (and lets it make motions in a millionth the time). This arm also builds complex things by putting parts together, one at a time, but the parts are reactive molecules and each assembly step is a chemical reaction.

The areas closest to nanotechnology today are microelectronics and biotechnology. Microelectronic engineers can design complex circuits, even computers on a chip, but they cannot arrange atoms as they please. As a result microcomputers are billions of times bulkier than the nanocomputers of the future. Biotechnologists can use the molecular machinery of life to build things atom by atom, but they can't (yet) design new molecular machines. They have only begun to learn the art of protein engineering.

With assemblers, molecular engineering will become easy, almost like building with Tinkertoys. Assemblers will be able to add carbon atoms to an object, a few at a time, building up a piece of diamond-fiber composite. Dozens of times stronger than ordinary metals, this material could be used to make almost anything—a car, a dome, a rocket—far stronger and lighter than anything we can build today. How would you cut and shape this material to make something? In general, you wouldn't. Just as a tree has no need to carve wood, so assembler-based production systems would have no need to carve their products: they would make them true to form in the first place, atomically perfect.

Life proves that systems of molecular machines can replicate themselves. One bacterium can copy itself in about twenty minutes, giving rise to many tons of duplicates in a few days. Other replicating molecular machines form complex patterns, such as crabgrass, redwood trees, blue whales, and human brains. Molecular machines have given earth coral reefs, an oxygen atmosphere, and a biosphere.

And just as the self-replicating molecular machinery of a seed can make a tree, so a properly programmed replicating assembler will be able to make a house, a computer, or a spaceship. This has implications for the cost of things: A tree uses sunlight and common materials to build itself, and a tree costs almost nothing to make—it takes no human effort, only room and time. With assembler-based technology, anything can be as inexpensive (pound for pound) as firewood, because it can be produced in essentially the same way, by growing it. (No, this analogy doesn't mean replacing forests!)

There is no new science in nanotechnology, only new engineering. The possibility of nanotechnology was implicit in the science known over thirty years ago, though no one saw it then. During the 1940s and 1950s biochemistry revealed more and more of the molecular machinery of the cell. In 1959 physicist Richard Feynman touched on a similar idea in a talk: he spoke of using small machines to build smaller machines (and these to build smaller machines, and so on). He suggested that the smallest machines would be able to "put atoms down where a chemist says" to make a "chemical substance." But Feynman didn't explain how these machines were to work, and said they "will really be useless," because chemists will be able to make whatever they want without them. Decades passed with little follow-up.

Molecular biology forged ahead, piling up ever more impressive examples of molecular machinery in viruses and cells. While I was studying at MIT in the winter of 1976, it became clear to me that advances in biology and chemistry would make possible what we now call assemblers. The consequences seemed large, then huge, then mind-boggling. Some were wonderful, some were terrifying. My initial optimism gave way to fear, and then to a more cautious hope. I gave talks on the idea at MIT, got criticism, wrote technical papers, and finally wrote a book, *Engines of Creation* (*WER* #53, p. 83), to help explain these consequences. Why has this effort devoured my time and attention? Simply because nanotechnology will bring dangers and opportunities on a grand scale. We need to understand it, or we may end up doing something stupid.

The Evolution of Nanotechnology

Several paths lead toward assemblers and nanotechnology: they include biotechnology, chemistry, and micromanipulators.

- *Biotechnology* Protein makes up most of the molecular machinery of life. Eventually, we will learn to design new molecular machines made of protein including protein machines for building better nonprotein machines. These in turn will be able to build almost anything.
- *Chemistry* Molecular machinery need not be made of protein. As our knowledge of chemistry advances, we will eventually learn to design and synthesize molecular machines like those made of protein, but different. These will include molecular machines for building better molecular machines, which in turn can build almost anything.
- *Micromanipulators* Physicists have made instruments for moving a sharp minute tip of matter with atomic precision. Eventually (perhaps) someone may learn how to use similar devices as manipula-

tors to build molecular machines for building better molecular machines, which in turn can build almost anything.

- *Combined approaches* The preceding techniques might be combined to build molecular machines able to build better molecular machines, which in turn . . .

As you can see, the starting point will make little difference. All roads lead to assemblers, and assemblers will let us make almost anything we are clever enough to design.

With many paths toward it open, nanotechnology will be a natural, evolutionary advance. Evolution occurs in ecologies, that is, in complex systems of cooperating, competing entities.

A biological ecology is a huge, multisided competition in which organisms compete to grab sunlight or each other's flesh; they sometimes form cooperative, symbiotic grabbing-teams. In a market ecology, on the other hand, the rules let individuals compete or cooperate but prohibit violent grabbing of flesh and possessions. This results in isolated patches of competition—industries—in which teams of cooperating people compete with other teams to see which can best cooperate with suppliers and buyers around the world.

The world political system is partly a pacifist, market ecology, and partly a militarist, biological ecology. But in one regard, both the competition to serve people and the competition to dominate them push in the same direction—toward more advanced technology based on better control over the structure of matter—in short, toward nanotechnology. Nanotechnology will make possible both better goods and better weapons.

In a world full of competing companies and governments, only global disaster or global domination could block the advance of technology. This seems to be a fundamental principle; if so, it must guide our plans.

We need to consider what is possible with advanced technology so we can better guide its emergence. A brief sketch is in order, covering what nanotechnology can do for us, what it can't do for us, and what it might be used to do to us.

This will set the stage for a look at what we might do to get ready. (If you want to believe in the sort of future that newspapers typically assume—where the next century is just a jazzed-up, run-down, or wrecked version of this one—then please keep a firm grip on your world view, stop reading, and don't consider the implications of what you've already read.)

New Options

What can nanotechnology do for us? Almost anything we want, in physical terms. Once we have the software to direct them, replicating assemblers can build almost anything, including more of themselves, without human labor. Because they will handle matter atom by atom, as trees do, they can be as clean as trees, or cleaner. They need not produce smoke or sludge or toxic chemical by-products.

Since shoes and ships and sealing wax—and spaceships and computers—can be made from elements common in air and rock, raw materials need cost little. And since sunlight is cheap and solar collectors can be as cheap as crabgrass, energy also need cost little. In general, almost any product imaginable can be made dirt cheap, even taking full account of energy costs and environmental side effects.

Assembler-based systems can replace modern manufacturing (and factories, and industrial corporations as we know them). Today, rich and poor countries alike shred and pollute the landscape with crude, inefficient technologies. With nanotechnology, all can be richer than the richest countries are now, yet do far less harm.

Just as trees do, systems of replicating assemblers can construct big, complicated structures, even extremely large pieces of hardware at low cost. By manufacturing tunneling machines cheaper, pound for pound, than firewood, nanotechnology will open up the underground frontier. There is as much room near the earth's surface below ground as there is above, yet we scar the landscape with highways and railroads. Why? Mostly because of costs. With cheap tunneling and construction, we could build a transportation system based on evacuated tunnels and magnetically levitated vehicles, giving transcontinental service in under an hour, and cross-town service in about two minutes. Fast, quiet, and efficient, this underground transport would leave more open land for earth's life, and for people who enjoy it.

By making spacecraft cheaper than firewood, nanotechnology

can open the space frontier. Space holds more room and resources than anyone can truly imagine. Near earth's sun are matter, energy, and room enough to build broad new lands with a million times the area of earth. We can bring life to the dead rocks of space, carrying forward the tradition of the plants and animals that brought life forth from the oceans to the land.

And by spreading life to space, we can ensure it against catastrophe. Earth is a small and fragile basket for something so precious as our only biosphere.

Nanotechnology can also provide stand-ins for life. Today people kill and eat both carrots and cattle. With advanced technology, there will be no need to kill animals in order to have meat. Cattle produce steaks (unintentionally, of course) by providing an environment for the growth of muscle tissue. Why not have a brainless beast that sits in a kitchen cabinet, one that grows fresh meat and vegetables as a tree grows apples? By growing meat outside animals, we could remove a major excuse for killing them.

So far, I've described how nanotechnology can be used to give us more of familiar things—goods, services, living space—in new and better ways. But these new ways of making things will also be able to make new kinds of things.

One important application will be the further miniaturization of computers. Detailed study shows that assemblers could build the equivalent of a large, modern computer in about 1/1,000 of the volume of a typical human cell. This would be a mechanical computer (they're easier to analyze than electronic computers), but moving parts on this scale can be small and fast enough to make the computer faster than today's electronic machines.

A desk-top machine could then have more raw computational power than any computer in the world today. In fact, it could have more raw power than all the computers in the world today combined. In these terms—which imply nothing about intelligence—such a machine would have the raw power of a million human brains.

Cooking this raw power into something useful is another matter—one of software and of designing or evolving patterns of computation that accomplish something valuable. The right patterns of activity will result in a fast-thinking artificial intelligence. (Otherwise all we have is a fancy adding machine.) This artificial intelligence applied to engineering will give us the ability to quickly design enormously complex systems.

With this sort of design ability, we will be able to build molecular machines able to repair living tissue. In the body we see molecular machines that enter tissue (white blood cells), that enter cells (viruses), that recognize molecules (antibodies), and that take apart and build all the parts of a cell (digestive enzymes, and the molecular machinery of cell reproduction). Build molecular machines with a similar range of abilities, place them under the control of sophisticated computer systems, and the result will be a system able to enter, diagnose, and repair a cell, tissue, or organ. The software seems a greater challenge than the hardware.

Since we live in bodies made of molecules, a technology able to rearrange molecules will mean a revolution in medicine. Today physicians cannot heal tissue; they can only provide the conditions for it to heal itself—if it can. And some tissues, like missing limbs or damaged spinal cords, can't. With cell repair systems based on nanotechnology, medicine will gain surgical control on the molecular level, healing tissue almost regardless of its condition.

Molecular-level repair can extend lifespan almost indefinitely (though the nature of the universe does not permit genuine immortality). And long life would not mean prolonged aging and deterioration. Wrinkled skin, clogged arteries, poor memory—these result not from some magic influence of the calender upon a life force, but from the disordering of patterns of cells and molecules. Restoring youthful patterns would restore youthful health.

The biosphere too has molecular problems. Many toxic waste molecules have entered the groundwater; too many carbon dioxide molecules have entered the air. With replicating assemblers (and enough knowledge) we can build plantlike devices with "roots" that reach far into the ground, capturing and destroying toxic molecules. We can even take on the biggest cleanup, reversing the greenhouse effect by converting carbon dioxide back into oil and coal and putting them back in the ground. (Remember, molecular machines made our atmosphere in the first place.) With nanotechnology, we can heal both our bodies and our biosphere.

In general, these applications of nanotechnology can mean wider choices for people. They allow independence from many of today's limits, escapes from poverty, from dependence on a global economy, and from short lifespans. They allow lives in wild futures like those dreamed of by some science fiction authors; equally, they allow lives in disciplined voluntary communities like those dreamed of by some new age writers.

Nanotechnology, being based on molecular machines, is compatible even with lives like those of our ancestors, and not only through its renunciation. Neolithic tribes made use of molecular machinery, packaged in yeast, seeds, and goats. It is only the hidden complexity of natural nanomachinery that makes possible agriculture on a human scale. Nanotechnology will follow this pattern, hiding immense complexity in easy-to-tend, self-reproducing systems.

With the coming of assemblers and nanotechnology, the world could be transformed for the better—if you prefer health to sickness, life to death, wealth to poverty, choice to limits, diversity to uniformity, and a clean environment to one burdened by the wastes of a coarse technology.

But if you prefer calm to change, watch out.

The Limits to Growth

Change exclusively for the better is far from guaranteed. Nanotechnology will not make everything possible.

The world—though so much larger than the earth—cannot yield infinite resources. Exponential growth of population, whether with long lives or short, would overrun available resources within a few thousand years, even allowing for interstellar flight. Nanotechnology offers a breathing space, but it cannot permanently cure the problem of population.

As there are limits to resource quantity, so there are also limits to hardware quality. No matter how you rearrange atoms, some things cannot be done. Natural law (whatever it may be) determines what matter is and what it can do. It will set bounds to the strength of materials, the speed of computers, and the rate of travel. And nanotechnology itself is but a vast extension of chemistry. It can no more change atomic nuclei than alchemists could transform lead into gold, and for the same reason: chemical technology is not nuclear technology.

Nanotechnology will push many limits far, far back, but it will not eliminate them. In many areas it will carry us forward to firm and lasting limits to growth, to an era (at last!) of stable technologies. Of course we also face practical limits set by our situation and ourselves. With our best efforts we cannot accomplish everything that is physically possible. Indeed, in a competitive world we sometimes can't even avoid things that are possible.

Dangers and Disruptions

Replicating assemblers will be a powerful technology, posing obvious dangers of accident and abuse. The issue here isn't whether assemblers and nanotechnology will raise problems on the scale of highway deaths or industrial accidents—they might or might not. The issue is whether they will smash civilization or destroy the biosphere.

An industrial base able to double in minutes, rather than decades, can serve as a potent basis for military power: Imagine an arms race in which one side can multiply its production a thousandfold in a single day. Further, nanotechnology will make possible potent new weapons, such as devices that act like programmable "germs" for germ warfare. These abilities make nuclear weapons look tame.

World-wrecking abuse of nanotechnology is a real threat. World-wrecking accidents are an unreal threat. True, a combination of ingenuity and criminal negligence could produce a replicator able to destroy the biosphere, but this is far from the ordinary meaning of "accident." It will be easy to make replicating assemblers that survive only with special help. For these limited replicators to run wild, they would have to do more than break down, they would have to gain special new abilities.

Genetically engineered organisms have stirred fears of accidental runaway plagues, and replicating assemblers may seem similar. Engineered organisms are generally modified with an eye to making them useful but incompetent (imagine a microscopic version of a modern domestic turkey), yet to survive in nature they would have to compete with the most effective organisms nature has evolved. As you'd expect, they tend to lose, and hence pose no threat. But engineered organisms are at least organisms—they are related to things that once evolved and survived in the natural world. Replicating assemblers, in contrast, need have no resemblance to anything that can survive unaided in nature. They would then be no more likely to accidentally escape and run wild than a can opener is to accidentally turn into a spaceship and fly to the moon.

Even if we avoid the abuse of nanotechnology—avoiding world war or the boot of technological dictatorship—we will still face disruptions from basically beneficial uses. Replicating assemblers will sweep away the foundations of our present economy. Mining, manufacturing, and shipping will wither in the face of machines able to make almost anything, anywhere, using just dirt, air, and sunlight. Hardware will matter little; software for making hardware will matter a lot.

Replicating assemblers will be able to make almost anything as cheaply as firewood, and in just as decentralized a way. Modern corporations chiefly coordinate people to work as parts of vast, complex machines; when complex machines become small, and homesteads can grow anything from tractors to supercomputers, who will need today's huge organizations? With its use in production and medicine, nanotechnology can enable us to eliminate poverty and disease, but it will also eliminate most corporations and industries. Will corporations try to stop it, and perpetuate misery to protect their position and profit? And will dictators then forge ahead, to seize absolute power?

What Is to Be Done?

Advances in many fields carry us toward nanotechnology, and all major powers are pursuing those advances. As a technology of tiny things made of ordinary materials, it cannot be monitored from satellites. I see no way to stop such a thing, short of invading every country and filling the world with incorruptible police. In a world armed with hydrogen bombs and full of real people, this somehow seems unlikely. If we can't stop nanotechnology, we must try to understand it, live with it, and use it as best we can.

Are good outcomes possible? It seems so, though I'm not offering any odds. This sketch of nanotechnology is too quick and rough to support any detail on this matter, but two points seem solid.

First, we need better understanding as individuals and as a society. Our survival may depend on our ability to tell sense from nonsense regarding a complex technology that doesn't exist yet. The nonsense will be abundant, no matter what we do: any field on the borders of science fiction, quantum mechanics, and biology is well positioned to import a lot of prefabricated crap; any field where experiments and experience aren't yet possible is going to have great trouble getting rid of that crap. When someone says "nanotechnology" and begins to expound, beware!

Second, a political movement to deal with nanotechnology must be a movement to guide an advance, not to stop it. I've already argued that attempts to stop it would be futile; here are some reasons for thinking such efforts would be socially irresponsible:

- Efforts to stop it would waste a tremendously valuable resource—the time and attention of that small minority that knows and cares enough to do anything at all.
- Efforts to stop it would cause further waste by consuming the efforts of the knowledgeable activists who would try to counter and redirect the stoppers.
- Useful efforts to guide this technology must bring together activists and technical experts—but efforts to stop it will tend to push these experts into a hostile camp, dividing potential allies to everyone's harm.
- What might seem like progress in stopping advances would instead drive research further and further from public scrutiny and control, first from universities and then from private companies, leaving secure, secretive military laboratories in the lead.
- The greatest possible success of a democratic movement to block this technology would merely drive it still further from public scrutiny and control, by ensuring that the democracies do not lead the way.

Since stopping it seems impossible, and attempts to stop it seem irresponsible, why not apply the time-tested sour-grapes principle and conclude that stopping it isn't even something to want? After all, blocking nanotechnology would condemn millions of people to misery and disease, and condemn the earth to further ravaging by desperate people seeking food and shelter with crude technologies. In contrast, success in guiding it could make many great dreams come true.

What is to be done? First, we need to become better informed about the basic facts of the coming revolution, about its real opportunities and dangers. Then, we need to build a broad movement, one with room for people who are pro-progress, but who take dangers seriously, and for people who are pro-caution, but who recognize the momentum of technological advance. With a broad movement—diverse but not fatally polarized—we will have a broad debate and will become yet better informed.

As nanotechnology comes near, it will seem more real. Concern will grow, and with it the sort of activism, knowledge, and organization that can make a difference in the world. If we hang together and hammer out a sound approach, we just might be ready when the breakthroughs arrive. We have years to prepare, to gather our wits and our strength, and to seek a measure of wisdom.

If we succeed, we could end up alive and free in a world worth living in.

Bibliography

General

John U. Nef. *The Conquest of the Material World: Essays on the Coming of Industrialism.* University of Chicago Press, IL, 1964. Meridian Books, Cleveland, Ohio, 1967.

W. O. Alexander. "The Competition of Materials." *Scientific American* 217 (September 1967), 255–266.

National Academy of Sciences, *Materials and Man's Needs.* Summary Report of the Commission on the Survey of Materials Science and Engineering (COSMAT). National Academy Press, Washington, DC, 1974. The COSMAT supplementary report was published in five volumes: Vol. 1: *The History, Scope and Nature of Materials Science and Engineering.* Vol. 2: *The Needs, Priorities and Opportunities for Materials Research.* Vol. 3: *The Institutional Framework for Materials Science and Engineering.* Vol. 4: *Materials Technology Abroad.* National Technical Information Service, Springfield, VA, 1975. Parts of these were reprinted in Morris Cohen (ed.), "Materials Science and Engineering: Its Evolution, Practice and Prospects." *Materials Science and Engineering* 37, 1 (January 1979).

Don Radford. *The Materials We Use.* Science in Today's World Series. Batsford/David and Charles, London, 1983.

Walter Hibbard Jr. "Materials Challenges of the 1980s." *Materials and Society* 8, 1 (1984).

Annual Report on High-Tech Materials: 1984—The Year That Was; 1985—The Year to Come. Technical Insights, Ft. Lee, NJ, 1985.

A Programme for the Wider Application of New and Improved Materials and Processes (NIMP). Report of the Materials Advisory Group, chaired by John Collyear ("Collyear committee"). Department of Trade and Industry, London (HMSO), 1985.

New Structural Materials Technologies: Opportunities for the Use of Advanced Ceramics and Composites. Office of Technology Assessment, U.S. Congress, Washington, DC, September 1986.

"Materials for Economic Growth." *Scientific American* 255, 4 (October 1986). Special issue with series of articles on new materials for information and communication, ground transportation, aerospace, energy, and medicine and on latest developments in electronic and magnetic materials, photonic materials, and advanced metals, ceramics, polymers, and composites. Contributors include Joel P. Clark, John S. Mayo, Morris A. Steinberg, Richard S. Claassen, Gerald L. Liedl, Praveen Chaudhari, and H. Kent Bowen.

Ronald F. Balazik and Barry W. Klein. "The Challenge of New Materials." *Materials and Society* 11, 2 (1987).

Thomas W. Eagar. "The Real Challenge in Materials Engineering." *Technology Review* (February–March 1987).

Peter A. Psaras and H. Dale Langford (eds.). *Advancing Materials Research*. National Academy Press, Washington, DC, 1987. Based on a symposium held by the National Academy of Engineering and the National Academy of Sciences with the participation of the National Materials Advisory Board and the Solid State Sciences Committee of the National Research Council, in Washington, DC, 28–29 October 1985.

Frost and Sullivan have done a series of reports on the U.S. and world markets for ceramics, composites, and other new materials, e.g., *Worldwide Advanced Composite Materials Market* (1985) and *U.S. Market for Technical Ceramics* (1985).

Materials Science

Rustum Roy (ed.). *Materials Science & Engineering in the United States*. Pennsylvania State University Press, 1970.

A. G. Guy. *Essentials of Materials Science*. McGraw-Hill, New York, 1976.

Richard S. Claassen and Alan G. Chynoweth. "Materials Science and Engineering as a Multidiscipline." In Morris Cohen (ed.), "Materials Science and Engineering: Its Evolution, Practice and Prospects." *Materials Science and Engineering* 37, 1 (January 1979).

Luke C. May. *The Nineteen Eighties—Payoff Decade for Advanced Materials*. The Science of Advanced Materials and Process Engineering (SAMPE) Series. Vol. 25. 1980.

Philip H. Abelson. "Materials Science and Engineering." *Science* 225, November 9, 1984.

Lawrence E. Murr. *Industrial Materials Science and Engineering*. Marcel Dekker, New York, 1984.

Advanced Materials Research: A Guide to R&D Centers. Technical Insights, Ft. Lee, NJ, 1984.

J. E. Gordon. *The New Science of Strong Materials, or Why You Don't Fall Through the Floor.* Princeton University Press, NJ, 1984.

B. H. Kear and B. C. Giessen (eds.), *Rapidly-Solidified Metastable Materials.* Proceedings of the Symposium held in Boston, MA. Elsevier, North-Holland, Amsterdam, 1984.

W. D. Callister. *Materials Science & Engineering: An Introduction.* Wiley, New York, 1985.

M. A. Meyers and O. Inal (eds.). *Frontiers in Materials Technologies,* Materials Science Monographs. Vol. 26. Elsevier, North-Holland, Amsterdam, 1985.

Advancing Technology in Materials & Processes. Proceedings of the Thirtieth National SAMPE Symposium. Anaheim, CA, 1985. (Plus other annual SAMPE conference papers)

Merton C. Flemings. "Materials Science and Engineering—Its Past and Its Future." *Transactions of the Iron and Steel Institute of Japan,* 26, 2 (1986).

W. Smith. *Principles of Materials Science and Engineering.* McGraw-Hill, New York, 1986.

W. F. Madden. "Future Developments in Materials and Composites." *International Journal of Materials and Product Technology* 1, 2 (1986).

Bernhard Ilschner. "The Influence of Materials Science on Metals Consumption." *Materials and Society* 10, 3 (1986).

Gerald L. Liedl. "The Science of Materials." *Scientific American* 255, 4 (October 1986).

Malcolm W. Browne. "New Era of Technology Seen in Diamond-Coating Process." *The New York Times,* Sunday, September 14, 1986, p. 1.

Larry Lewis. "Chemistry's New Workhorse: Organometallic Compounds Make Better Ceramics, Semiconductors and Catalysts." *High Technology* (July 1987).

Michael J. Bennett and Charles H. Kline. "Chemicals: An Industry Sheds Its Smokestack Image." *Technology Review* (July 1987).

Important journals are: *Progress in Materials Science, Current Topics in Materials Science, Annual Review of Materials Science, Treatise on Materials Science and Technology, International Journal of Materials and Product Technology, Materials and Society,* National SAMPE Technical Conference Series, Science of Advanced Materials and Process Engineering Series, Materials Research Society Symposium Series, and *The Encyclopedia of Materials Science and Engineering.*

The Superconductor Story

Gordon Graff. "Synthetic Metals Near Reality." *High Technology* (November 1985). *Comment:* Very interesting read in the light of later developments.

Arthur L. Robinson. "Record High-Temperature Superconductors Claimed." *Science* 235, January 30, 1987.

Michael D. Lemonick with J. Madeleine Nash and Richard Woodbury. "Superconductivity Heats Up." *Time,* March 2, 1987.

James Gleick. "Electricity Rushes into a New Era of Discovery: Momentous Advances in Superconductivity Set Off a Research Stampede." *The New York Times,* Tuesday, March 10, 1987, p. C1.

Emily T. Smith with Jo Ellen Davis and Evert Clark. "Our Life Has Changed: The Lightbulb, The Transistor—Now The Superconductor Revolution." And John W. Wilson with Otis Port. "The New World of Superconductivity." *Business Week,* April 6, 1987.

James Gleick. "New Superconductors Offer Chance to Do the Impossible." *The New York Times,* Thursday, April 9, 1987, p. 1.

Michael Kenward. "The Heat is On for Superconductors." *New Scientist,* May 7, 1987.

Michael D. Lemonick with Thomas McCarroll, J. Madeleine Nash, and Dennis Wyss, "Superconductors! The Startling Breakthrough That Could Change Our World." *Time,* May 11, 1987.

Emily T. Smith with Evert Clark, Michael Oneal and Randy Welch. "Putting Superconductors to Work—Superfast." *Business Week,* May 18, 1987.

"Not-So-Superconductors." *The Economist,* June 13, 1987.

Anthony Ramirez. "Superconductors Get into Business." *Fortune,* June 22, 1987.

T. A. Heppenheimer. "Superconducting: The New Billion Dollar Business." *High Technology* (July 1987).

John Greenwald with Dick Thompson and Dennis Wyss. "Frenzied Hunt for the Right Stuff." *Time,* August 10, 1987.

Kent Bowen. "The Superconducting Sprint: Trade Barriers Prompt Japan to Move Fast." *High Technology Business* (September 1987).

David R. Lampe. "High-Temperature Superconductivity: Exploring the Potential." *The MIT Report* (October 1987).

Alison Bass. "Superconductivity Research: Pace Slows, Reality Catches Up." *The Boston Globe,* November 9, 1987.

T. A. Heppenheimer. "1988's Hottest Superconductor Companies." *High Technology Business* (January 1988).

Simon Foner and Terry P. Orlando. "Superconductors: The Long Road Ahead." *Technology Review* (February–March 1988).

Emily T. Smith et al. "Superconductors: The All-Out Pursuit of Zero Resistance." *Business Week,* March 14, 1988.

Gunter K. Muecke and Peter Möller. "The Not-So-Rare Earths." *Scientific American* (January 1988).

The New Materials

Ceramics

W. W. Kriegel and H. Palmer. *Ceramics in Severe Environments.* Materials Science Research Series, Vol. 5. Plenum Press, New York, 1971.

W. D. Kingery. *Introduction to Ceramics.* John Wiley, New York, 1976.

David W. Richerson. "Applications of Modern Ceramic Engineering." *Mechanical Engineering* (December 1982).

David W. Richerson. *Modern Ceramic Engineering.* Marcel Dekker Inc., New York, 1982.

John W. Dizard. "The Amazing Ceramic Engine Draws Closer." *Fortune,* July 25, 1983.

J. Hunt. "Engineering with Ceramic Materials." *Engineering Materials and Design* (November 1983).

Gordon Graff. "Ceramics Take on Tough Tasks." *High Technology* (December 1983).

John Bell. "The Ceramics Age Dawns." *New Scientist,* January 26, 1984.

William J. Broad. "New Ceramic Compounds Set for Major Technology Role." *The New York Times,* Tuesday, November 13, 1984, p. C1.

Malcolm Brown. "New Tech . . . from Old Crocks," *The Sunday Times,* London, January 20, 1985.

H. Garrett De Young. "Marching Into the New Stone Age." Special report on "Japan's Technology Agenda," *High Technology* (August 1985).

Bob Johnstone. "Ceramics Faces Its Big Test." *New Scientist,* September 12, 1985.

Jan Skalny, "Ceramics in the 1990s: Will They Still Be Brittle?" *Materials and Society* 9, 2 (1985).

J. Perry. "Ceramics as Everyday Engineering Components." *Ceramics Industries Journal* (December 1985).

Noboru Ichinose. "Electronic Ceramics for Sensors." *American Ceramic Society Bulletin* 64, 12 (December 1985).

Ernest Raia. "New Ceramic Makes Better Cutting Tool." *High Technology* (December 1985).

Peter Marsh. "Science Starts to Gel in World of Ceramics." *Financial Times,* London, December 20, 1985.

L. M. Sheppard. "New Generation Ceramics." *Advanced Ceramics and Processes* (January 1986).

John D. Moteff and Paul Wynblatt. "A Measure of the Cost-Effectiveness of Developing Ceramic-Based Automotive Engines." *Materials and Society* 10, 4 (1986).

Michael Rogers. "Feats of Clay." *Fortune,* February 3, 1986.

R. C. Buchanan (ed.). *Ceramic Materials for Electronics* Marcel Dekker, New York, 1986.

Evaluation of Technological Priorities: Engineering Ceramics. PREST. University of Manchester, UK, January 1987.

Earl E. Conabee. "Developments in the Advanced Ceramics Industry." *Materials and Society* 11, 2 (1987).

Vivien Mitchell. *Advanced Ceramics.* Financial Times Business Information, London, 1987.

Tim Brady. *Advanced Ceramics: Research, Innovation and the Implications for Skills and Training.* Manpower Services Commission, London, 1988.

Tim Brady. *Advanced Ceramics: Research, Innovation and the Implications for Skills and Training.* Manpower Services Commission, London, 1988.

Fiber Optics

S. E. Miller and A. G. Chynoweth (eds.). *Optical Fiber Telecommunications.* Academic Press, Orlando, FL, 1979.

Les C. Gunderson and Donald B. Keck. "Optical Fibers: Where Light Outperforms Electrons. "*Technology Review* (May–June 1983).

Jeff Hecht. "Wiring up the Big Country." *New Scientist,* January 6, 1983. And "Fibre Optics Call up the Past." *New Scientist,* January 12, 1984.

Herwig Kogelnik. "High-Speed Lightwave Transmission in Optical Fibers." *Science,* 228, May 31, 1985.

Martin Pyykkonen. "Networking with Light." *High Technology* (February 1986).

Jeff Hecht. "Optical Cables Take to Sea." *New Scientist,* February 6, 1986.

Stephen Koepp. "And Now, the Age of Light." *Time,* October 6, 1986. And "London Calling on a Beam of Light." *Time,* January 19, 1987.

Tom Forester. *High-Tech Society: The Story of the Information Technology Revolution.* Basil Blackwell, Oxford, England, and MIT Press, Cambridge, MA, 1987, pp 97–102.

Jeff Hecht. *Understanding Fiber Optics.* Howard Sams & Co., New York, 1987.

C. David Chaffee. *The Rewiring of America.* Academic Press, New York, 1988.

Plastics and Composites

Jack R. Vinson and Tsu-Wei Chou. *Composite Materials and Their Use in Structures.* Wiley, New York, 1975.

Stephen W. Tsai and M. Thomas Hahn. *Introduction to Composite Materials*. Technomic Publishing, Lancaster, PA, 1980.

John A. Manson and Leslie H. Sperling. *Polymer Blends and Composites*. Plenum Press, New York, 1981.

Ferdinand Rodriquez. *Principles of Polymer Systems*. McGraw-Hill, New York, 1982.

H. Garrett De Young. "Plastic Composites Fight for Status." *High Technology* (October 1983).

Gordon Graff. "Superstrong Plastics Challenge Metals." *High Technology* (February 1985).

Replacement of Metals with Plastics. Report of the Production Engineering Research Association. National Economic Development Office, London, 1985.

Peter Marsh. "High-Fibre Diet for Tomorrow's Materials." *Financial Times*, London, December 31, 1985.

Kenneth R. Sheets. "The New High-Tech World of Plastics." *US News and World Report*, February 24, 1986, pp. 52–53.

Gordon Graff. "Piezopolymers: Good Vibrations." *High Technology* (June 1986).

H. Garrett De Young. "What We Need Is a Breakthrough." Special report "Japanwatch '86," *High Technology* (August 1986).

Tony Jackson. "A Game for Big Players." *Financial Times*, London, August 15, 1986.

Gordon Graff. "High-Performance Plastics." *High Technology (October 1986)*.

Business Opportunities in Resin-Based Advanced Composites. SRI International, Menlo Park, CA, 1986.

Jeffrey L. Meikle. "Plastic, Material of a Thousand Uses." In Joseph J. Corn (ed.). *Imagining Tomorrow*. MIT Press, Cambridge, MA, 1986. *Comment:* Fascinating history of plastic and its role in industrial design. Especially noteworthy is the plastics utopianism of 1920–1950.

Joseph K. Lees. "Advanced Composite Materials: A Strong Growth Industry." *Materials and Society* 11, 2 (1987).

Richard B. Kaner and Alan G. MacDiarmid. "Plastics That Conduct Electricity." *Scientific American* (February 1988).

Alloys, Semiconductors, Cements, and Superglues

Deon Kruger. "Recent Developments in the Use of Polymer Concrete." *Materials and Society*, 9, 3 (1985).

Gene Bylinsky. "What's Sexier and Speedier Than Silicon." *Fortune*, June 24, 1985.

T. A. Heppenheimer. "Making Planes from Powder." *High Technology* (September 1986).

Nancy W. Stauffer. "New-Fangled Cement," *Technology Review* (July 1986).

Della M. Roy. "New Strong Cement Materials" Chemically-Bonded Ceramics." *Science* 235, February 6, 1987.

John Kerr. "Superglues Will Transform Manufacturing Methods." *Financial Times,* London, April 8, 1983, and subsequent reports, March 4, 1985, and July 24, 1985.

Ron M. Haeberle. "Development of Mechanically Alloyed Products for Advanced Materials Applications." *Materials and Society* 11, 2 (1987).

Eric L. Hyman. "The Comparative Merits of Ferrocement as a Substitute for Wood in Fishing Boats." *Materials and Society* 11, 2 (1987).

T. A. Heppenheimer. "Industrial Adhesives Start to Spread." *High Technology* (June 1987).

Application Areas

Computers and Telecommunications

M. B. Panish. "Molecular Beam Epitaxy." *Science* 228, May 23, 1980.

"Fiber Optics: The Big Move in Communications—and Beyond." *Business Week,* May 21, 1984.

"Frontiers in Computers." Special issue of *Science* 231, February 28, 1986.

Malcolm Brown. "Ions Are in the Firing Line. "*The Sunday Times,* London, January 20, 1985.

"Advanced Computing." Special issue of *Scientific American* 257, 4 (October 1987).

Manufacturing, Transport

Carl Hain. "Future Auto Engines: Competition Heats Up." *High Technology* (May–June 1982).

Steel and the Automotive Sector. International Iron and Steel Institute, Brussels, Belgium, 1983.

E. Paul De Garmo and Ronald A. Kohser. *Materials and Processes in Manufacturing.* 6th ed. Macmillan, New York, 1984.

John Bell. "Car Makers Shy from the Polymer Gamble." *New Scientist,* June 7, 1984.

Emmett J. Horton and W. Dale Compton. "Technological Trends in Automobiles." *Science* 225, August 10, 1984.

William J. Hampton. "The Car of the Future Will Have Skin and Bones." *Business Week,* July 29, 1985.

Jeffrey Zygmont. "Plastic Cars Hit the Road." *High Technology* (December 1985).

N. A. Gjostein. "Automotive Materials Usage Trends." *Materials and Society* 10, 3 (1986).

Robert A. Bonewitz. "Advanced Materials, Products, and Manufacturing Processes—A Challenge to the Metals Industry." *Materials and Society* 11, 2 (1987).

G. L. Bata and G. Salloum. "Computer-Integrated Materials Processing." *International Journal of Materials and Product Technology* 2, 2 (1987).

Construction

Fred Moavenzadeh. "Construction's High-Technology Revolution." *Technology Review* (October 1985).

Shannon A. Horst. "Adobe: New Look at a Centuries-Old Building Material." *The Christian Science Monitor,* April 17 and April 18, 1986, pp. 20–21 and 16–17.

Doug Stewart. "Skylines of Fabric." Technology Review (January 1987).

Sue Armstrong. "Pied à terre." New Society, March 10, 1988.

Aerospace

Aeronautical Technology 2000. Report by the Aeronautics and Space Engineering Board, the National Research Council, and the National Academy of Engineering. National Academy Press, Washington, DC, 1984.

Paul Kinnucan. "Superfighters." *High Technology* (April 1984). *Comment:* Use of composites in new fighter aircraft.

Jerome Persh. "Materials and Structures: Science and Technology Requirements for the DOD Strategic Defense Initiative." *American Ceramic Society Bulletin* 64, 4 (April 1985).

Mark Drela and John S. Langford. "Human-Powered Flight." *Scientific American* (November 1984). *Comment:* Use of new lightweight materials makes this possible.

T. A. Heppenheimer. "The Light Stuff: Burt Rutan Transforms Aircraft Design." *High Technology,* (December 1986). *Comment:* Use of composites, fibers and foams in *Voyager,* the remarkable craft that went round the world nonstop in 1987.

Energy

Materials Technology in the Near-Term Energy Program. Report by the National Research Council's Commission on Sociotechnical Systems. National Academy Press, Washington, DC, 1974.

Lawrence E. Murr (ed.). *Solar Materials Science*. Academic Press, Orlando, FL, 1980.

B. L. Butler and R. S. Claassen. "Survey of Solar Materials." *Journal of Solar Energy Engineering* 102, 3 (August 1980).

Energy in Transition: 1985–2010. Report by the National Research Council. National Academy Press, Washington, DC, 1982.

Medicine

Larry L. Hench and June Wilson. "Surface-Active Biomaterials." *Science* 226, November 9, 1984.

J. Russell Parsons. "Resorbable Materials and Composites: New Concepts in Orthopedic Biomaterials." *Orthopedics* 8, 7 (July 1985).

K. J. Quinn, J. M. Courtney, J. H. Evans, J. D. Gaylor, and W. H. Reid. "Principles of Burn Dressings." *Biomaterials* 6, 6 (November 1985).

Helen E. Kambic, Shun Murabayashi, and Yukihiko Nose. "Biomaterials in Artificial Organs." *Chemical and Engineering News* 64, 15 (April 14, 1986).

Richard Fifield. "New Lives for Painful Hips." *New Scientist*, December 10, 1987.

Space Processing

Tom Forester. "The High Frontier." *New Society*, August 10, 1978.

Richard S. Lewis. "Will Space Processing Get into Orbit?" *New Scientist*, November 9, 1978.

Extraterrestrial Materials Processing and Construction. Report by the Lunar and Planetary Institute, Houston, TX, 1980.

Robert Naumann and Harvey Herring. *Materials Processing in Space: Early Experiments*. National Aeronautics and Space Administration, 1980.

A. Brewersdorff (ed.). *Materials Science in Space*. Pergamon, Elmsford, New York, 1981.

David Criswell and Peter Marsh. "Working on the Moon." *New Scientist*, October 1, 1981.

A. Rindone. *Materials Processing in the Reduced Gravity Environment of Space*. Elsevier, North-Holland, Amsterdam. 1982.

David Marsh. "Future Factories May Orbit in Space." *Financial Times*, London, March 14, 1983.

Materials Sciences under Microgravity. Proceedings of an ESA Symposium. European Space Agency, 1983.

"For Industry It's Almost Lift-off Time." *Business Week*, June 20, 1983.

Peter Marsh. "Scientists Gain a Foothold in Space." *New Scientist*, September 22, 1983.

"Perfect Spheres: The First Product from Space." *Business Week,* January 30, 1984.

John S. De Mott. "Business Heads for Zero Gravity." *Time,* November 26, 1984.

Malcolm Brown. "1992: Factory in the Firmament." *The Sunday Times,* December 23, 1984.

Seth Payne and Alan Hall. "Business Is Starting to Get Serious about Space." *Business Week,* September 23, 1985.

Teresa Foley. "Commerce in Space Fails to Take Off." *New Scientist,* October 31, 1985. *Comment:* A valuable corrective to all the hype.

Brian Dumaine. "Still A-OK: The Promise of Factories in Space," *Fortune,* March 3, 1986.

Economic and Social Implications

The Global Battle (See also "General" listings)

G. B. Kenny and H. K. Bowell. "High-Tech Ceramics in Japan: Current and Future Markets." *Ceramic Bulletin* 62, 5 (1983).

Gene Gregory. "Fine Ceramics: Basic Materials for Japan's Next Industrial Structure." *Materials and Society* 8, 3 (1984).

The Competitive Status of the U.S. Steel Industry. Report by the National Research Council. National Academy Press, Washington, DC, 1985.

High Technology Ceramics in Japan. National Research Council, Washington, DC, 1985.

Gene Bylinsky. "Advanced Materials" and "Optoelectronics." In "Where the US Stands vs. Japan, Europe and Russia." Part of special report, "The High-Tech Race," *Fortune,* October 13, 1986.

Gene Gregory. "New Materials Technology in Japan." *International Journal of Materials and Product Technology* 2, 1 (1987).

Beyond Materials

D. H. Meadows, D. L. Meadows, J. Randers, and W. W. W. Behrens III. *The Limits to Growth* Universe Books, New York, 1972.

Wilfred Malenbaum. *World Demand for Raw Materials in 1985 and 2000.* McGraw-Hill, New York, 1978.

George F. Ray. "The Contribution of Science and Technology to the Supply of Industrial Materials." *National Institute Economic Review,* no. 92 (National Institute of Economic and Social Research, London, May 1980).

D. G. Altenpohl et al. *Materials in World Perspective.* Materials Research and Engineering Series. Vol. 1. Springer-Verlag, Berlin, 1980.

Paul Hawken. *The Next Economy.* Holt, Rinehart and Winston, New York, 1983.

John E. Tilton. "Materials Substitution: Lessons from the Tin-Using Industries." In John E. Tilton (ed.), *Materials Substitution: Lessons from the Tin-Using Industries.* Resources for the Future Inc., Washington, DC, 1983.

Reports of annual Mineral Economics symposiums in *Materials and Society* 7, 1 (1983); 8, 2 (1984); 9, 2 (1985); 10, 2 (1986); and 11, 2 (1987).

George F. Ray. "Mineral Reserves: Projected Lifetimes and Security of Supply." *Resources Policy* 10, 2, June, 1984.

Strategic Materials—Technologies to Reduce U.S. Import Vulnerability. Office of Technology Assessment. U.S. Congress. Washington, DC, 1985.

E. D. Larson, R. H. Williams, and A. Bienkowski. *Material Consumption Patterns and Industrial Energy Demand in Industrialized Countries.* Center for Energy and Environmental Studies, Princeton University, 1984.

Raymond F. Mikesell. "Economic Stockpiles For Dealing With Vulnerability to Disruption of Foreign Supplies of Minerals." *Materials and Society* 9, 1 (1985).

Raymond F. Mikesell. *Nonfuel Minerals: Foreign Dependence and National Security.* University of Michigan Press, Ann Arbor, MI, 1985.

Joel P. Clark and Frank R. Field III with John V. Busch, Thomas B. King, Barbara Poggiali, and Elaine P. Rothman. "How Critical Are Critical Materials?" *Technology Review* (August–September 1985).

Alan M. Strout. "Energy-Intensive Materials and the Developing Countries." *Materials and Society* 9, 3 (1985).

George F. Ray. "Innovation in Materials." In Roy M. Macleod (ed.), *Technology and the Human Prospect.* Frances Pinter, London, 1986.

Eric D. Larson, Marc H. Ross, and Robert H. Williams. "Beyond the Era of Materials." *Scientific American* 254, 6 (June 1986).

E. D. Hondros. "Materials, Year 2000—A Perspective." *International Journal of Materials and Product Technology* 1, 1 (1986).

James M. Broadus. "Seabed Materials." *Science,* February 20, 1987.

K. Eric Drexler. *Engines of Creation: The Coming Era of Nanotechnology.* Anchor Press, Doubleday, New York, 1986. See also K. Eric Drexler. "A Technology of Tiny Things: Nanotechnics and Civilization." *Whole Earth Review,* no. 54 (Spring 1987), pp. 8–14.

Robert C. Cowen. "Mapping the Nation's Underseas Wealth." *Technology Review* (November–December 1987).

Ian Steele. "Materials Technology and the Third World." *Technology Review* (January 1988).

Index